Engine Testing
The Design, Building, Modification and Use of Powertrain Test Facilities

Engine Testing

The Design, Building, Modification and Use of Powertrain Test Facilities

A. J. Martyr
M. A. Plint

ELSEVIER

AMSTERDAM • BOSTON • HEIDELBERG • LONDON
NEW YORK • OXFORD • PARIS • SAN DIEGO
SAN FRANISCO • SINGAPORE • SYDNEY • TOKYO

Butterworth-Heinemann is an imprint of Elsevier

Butterworth-Heinemann is an imprint of Elsevier
The Boulevard, Langford Lane, Kidlington, Oxford OX5 1GB, UK
225 Wyman Street, Waltham, MA 02451, USA
525 B Street, Suite 1900, San Diego, CA 92101-4495, USA

First edition 1995
Second edition 1999
Third edition 2007
Fourth edition 2012

Notice
No responsibility is assumed by the publisher for any injury and/or damage to persons or
property as a matter of products liability, negligence or otherwise, or from any use or operation of any methods, products, instructions or ideas contained in the material herein. Because
of rapid advances in the medical sciences, in particular, independent verification of diagnoses
and drug dosages should be made

British Library Cataloguing in Publication Data
A catalogue record for this book is available from the British Library

Library of Congress Cataloging-in-Publication Data
A catalog record for this book is availabe from the Library of Congress

ISBN–13: 978-0-08-096949-7

For information on all Butterworth-Heinemann publications
visit our web site at books.elsevier.com

Printed and bound in the US

12 13 14 15 16   10 9 8 7 6 5 4 3 2 1

Working together to grow
libraries in developing countries

www.elsevier.com | www.bookaid.org | www.sabre.org

ELSEVIER   BOOK AID International   Sabre Foundation

Contents

5.   Electrical Design Requirements of Test Facilities

6. Ventilation and Air-Conditioning in Powertrain Test Facilities

7. Test Cell Cooling Water and Exhaust Gas Systems

10. Dynamometers: The Measurement of Torque, Speed, and Power

11. Rigging the Engine and Shaft Selection

12. Test Cell Safety, Control, and Data Acquisition

13. Data Handling, the Use of Modeling, and Post-Test Processing

17.   Chassis or Rolling-Road Dynamometers

21. Thermal Efficiency, Measurement of Heat, and Mechanical Losses

The original intention of myself and my late co-author of the first two editions, Mike Plint, was to pass on to younger engineers our wide, but nonspecialist, knowledge of powertrain testing and the construction of the cells in which it takes place.

I am a product of what is probably the last generation of mechanical engineers to have benefitted from a five-year apprenticeship with a UK-based engineering company who was able to give its trainees hands-on experience of almost every engineering trade, from hand-forging and pattern making, through machine-shop practice and fitting, to running and testing of steam and gas turbines and medium-speed diesel engines.

After 50 years of involvement in the testing and commissioning engines and transmissions, of designing and project managing the construction of the test equipment and facilities required, this will be the last edition of this book in which I play a part.

The specialist engineer of today is surrounded by sources of information on every subject he or she may be required to learn in the course of their career. Should they be asked to carry out, or report on the task, for example, of converting a diesel engine test cell to also run gasoline engines, the immediate reaction of many will be to sit in front of a computer and type the problem into a search engine. In less than one second they will be confronted with over four million search results, the majority of which will be irrelevant to their problem and a few will be dangerously misleading. It is my hope that occasionally those searches might find this book and that not only the section related to a problem will be read.

My own research and reader feedback has led me to define three general types of readership.

The first, and for any author the most rewarding, is the student engineers who have been given the book by their employers at the start of their career and who have read most of it, from start to finish, as it was written. To those readers I apologize for repeating myself on certain subjects; such repetition is to benefit those who only look at the book to gain specific, rather than general, knowledge. The least rewarding is those specialist engineers who, as an exercise in self-reassurance, read only those sections in which they have more expertise than myself and who might have found benefit in reading sections outside their specialization. Of the remaining readership the most irritating are those who obtain the book in order to resolve some operational or constructional problem within a test facility, that would have been avoided had the relevant section been read before the work was done. The most frequent problems faced by the latter group, much to my irritation and their expense, are those dealing with some form of cell ventilation problem or those who "have always used this type of shaft and never had any problems before".

We all face the problems of working in an increasingly risk-averse world where many officials, representing some responsible authority, seem to consider the operation of an engine test cell to be a risk akin to some experimental explosives research institute, an opinion confirmed if they are allowed to witness a modern motor-sport engine running at full power before they drive away, safely, in their own cars.

The subjects covered in this book now exceed the expertise of any one engineer and I have benefitted greatly from the knowledge and experience of many talented colleagues.

Because of the risk of unforgivably forgetting someone, I hesitate to name all those who have unstintingly answered my questions and commented on some aspect of my work. However, I want to record my particular thanks to the following:

To Stuart Brown, Craig Andrews, Colin Freeman, David Moore, and John Holden, with whom I have had the honor of working for some years and whose support has been invaluable, not only in the production of this book but in my working life. To Hugh Freeman for his cheerfully given help concerning modern automotive transmission testing and Ken Barnes for his guidance on the American view on the subjects covered. To George Gillespie and his team at MIRA, and to engineers from specialist companies (mentioned in the relevant chapters) who have responded to my requests for information or the use of graphics.

My colleagues at the School of Engineering at the University of Bradford, Professor Ebrahimi and Byron Mason, have allowed me to keep up to date with engine research and the operation of the latest instrumentation. Of my past colleagues based in Graz special mention must be made of electrical engineer Gerhard Müeller. Finally, particular thanks to Antonios Pezouvanis of the University of Bradford, who has supplied both assistance and illustrations.

Writing a book is an act of arrogance, for which the author pays dearly by hours and hours of lonely typing. Thanks must be given to my neighbor and friend David Ballard for proofreading those chapters that had become so agonized over that I was incapable of judging their syntax. Finally, Hayley Salter and Charlotte Kent of Elsevier, who have been my "help of last resort", and to my family for their tolerance concerning the hours spent locked away on "the bloody book".

Tony Martyr
Inkberrow
July 2011

A. J. Martyr has held senior technical positions with several of the major test plant manufacturers and consultancy firms over the last 45 years. He is now Honorary Visiting Professor of Powertrain Engineering at Bradford University.

M. A. Plint died in November 1998, four days after the publication of the second edition and after a long and distinguished career in engineering and authorship.

Introduction

This book is not intended to be exclusively of interest to automotive engineers, either in training or in post, although they have formed the majority of the readership of previous editions. It is intended to be of assistance to those involved not only with the actual testing of engines, powertrains and vehicles, but also with all aspects of projects that involve the design, planning, building, and major modernization of engine and powertrain test facilities.

We are today (2011) at a significant break in the continuity of automotive engine and powertrain development. Such is the degree of system integration within the modern vehicle, marine, and generating machinery installations that the word "engine" is now frequently replaced in the automotive industries by the more general term "powertrain".

So, while much of this book is concerned with the design, construction, and use of facilities that test internal combustion engines, the boundaries of what exactly constitutes the primary automotive IC power source is becoming increasingly indistinct as hybridization, integration of electrical drives, and fuel cell systems are developed.

The unit under test (UUT) in most cells today, running automotive engines, has to either include actual or simulated vehicle parts and controllers, not previously thought of as engine components. This volume covers the testing of these evolving powertrain technologies, including transmission modules, in so far as they affect the design and use of automotive test facilities.

Drivers' perception of their vehicle's performance and its drivability is now determined less by its mechanical properties and more by the various software models residing in control systems interposed between the driver and the vehicle's actuating hardware. Most drivers are unaware of the degree to which their vehicles have become "drive by wire", making them, the driver, more of a vehicle commander than a controller. In the latter role the human uses the vehicle controls, including the accelerator pedal, to communicate his or her intention, but it is the engine control unit (ECU), calibrated and mapped in the test cell, that determines how and if the intention is carried out. In the lifetime of this volume this trend will develop to the point, perhaps, where driver behavior is regionally constrained.

Twenty years ago drivability attributes were largely the direct result of the mechanical configuration of the powertrain and vehicle. Drivability and performance would be tuned by changing that configuration, but today it is the test engineers and software developers that select and enforce, through control "maps", the powertrain and vehicle characteristics.

In all but motor sport applications the primary criteria for the selected performance maps are those of meeting the requirements of legislative tests, and only secondarily the needs of user profiles within their target market.

Both US and European legislation is now requiring the installation, in new light vehicles, of vehicle stability systems that, in a predetermined set of circumstances, judge that the driver is about to lose control or, in conditions that are outside a pre-programmed norm, intervenes and, depending on one's view, either takes over powertrain control and attempts to "correct" the driver's actions, or assists the driver to keep a conventional model of vehicle control.

A potential problem with these manufacturer-specific, driver assistance systems is their performance in abnormal conditions, such as deep snow or corrugated sand, when drivers, few of whom ever read the vehicle user manual, may be unaware of how or if the systems should be switched on or off.

Similarly, on-board diagnostic (OBD) systems are becoming mandatory worldwide but their capabilities and roles are far exceeding the legislatively required OBD-11 monitoring of the performance of the exhaust emission control system. Such systems have the potential to cause considerable problems to the test engineer rigging and running any part of an automotive powertrain in the test cell (see Chapter 11).

The task of powertrain and vehicle control system optimization known as *powertrain and vehicle calibration* has led to the development of a key new role of the engine test cell, a generation of specially trained engineers, test techniques, and specialized software tools.

The task of the automotive calibration engineer is to optimize the performance of the engine and its transmission for a range of vehicle models and drivers, within the constraints of a range of legislation. While engines can be optimized against legislation in the test cell, provided they are fitted with their vehicle exhaust systems, vehicle optimization is not such a precise process. Vehicle optimization requires both human and terrain interfaces, which introduces another layer of integration to the powertrain engineer. The same "world engine" may need to satisfy the quite different requirements of, for example, a German in Bavaria and an American in Denver, which means much powertrain calibration work is specific to a vehicle model defined by chosen national terrain and driver profiles.

This raises the subject of *drivability*, how it is specified and tested. In this book the author has, rather too wordily, defined drivability as follows:

For a vehicle to have good drivability requires that any driver and passengers, providing they are within the user group for which the vehicle was designed, should feel safe and confident, through all their physical senses, that the vehicle's reactions to any driver input, during all driving situations, are commensurate to that input, immediate, yet sufficiently damped and, above all, predictable.

Testing this drivability requirement in an engine or powertrain test bed is difficult, yet the development work done therein can greatly affect the character of the resulting vehicle(s); therefore, the engine test engineer must not work in organizational or developmental isolation from the user groups.

A proxy for drivability of IC engine-powered vehicles that is currently used is a set of constraints on the rate of change of state of engine actuators. Thus, within the vehicle's regions of operation covered by emission legislation, "smoothness" of powertrain actuator operation may be equated with acceptable drivability.

The coming generation of electric vehicles will have drivability characteristics almost entirely determined by their control systems and the storage capacity of their batteries. The whole responsibility for specification, development, and testing this "artificial" control and drivability model, for every combination of vehicle and driver type, will fall upon the automotive engineer.

Most drivability testing known to the author is based on a combination of subjective judgment and/or statistically compiled software models based on data from instrumented vehicles; this area of modeling and testing will be an interesting and demanding area of development in the coming years.

Fortunately for both the author and readers of this book, those laws of chemistry and thermodynamics relevant to the internal combustion engine and its associated plant have not been subject to change since the publication of the first edition over 17 years ago. This means that, with the exception of clarifications based on reader feedback, the text within chapters dealing with the basic physics of test facility design has remained little changed since the third edition.

Unfortunately for us all, the laws made by man have not remained unchanging over the lifetime of any one of the previous editions. The evolution of these laws continues to modify both the physical layout of automotive test cells and the working life of many automotive test engineers. Where possible, this volume gives references or links to sources of up-to-date information concerning worldwide legislation.

Legislation both drives and distorts development. This is as true of tax legislation as it is for safety or exhaust emission legislation. A concentration on CO_2 emission, enforced via tax in the UK, has distorted both the development of engines and their test regimes. Legislation avoidance strategies tend to be developed, such as those that allow vehicles to meet "drive-by" noise tests at legislative dictated accelerations but to automatically bypass some silencing (muffling) components at higher accelerations.

From many site visits and discussions with managers and engineers, it has been noticeable to the author that the latest generation of both test facility users and the commissioning staff of the test instrumentation tend to be specialists, trained and highly competent in the digital technologies. In this increasingly software-dependent world of automotive engineering, this expertise is vital, but it can be lacking in an appreciation of the mechanics, physics, and established best practices of powertrain test processes and facility requirements. Narrowing specialization, in the author's recent experience, has led to operational problems in both specification and operation of test facilities, so no apology is offered for repeating in this edition some fundamental advice based on experience. Many of the recommendations based on experience within this book have stories behind them worthy of a quite different type of volume.

All test engineers live in a world that is increasingly dominated by digital technology and legal, objective, audited "box-ticking" requirements, yet the outcome of most automotive testing remains stubbornly analog and subjective.

A typical requirement placed upon a powertrain test department could be:

Carry out such testing that allows us to guarantee that the unit or component will work without failure for 150,000 miles (240,000 km).

Such a task may be formalized through the use of a "development sign-off form".

If and when the prescribed test stages are concluded and without failure, such a procedure allows that the required box be ticked to acknowledge that the specified requirement can be guaranteed.

But the true response is that we have simply increased our confidence in the unit being sufficiently durable to survive its design life.

This not so subtle difference in approach to test results appears to the author to be one of the defining differences between the present generation, brought up in a world dominated by digital states and numbers, and a, usually older, generation whose world view is much more analog—successful test operations will have a well-managed mixture of both approaches.

In designing and running tests it is a fundamental requirement to ensure that the test life so far as is possible represents real life.

Powertrain test cells had to become physically larger in order to accommodate the various full vehicle exhaust systems, without which the total engine performance cannot be tested. Similarly, cell roof and corridor space has had to be expanded to house exhaust gas emission analyzers and their support systems (Chapter 16), combustion air treatment equipment, large electrical drives, and battery simulator cabinets (Chapter 5).

Completely new types of test facilities have been developed, in parallel with the development of legislative requirements, to test the electromagnetic emission and vulnerability of whole vehicles, their embedded modules, wiring harnesses, and transducers (Chapter 18).

The testers of medium-speed and large diesels have not been entirely forgotten in this edition and information covering their special area of work is referenced in the index.

The final testers of a powertrain, and the vehicle system in which it is installed, are the drivers, the operators, and the owners. The commercial success of the engine manufacturer depends on meeting the range of expectations of this user group while running a huge variety of journeys; therefore, it has always been, and still remains, a fundamental part of the engine test engineer's role to anticipate, find, and ensure correction of any performance faults before the user group finds them.

The owner/driver of the latest generation of vehicles may consider that the majority of the new additions to the powertrain and vehicle are secondary to its prime function as a reliable means of locomotion. It can be argued that the increased complexity may reduce vehicle reliability and increase the cost of fault-finding and after-market repair; OBD systems need to become a great deal smarter and more akin to "expert systems". The author cannot be alone in wondering about the long-term viability of this new generation of vehicles in the developing world, where rugged simplicity and tolerance to every sort of abuse is the true test of suitability.

Thus, new problems related to the function, interaction, reliability, vulnerability, and predictability of an increasingly complex "sum of the parts" arise to test the automotive test engineer and developer.

Unfortunately it is often the end user that discovers the vulnerability of the technologies embedded in the latest, legislatively approved, vehicles to "misuse".

This may be because the test engineer may, consciously or unconsciously, avoid test conditions that could cause malfunction; indeed, the first indication of such conditions represent the operational boundaries in a device's control map during its development.

The ever increasing time pressure on vehicle development has for many years forced testing of powertrain and vehicle modules to be done in parallel rather than in series. In modern systems this has necessitated increased module testing using hardware-in-the-loop (HIL) and software-in-the-loop (SIL) techniques, all of which rely on the use of software-based models of the missing components. Using modeling when the device being modeled is available, cheaper and easier to calibrate than the model generator is just one of the developments that raise some fundamental questions about the role of the test engineer, the test sequences used, and the criteria used to judge good results from poor ones.

Chapter 1

Test Facility Specification, System Integration, and Project Organization

INTRODUCTION: THE ROLE OF THE TEST FACILITY

If a "catch-all" task description of automotive test facilities was required it might be "to gain automotive type approval for the products under test, in order for them to enter the international marketplace".

Engine Testing. DOI: 10.1016/B978-0-08-096949-7.00001-7

The European Union's Framework Directive 2007/46/EC covers over 50 topics (see Figure 2.3) for whole vehicle approval in the categories M (passenger cars), N (light goods) and O (trucks), and there are similar directives covering motorcycles and many types of off-road vehicles. Each EU member state has to police the type approval certification process and have their own government organization so to do. In the UK, the government agency is the Vehicle Certification Agency (VCA) [1].

The VCA, like its European counterparts, appoints technical services organizations to carry out testing of separate approval topics and each of these organizations requires ISO 17025 accreditation for the specific topic in order to demonstrate competency.

PART 1. THE SPECIFICATION OF TEST POWERTRAIN FACILITIES

An engine or powertrain test facility is a complex of machinery, instrumentation, and support services, housed in a building adapted or built for its purpose. For such a facility to function correctly and cost-effectively, its many parts must be matched to each other while meeting the operational requirements of the user and being compliant with relevant regulations.

Engine, powertrain, and vehicle developers now need to measure improvements in performance that are frequently so small as to be in the noise band of their instrumentation. This level of measurement requires that every device in the measurement chain is integrated with each other and within the total facility, such that their performance and the data they produce is not compromised by the environment in which they operate, or services to which they are connected.

Powertrain test facilities vary considerably in layout, in power rating, performance, and the markets they serve. While most engine test cells built in the last 20 years have many common features, all of which are covered in the following chapters, there are types of cells designed for very specific and limited functions that have their own sections in this book.

The common product of all these cells is *data*, which will be used to identify, modify, homologate, or develop performance criteria of all or part of the unit under test (UUT).

All post-test work will rely on the relevance and veracity of the test data; the quality audit trail starts in the test cell.

To build, or substantially modify, a modern powertrain test facility requires the coordination of a wide range of specialized engineering skills; many technical managers have found it to be an unexpectedly wide-ranging complex project.

The task of putting together test cell systems from their many component parts has given rise, particularly in the USA, to a specialized industrial role known as "system integration". In this industrial model a company, more rarely a consultant, having relevant experience of one or more of the core technologies required, takes contractual responsibility for the integration of all the test facility components from various sources. Commonly the integrator role has been carried out by the supplier of test cell control systems and the contractual responsibility may, ill-advisedly, be restricted to the integration of the dynamometer and control room instrumentation.

In Europe the model was somewhat different because the long-term development of the dynamometry industry has led to a very few large test plant contracting companies. Now in 2012, new technologies are being used, such as those using isotopic tracers in tribology and wireless communication in transducers; this has meant that the number of individual suppliers of test instrumentation has increased, making the task of system integration ever more difficult. Thus, for every facility build or modification project it is important to nominate the role of systems integrator, so that one person or company takes the contractual responsibility for the final functionality of the total test facility.

Levels of Test Facility Specification

Without a clear and unambiguous specification no complex project should be allowed to proceed.[1]

This book suggests the use of three levels of specification:

1. *Operational specification*, describing "what is it for", created and agreed within the user group, prior to a request for quotation (RFQ) being issued. This may sound obvious and straightforward, but experience shows that different groups and individuals, within an industrial or academic organization, can have quite different and often mutually incompatible views as to the main purpose of a major capital expenditure.
2. *Functional specification*, describing "what it consists of and where it goes", created by a user group, when having or employing the necessary skills. It might also be created as part of a feasibility study by a third party, or by a nominated main contractor as part of a design study contract.
3. *Detailed functional specification*, describing "how it all works", created by the project design authority within the supply contract.

Note Concerning Quality Management Certification

Most medium and large test facilities will be part of organizations certified to a Quality Management System equivalent to ISO 9001 and an Environmental Management System equivalent to ISO 14000 series. Some of the management implications of this are covered in Chapter 2 but it should be understood that such certification has considerable bearing on the methods of compilation and the final content of the Operational and Functional specifications.

---

1. Martyr's First Law of Project Management: see Appendix 1.

Creation of an Operational Specification

This chapter will tend to concentrate on the operational specification, which is a *user-generated document*, leaving some aspects of the more detailed levels of functional specification to subsequent chapters covering the design process.

The operational specification should contain within its first page a clear description of the task for which the facility is being created; too many "forget to describe the wood and concentrate on the trees".

Its creation will be an iterative task and in its first draft it need not specify in detail the instruments required, nor does it have to be based on a particular site layout. Its first role will normally be to support the application for budgetary support and outline planning; subsequently it remains the core document on which all other detailed specifications and any requests for quotations (RFQ) are based.

It is sensible to consider inclusion of a brief description of envisaged facility acceptance tests within the operational specification document. When considering what form any acceptance tests should take it is vital they be based on one or more test objects that will be available on the project program. It is also sensible for initial "shake-down" tests to use a test piece whose performance is well known and that, together with its rigging kit, is readily available.

During the early stages of developing a specification it is always sound policy to find out what instrumentation and service modules are available on the market and to reconsider carefully any part of the operational specification that makes demands that may unnecessarily exceed the operational range that exists.

A general cost consciousness at this stage can have a permanent effect on capital and subsequent running costs.

Because of the range of skills required in the design and building of a "greenfield" test laboratory, it is remarkably difficult to produce a succinct specification that is entirely satisfactory to all stakeholders, or even one that is mutually comprehensible to all specialist participants.

Producing a preliminary cost estimate is made more difficult by the need for some of the building design details, such as floor loadings and electrical power demand, to be determined before the detailed design of the internal plant has been finalized.

The specification must include pre-existing site conditions or imposed restrictions that may impact on the facility layout or construction. In the UK this requirement is specifically covered by law, since all but the smallest contracts involving construction or modification of test facilities will fall under the control of a section of health and safety legislation known as *Construction Design and Management Regulations 1994 (CDM)* [2]. Not to list site conditions that might affect subsequent work, such as the presence of contaminated ground or flood risk, can jeopardize any building project and risk legal disputes.

The specification should list any prescribed or existing equipment that has to be integrated within the new facility, the level of staffing intended, and any special industrial standards the facility is required to meet. It is also appropriate that the operational specification document contains statements concerning the general "look and feel".

Note that the certification or accreditation of any test laboratory by an external authority such as the United Kingdom Accreditation Service (UKAS) or the International Organization for Standardization (ISO) has to be the responsibility of the operator, since it is based on approved management procedures as much as the equipment. External accreditation cannot realistically be made a contractual condition placed upon the main contractor.

In summary, the operational specification should, at least, address the following questions:

- What are the primary and secondary purposes for which the facility is intended and can these functions be condensed into a sensible set of Acceptance Procedures to prove the purposes may be achieved?
- What is the geographical location, altitude, proximity to sensitive or hostile neighbors (industrial processes or residential), and seasonal range of climatic conditions?
- What is the realistic range of units under test (UUT)? How are test data (the product of the facility) to be displayed, distributed, stored, and post-processed?
- How many individual cells have been specified, and is the number and type supported by a sensible workflow and business plan?
- What possible extension of specification or further purposes should be provided for in the initial design?
- May there be a future requirement to install additional equipment and how will this affect space requirement?
- How often will the UUT be changed and what arrangements are made for transport into and from the cells, and where will the UUT be prepared for test?
- How many different fuels are required and are arrangements made for quantities of special or reference fuels?
- What up-rating, if any, will be required of the site electrical supply and distribution system? Be aware that modern AC dynamometers may require a significant investment in electrical supply up-rating and specialized transformers.
- To what degree must engine vibration and exhaust noise be attenuated within the building and at the property border?
- Have all local regulations (fire, safety, environment, working practices, etc.) been studied and considered within the specification? (See below.)
- Have the site insurers been consulted, particularly if insured risk has changed or a change of site use is being planned?

Feasibility Studies and Outline Planning Permission

The investigatory work required to produce a site-specific operational specification may produce a number of alternative layouts, each with possible first-cost or operational problems. Part of the investigation should be an environmental impact report, covering both the facility's impact of its surroundings and the locality's possible impact on the facility.

Complex techno-commercial investigatory work may be needed, in which case a formal "feasibility study", produced by an expert third party, might be considered. In the USA, this type of work is often referred to as a "proof design" contract. Typically it would cover the total planned facility, but may only be concerned with that part that gives rise to techno-commercial doubt or is the subject of radically differing possible strategies.

The secret of success of such studies is the correct definition of the required "deliverable". An answer to the technical and budgetary dilemmas is required, giving clear and costed recommendations, rather than a restatement of the alternatives; so far as is possible the study should be supplier neutral.

A feasibility study will invariably be site specific and, providing appropriate expertise is used, should prove supportive to gaining budgetary and outline planning permission. The inclusion within any feasibility study or preliminary specification of a site layout drawing and graphical representation of the final building works will be extremely useful in subsequent planning discussions. Finally, the text should be capable of easy division and incorporation into the final functional specification documents.

Benchmarking

Cross-referencing with other test facilities or test procedures is always useful when specifying your own. Benchmarking is merely a modern term for an activity that has long been practiced by makers of products intended for sale. It is the act of comparing your product with competing products and your production and testing methods with those of your competitors. Once it is on the market any vehicle or component thereof can be bought and tested by the manufacturer's competitors, with a view to copying any features that are clearly in advance of the competitor's own products. There are test facilities built and run specifically for benchmarking competitor's products.

Maintenance of confidentiality by the restriction access, without hindrance of work, needs to have been built into the facility design rather than added as an afterthought.

Regulations, Planning Permits, and Safety Discussions Covering Test Cells

In addition to being technically and commercial for viable, it is necessary the new or altered test laboratory to be permitted by various civil authorities.

Therefore, the responsible project planner should consider discussion at an early stage with the following agencies:

- Local planning authority
- Local petroleum officer and fire department
- Local environmental officer
- Building insurers
- Local electrical supply authority
- Site utility providers.

Note the use of the word "local". There are very few regulations specifically mentioning engine test cells; much of the European and American legislation is generic and frequently has unintended consequences for the automotive test industry. Most legislation is interpreted locally and the nature of that interpretation will depend on the highly variable industrial experience of the officials concerned. There is always a danger that inexperienced officials will overreact to applications for engine test facilities and impose unrealistic restraints on the design or function. It may be useful to keep in mind one basic rule that has had to be restated over many years:

An engine test cell, using volatile fuels, is a "zone 2" hazard containment box. While it is possible and necessary to maintain a non-explosive environment, it is not possible to make its interior inherently safe since the unit under test is not inherently safe; therefore, the cell's function is to minimize and contain the hazards by design and function and to inhibit human access when hazards may be present.

It may also be useful to remind participants in safety-related discussions that their everyday driving experiences take them far closer to a running engine than is ever experienced by anyone sitting at a test cell control desk.

Most of the operational processes carried out within a typical engine or powertrain test cell are generally less potentially hazardous than those experienced by garage mechanics, motor sport pit staff, or marine engineers in their normal working life. The major difference is that in a cell the running automotive powertrain module is stationary in a space and humans could have, unless prevented by safety mechanisms, potentially dangerous access to it.

It is more sensible to interlock the cell doors to prevent access to an engine running above "idle" state, than to attempt to make the rotating elements "safe" by the use of close-fitting and complex guarding that will inhibit operations and inevitably fall into operational disuse.

The authors of the high-level operational specification would be ill-advised to concern themselves with some of these minutiae, but should simply state that industrial best practice and compliance with current legislation is required.

The arbitrary imposition of existing operational practices on a new test facility should be avoided until confirmed as appropriate, since they may restrict the inherent benefits of the technological developments available.

One of the restraints commonly imposed on the facility buildings by planning authorities concerns the number and nature of chimney stacks or ventilation ducts; this is often a cause of tension between the architect, planning authority, and facility designers.

With some ingenuity and extra cost these essential items can be disguised, but the resulting designs will inevitably require more upper building space than the basic vertical inlet and outlet ducts. Similarly, noise breakout through such ducting may, as part of the planning approval, have be reduced to the pre-existing background levels at the facility border. This can be achieved in most cases but the space required for attenuation will complicate the plant room layout (see Chapter 6 concerning ventilation).

The use of gaseous fuels such as LPG, stored in bulk tanks, or natural gas supplied through an external utility company will impose special restrictions on the design of test facilities and if included in the operational specification the relevant authorities and specialist contractors must be involved from the planning stage. Modifications may include blast pressure relief panels in the cell structure and exhaust ducting, all of which needs to be included from design inception.

The use of bulk hydrogen, required for the testing of fuel-cell-powered powertrains, will require building design features such as roof-mounted gas detectors and automatic release ventilators.

Specification of Control and Data Acquisition Systems

The choice of test automation supplier need not be part of the first draft operation specification. However, since test automation will form part of the functional specification, and since the choice of test cell software may be the singularly most important techno-commercial decision in placing a contract for a modern test facility, it would seem sensible to consider the factors that should be addressed in making that choice.

The test cell automation software lies at the core of the facility operation; therefore, its supplier will play an important role within the final system integration. The choice therefore is not simply one of a software suite but of a key support role in the design and ongoing development of the new facility.

Project designers of laboratories, when considering the competing auto-mation suppliers, should consider detailed points covered in Chapter 12 and the following strategic points:

- The installed base, relevant to their own industrial sector.
- Does one or more of their major customers exclusively use a particular control system? (Commonality of systems may give a significant advantage in exchange of data and test sequences.)
- Level of operator training and support required.
- Has the control system been proven to work with any or all of intended third-party hardware?

- Is communication with the control modules of the units under test required and is it possible via the designated "comms bus"?
- How much of the core system is based on industrial standard systems and what is the viability and cost of both hardware and software upgrades? (Do not assume that a "system X lite" may be upgraded to a full "system X".)
- Requirements to use pre-existing data or to export data from the new facility to existing databases.
- Ease of creating your test sequences.
- Ease of channel calibration and configuration.
- Flexibility of data display, post-processing, and exporting options.

A methodical approach requires a "scoring matrix" to be drawn up whereby competing systems may be objectively judged.

Anyone charged with producing specifications is well advised to carefully consider the role of the test cell operators, since significant upgrades in test control and data handling will totally change their working environment. There are many cases of systems being imposed on users and that never reach their full potential because of inadequacy of training or a level of system complexity that was inappropriate to the task or the grade of staff employed.

Use of Supplier's Specifications

It is all too easy for us to be influenced by headline speed and accuracy numbers in the specification sheets for computerized systems.

The effective time constants of many powertrain test processes are not limited by the data handling rates of the computer system, but rather by the physical process being measured and controlled. Thus, the speed at which an eddy-current dynamometer can make a change in torque absorption is governed more by the rate of magnetic flux generation in its coils, or the rate at which it can change the mass of water in a water-brake's internals, rather than the speed at which its control algorithm is being recalculated. The skill in using such information is to identify the numbers that are relevant to the task for which the item is required.

Faster is not necessarily better, but it is often more expensive.

Functional Specifications: Some Common Difficulties

Building on the operational specification, which describes what the facility has to do, the functional specification describes how the facility is to perform its defined tasks and what it will need to contain. If the functional specification is to be used as the basis for competitive tendering then it should avoid being unnecessarily prescriptive.

Overprescriptive specifications, or those including sections that are technically incompetent, are not rare and create a problem for specialist contractors. Overprescription may prevent a better or more cost-effective solution being quoted, while technical errors mean that a company who, through lack of experience, claims compliance and wins the contract will then inevitably fail to meet the customer's expectations.

Examples of overprescription range from choice of ill-matching of instrumentation to an unrealistically wide range of operation of subsystems.

A classic problem in facility specification concerns the range of engines that can be tested in one test cell using common equipment and a single shaft system. Clearly there is an operational cost advantage for the whole production range of a manufacturer's engines to be tested in any one cell. However, the detailed design problems and subsequent maintenance implications that such a specification may impose can be far greater than the cost of creating two or more cell types that are optimized for a narrower range of engines. Not only is this a problem inherent in the "turn-down" ratio of fluid services and instruments having to measure the performance of a range of engines from, say 450 to 60 kW, but the range of vibratory models produced may exceed the capability of any one shaft system.

This issue of dealing with a range of torsional vibration models may require that cells be dedicated to particular types or that alternative shaft systems are provided for particular engine types. Errors in this part of the specification and the subsequent design strategy are often expensive to resolve post-commissioning.

Not even the most demanding customer or specialized software supplier can attempt to break the laws of physics with impunity.

Before and during the specification and planning stage of any test facility, all participating parties should keep in mind the vital question:

By what cost and time-effective means do we prove that this complex facility meets the requirement and specification of the user?

At the risk of over-repetition it must be stated that it is never too early to consider the form and content of acceptance tests, since from them the designer can infer much of the detailed functional specification.

Failure to incorporate these into contract specifications from the start can lead to delays and disputes at the end.

Interpretation of Specifications by Third-Party Stakeholders

Employment of contractors with the relevant industrial experience is the best safeguard against overblown contingencies or significant omissions in quotations arising from user-generated specifications.

Provided with a well-written operational and functional specification, any competent subcontractor, experienced in the relevant area of the powertrain or vehicle test industry, should be able to provide a detailed quote and specification for their module or service within the total project.

Subcontractors who do not have experience in the industry will not be able to appreciate the special, sometimes subtle, requirements imposed upon their designs by the transient conditions, operational practices, and possible system interactions inherent in our industry. In the absence of a full appreciation of the project based on previous experience, inexperienced sales staff will search the specification for "hooks" on which to hang their standard products or designs, and quote accordingly. This is particularly true of air- or fluid-conditioning plant, where the bare parameters of temperature range and heat load can lead the inexperienced to equate test cell conditioning with that of a chilled warehouse. An escorted visit to an existing test facility should be the absolute minimum experience for subcontractors quoting for systems such as chilled water, electrical installation, and HVAC.

PART 2. MULTIDISCIPLINARY PROJECT ORGANIZATION AND ROLES

In all but the smallest test facility projects, there will be three generic types of contractor with whom the customer's project manager has to deal. They are:

- Civil contractor
- Building services contractors
- Test instrumentation contractor.

How the customer decides to deal with these three industrial groups and integrate their work will depend on the availability of in-house skills and the skills and experience of any preferred contractors.

The normal variations in project organization, in ascending order of customer involvement in the process, are:

- A consortium working within a design and build or "turnkey"[2] contract based on the customer's operational specification and working to the detailed functional specification and fixed price produced by the consortium.
- Guaranteed maximum price (GMP) contracts, where a complex project management system, having an "open" cost-accounting system, is set up with the mutual intent to keep the project within an agreed maximum value. This requires joint project team cohesion of a high order.
- A customer-appointed main contractor employing a supplier chain working to the customer's full functional specification.
- A customer-appointed civil contractor followed by services and system integrator contractor each appointing specialist subcontractors, working with the customer's functional specification and under the customer's project management and budgetary control.

---

2. The term "turnkey" is now widely misused by clients in our industrial sector. The original concept of a turnkey contract was of one carried out to an agreed and fixed specification by a contractor taking total responsibility for the site and all associated works, with virtually no involvement by the end user until the keys to the facility were handed over so that witnessed acceptance tests could be performed.

- A customer-controlled series of subcontract chains working to the customer's detailed functional specification, project engineering, site and project management.

Whichever model is chosen, the two vital roles of project manager and design authority (systems integrator) have to be clear to all and provided with the financial and contractual authority to carry out their allotted roles.

Project Roles and Management

The key role of the client, or user, is to invest great care and effort into the creation of a good operational and functional specification. Once permission to proceed has been given, based on this specification and budget, the client has to invest the same care in choosing the main contractor.

When the main contractor has been appointed, the day-to-day role of the client user group should, ideally, reduce to that of attendance at review meetings and being "on call".

Nothing is more guaranteed to cause project delays and cost escalation than ill-considered or informal changes of specification detail by the client's representatives.

Whatever the project model, the project management system should have a formal system of "notification of change" and an empowered group within both the customer and contractor's organization to deal with such requests quickly. The type of form shown in Figure 1.1 allows individual requests for project change to be recorded and the implications of the change to be discussed and quantified. Change can have either a negative or positive effect on project costs and may be requested by either the client or contractor(s).

All projects have to operate within the three restraints of time, cost, and quality (content). The relative importance of these three criteria to the specific project has to be understood by the whole project management group. The model is different for each client and for each project, and however much a client may protest that all three criteria have equal weighting and are fixed, if change is introduced, one has to be a variable (Figure 1.2).

The later in the program that change, within the civil or service systems, is required, the greater the consequential effect. The effect of late changes within the control and data acquisition systems are much more difficult to predict; they may range from trivial to those requiring a significant upgrade in hardware and software, which is why a formal "change request" process is so important.

One often repeated error, which is forced by time pressure on the overall program, is to deliver instrumentation and other electromechanical equipment into a facility building before the internal environmental is suitable. In the experience of the author, it is always better to deliver such plant late and into a suitable environment, then make up time by increasing installation man hours, than it is to have incompatible trades working in the same building space and suffering the almost inevitable damage to expensive equipment.

Customer:	Variation No:
Project name:	Project No:

Details of proposed change:
(include any reference to supporting documentation)

Requirement (tick and initial as required)
Design authority required design change
Customer instruction:
Customer request:
Contractor request:
Urgent quotation required:
Customer agreed to proceed at risk:
Work to cease until variation agreed:
Requested to review scope of supply:
Other (specify):

	Name	Email/Phone No.
Contractor representative		
Authorized		

Customer representative	Name	Email/Phone No.
Authorized		

Actions:
Sales quote submitted: (date and initial)
Authorized for action: (date and initial)
Implemented: (date and initial)

FIGURE 1.1 A sample contract variation record sheet.

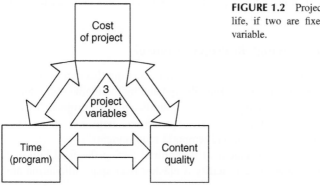

FIGURE 1.2 Project constraints: in real life, if two are fixed the third will be variable.

Cost of project

3 project variables

Time (program)

Content quality

Project Management Tools: Communications and Responsibility Matrix

Any multicontractor and multidisciplinary project creates a complex network of communications. Networks and informal subnetworks, between suppliers, contractors, and personnel within the customer's organization, may pre-exist or be created during the project; the danger is that informal communications may cause unauthorized variations in project content or timing.

Good project management is only possible with a disciplined communication system and this should be designed into, and maintained during, the project.

The arrival of email as the standard communication method has increased the need for communication discipline and introduced the need, within project teams, of creating standardized computer-based filing systems.

Web-Based Control and Communications

The proliferation of informal SMS (Short Message Service) messaging, web-based communication, and social networking tools is potentially disastrous when used for project communications. Not only does the use of such systems, on company computers or mobile phones, have the potential for confusion, but also confidentially is endangered.

There are a number of powerful Document Control software packages available to use in large multidisciplinary projects, such as those developed by NextPage® Inc. or BIW technologies Ltd.

Some corporate customers prefer to create and maintain a project-specific intranet or internet website by which the project manager has an effective means of maintaining control over formal communications. Such a network can give access permission, such as "read only", "submit", and "modify", as appropriate to individuals' roles and the nominated staff having any commercial or technical interest in the project.

The creation of a responsibility matrix is most useful when it covers the important minutiae of project work—that is, not only who supplies a given module, but who insures, delivers, offloads, connects, and commissions the module.

Use of "Master Drawing" in Project Control

The use of a common facility layout or schematic drawing that can be used by all tendering contractors, and is continually updated by the main contractor or design authority, can be a vital tool in any multidisciplinary project. In such projects there may be little detailed appreciation between specialized contractors of each other's spatial and temporal requirements.

Constant, vigilant site management is required during the final building "fit-out" phase of a complex test facility if clashes over space allocation are to be avoided, but good contractor briefing while using a common layout can reduce the inherent problem. If the systems integrator or main contractor takes ownership of project floor layout plans and these plans are used at every subcontractor meeting, and kept up to date to record the layout of all services and major modules, then most space utilization, service route, and building penetration problems will be resolved before work starts. Where possible and appropriate, contractors method statements[3] should use the common project general layout drawing to show the area of their own installation in relation to the building and installations of others.

Project Timing Chart

Most staff involved with a project will recognize a classic Gantt chart, but not all will understand the relevance of their role or the interactions of their tasks within that plan. It is the task of the project manager to ensure that each contractor and all key personnel work within the project plan structure. This is not served by sending repeatedly updated, electronic versions of a large and complex Gantt chart to all participants, but by early contract briefing and pre-installation progress meetings.

There are some key events in every project that are absolutely time critical and these have to be given special attention by both client and project manager. Consider, for example, the site implications of the arrival of a chassis dynamometer for a climatic cell:

- Although the shell building must be weather-tight, access into the chassis dynamometer pit area will have to be kept clear for special heavy handling equipment, by deliberately delayed building work, until the unit is installed; the access thereafter will be closed up.
- One or more large trucks will have to arrive on the client's site, in the correct order, and require suitable site access, external to the building, for maneuvering.
- The chassis dynamometer sections will require a large crane to offload, and probably a special lifting frame to maneuver them in place. To minimize hire costs, the crane's arrival and site positioning will have to be coordinated some hours only before the trucks' arrival.
- Other contractors will have to be kept out of the affected work and access areas for the duration, as will client's and contractor's vehicles and equipment.

---

3. There exist in the UK, and elsewhere, international trade associations that produce for their members standardized method statements that can be reused with the minimum site-specific alteration. Such purely bureaucratic exercises should be rejected by the client unless they address the key questions relating to the specifics of project timing, the site logistics, and site work supervision, as a minimum.

Preparation for such an event takes detailed planning, good communications, and authoritative site management. The non-arrival, or late arrival, of one of the key players because "they did not understand the importance" clearly causes acute problems in the example above. The same ignorance or disregard of programmed roles can cause delays and overspends that are less obvious than the above example throughout any project where detailed planning and communications are left to take care of themselves.

A Note on Documentation

Complex fluid services and electrical systems, particularly those under the control of programmable devices, are, in the nature of things, subject to detailed modification during the build and commissioning process. The final documentation, representing the "as commissioned" state of the facility, must be of a high standard and easily accessible, post-handover, to maintenance staff and subcontractors. The form and due delivery of documentation should be specified within the functional specification and form part of the acceptance criteria. Subsequent responsibility for keeping records and schematics up to date within the operator's organization must be clearly defined and controlled.

SUMMARY

The project management techniques required to build a modern test facility are the same as those for any multidisciplinary laboratory construction, but require knowledge of the core testing process so that the many subtasks are integrated appropriately.

The statement made early in this chapter, "Without a clear and unambiguous specification no complex project should be allowed to proceed", seems self-evident, yet many companies and government organizations, within and outside our industry, continue either to allocate the task inappropriately or underestimate its importance, and consequently subject it to post-order change. The result is that project times are extended by an iterative quotation period or there develops a disputatious period of modification from the point at which the users realize, usually during commissioning, that their (unstated or misunderstood) expectations are not being met.

REFERENCES

[1] UK Government Vehicle Certification Agency website: http://www.dft.gov.uk/vca/
[2] CDM Regulations (UK). Available at: http://www.hse.gov.uk/construction/cdm.htm

Quality and H&S Legislation and Management, Type Approval, Test Correlation, and Reporting of Results

Engine Testing. DOI: 10.1016/B978-0-08-096949-7.00002-9

TEST FACILITY EFFICIENCY AND QUALITY CERTIFICATION

The prime task of technical management of any test laboratory is to ensure that the test equipment is chosen, maintained, and used to its optimum efficiency in order to produce data of the quality required to fulfill its specified tasks. In many cases "efficiency" is interpreted by management as ensuring plant achieves maximum "up-time" or "shaft rotation time" figures. However, any test facility, like any individual, can work as a "busy fool" if or when the tests are badly designed or undertaken in a way that is not time and cost efficient or, much worse, produce data corrupted by some systematic mishandling or post-processing.

An important line management task in ensuring cost-effective cell use is to decide how detected faults in the unit under test (UUT) are treated. If test work is queuing up, do you attempt to resolve the problem in the cell, turning the space into an expensive workshop, or does the UUT get removed and the next test scheduled take its place? Cell use and scheduling can pose complex techno-commercial problems and the decisions taken will be substantially affected by the design of the facility (see Chapter 4).

With the possible exception of academic organizations, all test facilities carrying out work for, or within, original equipment manufacturer (OEM) organizations or for government agencies will need to be certified to the International Standards Organization (ISO) 9001 or an equivalent quality standard.

Independent confirmation that organizations meet the requirements of ISO 9001 are obtained from third party, national or international certification bodies. Such certification does not impose a standard model of organization or management, but all certified test facilities will be required to create and maintain documented processes and have the organizational positions to support them.

ISO 9001 requires a quality policy and quality manual that would usually contain the following compulsory documents defining the organizations systems for:

1. Control of documents
2. Control of records, including test results, calibration, etc.
3. Internal audits, including risk analysis, calibration certification, etc.
4. Control of nonconforming product/service, including customer contract, feedback, etc.
5. Corrective action
6. Preventive action, including training.

Small uncertified test organizations should use such a framework in their development, while directors of certified organizations have to understand that the role of quality management within their organization is not that of simply feeding a bureaucratic monster but of continuous improvement of company products and services.

MANAGEMENT ROLES

Although the organizational arrangements may differ, a medium to large test facility will employ staff having three distinctly different roles:

- Facility staff and management charged with building, maintaining, and developing the installed plant, its support services, and the building fabric.
- The internal user group charged with designing and conducting tests, collecting data, and disseminating information.
- A quality group charged with ISO 9001 certification, if applicable, internal audit, and management of instrument calibration system.

Each group will have some responsibility for two funding streams, operational and project specific. The ever narrower specialization in the user group and increasing scarcity of multidisciplinary facility or project staff were the original justification for the first edition of this book. It is noteworthy that one of the important tasks that may tend to float between groups is that of specification of new or modified facilities (see Chapter 1).

It is not unusual for test industry suppliers to negotiate and agree specifications with a purchase or facility "customer", yet deliver to a "user" with different detailed requirements. The responsibility for avoiding such wasteful practices and having a common specified requirement is that of the first level of common management.

The listed tasks are not mutually exclusive and in a small test shop may be merged, although in all but the smallest department management of the QA task, which includes the all-important responsibility for calibration and accuracy of instrumentation, should be kept distinct from line management of the users of the facility.

Figure 2.1 shows the allocation of tasks in a large test department and indicates the various areas of overlapping responsibility, along with the impact of "quality management". Periodical calibration of dynamometers, instruments, tools, and the maintenance of calibration records may be directly in the hands of the quality manager or carried out by the facilities department under his supervision; either way the procedure or computerized process must be clearly documented and imposed.

WORK SCHEDULING

A test facility producing high-quality data needs the same management controls as any other production facility of the same size and complexity producing "widgets". The cost of production will depend critically on ensuring staff, at all stages, do not reinvent test methodologies, use common I/O configurations for similar tasks, and identical calibration data. Best practices have to be developed and imposed on the process from the beginning of the

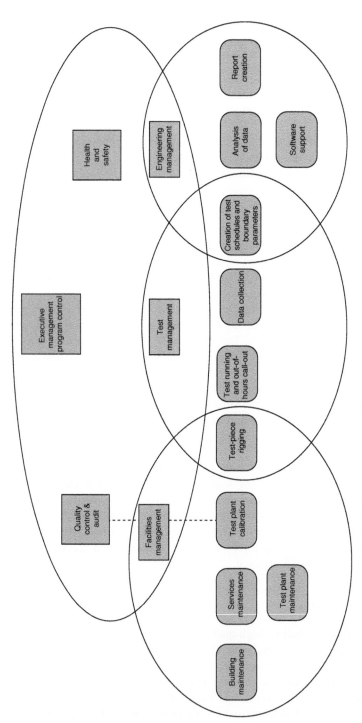

FIGURE 2.1   Allocation of tasks within a management structure of a medium/large test facility.

process (test requests or RFQ) to the end of the process (test results format and delivery).

A particular problem experienced by all test facilities, when good multi-departmental project management is lacking, is the on-schedule availability of the unit to be tested, its required rigging attachments, and its key data (ECU settings, performance limits, inertia data for shaft selection, etc.) required to set the test parameters.

There is always a need to ensure that the data produced and lessons learnt in its production are easily recalled and appropriately reused. This is not a major problem for very small test organizations with a stable workforce, because the experience acquired is held and applied by the few individual operatives, whereas in large organizations experience is very widely diffused or jealously guarded so requires a capture and dispersal process.

To solve the problem of UUT scheduling it is vital to have a test request quality gateway wherein the specifications of the test are defined, preferably using an information template. The appropriate reuse of test schedules and configurations saves time and improves the repeatable quality of the data. The management system must be capable of auditing the quality of the test request and converting it into a test schedule including cell allocation, equipment and transducer configuration, and data reporting format, all based, but not always rigidly subject to, previously used templates.

Major OEMs will use enterprise resource planning (ERP) systems, such as those developed by SAP AG for controlling their entire operations, but specialist tools developed specifically for large powertrain test facilities are also used within a global business management system, of which the Test "Factory Management Suite" developed by AVL is an example.

HEALTH & SAFETY (H&S) LEGISLATION, MANAGEMENT, AND RISK ASSESSMENT

There are very few, if any, H&S regulations that have been developed exclusively to cover powertrain test facilities; worldwide they come under general laws related to safety at work and environmental protection. Yet the application of these general industrial rules sometimes has unintended consequences and causes operational complications, as in the case of the European ATEX regulations (see Chapter 4), the New Machinery Directive (EN ISO 13849-1), and EN 62061 [1,2]. The demands of EN ISO 13849-1 regulations relevant to the automotive powertrain industry are, at the time of writing (mid 2011), under active discussion and some confusion. Many authorities believe them to require the treatment of the cell structure as a "machine guard" and therefore to require a dual-processor, "safety" PLC-based system to prevent access to the cell, unless under very specific conditions. In the numerical Safety Integrity Level (SIL) scoring required within EN 62061, typical powertrain test cells have been graded as SIL level 2 and negotiations with accredited national organizations such as TÜV are ongoing in an attempt to arrive at a mutually acceptable level of integration, practices, and the required SIL status of various levels of test facility upgrade. The legislation has not so far been applied retrospectively so existing, well-run facilities, compliant with earlier legislation, may remain unchanged.

To allow engine testing to be performed without being made impractical or prohibitively expensive, and in order to maintain our good record of safety, the industrial procedures have tended to be based on established and generally understood best practices. It is to be hoped that the new regulations can be translated so as to support successful precedent. However, where precedent does not exist, as in the use of the new technologies in hybrid and electrical powertrains and vehicles involving large batteries and battery simulation, then renewed vigilance and specific risk analysis is required.

It should be noted that the New Machinery Directive requires the good electrical practices recommended in Chapter 5 of this book, particularly those relating to equipotential bonding of electrical earth (ground) and practices aimed at the avoidance of electromagnetic interference (EMI) problems. It does, however, contain sentences such as "an emergency stop is a safety measure but not a protective device", which may confuse experienced operators familiar with using lock-out EM stop buttons.

The author recommends involvement with the forums and their websites of your national machinery manufacturing trade associations, many of which tend to give up-to-date advice on detailed compliance with these regulations.

Formal responsibility for H&S within a large organization will be that of a manager trained to ensure that policies of the company and legal requirements are adhered to by the supervisory organization. However, everyone employed by or visiting a test facility has responsibilities (under the law in the UK) in this regard.

COMMON HAZARDS IN ALL POWERTRAIN FACILITIES

The vast majority of accidents in engine test facilities do not result in human injury because of compliance with the "rule" relating to the test cell having to form a hazard containment box. Reported injuries are very largely confined to those caused by poor housekeeping, such as slipping on fluid-slicked surfaces, tripping on cables or pipes, and accidental contact with hot surfaces.

The two most common, serious malfunction incidents experienced in the last 20 years are:

1. *Shaft failures*, caused by inappropriate system design and/or poor assembly.
2. *Fire initiated in the engine unit*, in the last 10 years more commonly caused by fuel leakages from high-pressure (common rail) engine systems, probably the result of poor system assembly or modification.

It follows, therefore, that a high standard of test assembly and checking procedures together with the design and containment of shafts, plus staff training in the correct actions to take in the case of fire, are of prime importance.

An explosive release into the cell of parts of rotating machinery, other than those resulting from a shaft failure, is rarer than many suppose, but internal combustion (IC) engines do occasionally throw connecting rods and ancillary units do vibrate loose and throw off drive-belts. In these cases the debris and the consequent oil spillage should be contained by the cell structure and drainage system, and humans should, through robust interlocks, good operating procedure, and common sense, be kept outside the cell when running, above idle speed, is taking place.

Incidents of *electric shock* in well-maintained test facilities have been rare, but with the increasing development of hybrid and electric vehicle power-trains there must be an increasing danger of electric shock and electrical burns. Large battery banks, battery simulations, and super-capacitors, like all energy storage devices, must be treated with great caution and the testing of, or with, any such device should be subject to risk analysis and appropriate training.

RISK ANALYSIS

Risk can be defined as the danger of, or potential for, injury, financial loss, or technical failure. While H&S managers will concentrate on the first of these, senior managers have to consider all three at the commencement of every new testing enterprise or task.

The legislatively approved manner of dealing with risk management is to impose a process by which, before commencement, a responsible person has to carry out and record a risk assessment. The requirements, relating to judgment of risk level, of the Machinery Directive EN ISO 13849-1, which replaces EN 954-1, are shown in Figure 2.2.

Risk assessment is not just a "one-off" paper exercise, required by a change in work circumstances; it is a continuous task, particularly during complex projects where some risks may change by the minute before disappearing on task completion.

Staff involved in carrying out risk assessments need to understand that the object of the exercise is less about describing and scoring the risk but much more about recognizing and putting in place realistic actions and procedures that eliminate or reduce the potential effects of the hazard.

Both risks of injury (acute), such as falling from a ladder, and risks to health (chronic), such as exposure to carcinogenic materials, should be considered in the risk assessments, as should the risks to the environment, such as fluid leaks, resulting from incidents that have no risk to human well-being.

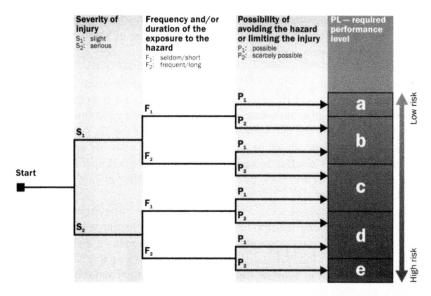

FIGURE 2.2 Risk assessment performance levels as per EN ISO 13849-1. Refer to directive documentation for details of types of action required for each level of PL in particular applications.

There are important events within the life cycle of test facilities when H&S processes and risk assessment should be applied:

- Planning and pre-start[1] stages of a new or modified test facility, both project specific and operational.
- At the change of any legislation explicitly or implicitly covering the facility.
- Service, repair, and calibration periods by internal or subcontract staff.[2]
- A significantly different test object or test routine such as those requiring unmanned running or new fuels.
- Addition of new instrumentation.

The formal induction of new staff joining a test facility workforce and the regular review of the levels of training required with its development are important parts of a comprehensive H&S and environmental policy. In any company that carries out an annual appraisal of staff, the subject of training will be under review by both management and staff member; where no such policy exists, training should be the formal responsibility of line management.

---

1. It is sound common sense, and in the UK required by CDM regulations, that the customer is responsible for advising the contractors of any pre-existing conditions at the site that may affect any work, and therefore any risk assessment, they carry out.

2. The provision of a risk assessment by a subcontractor does not abrogate the responsibility of the site management for H&S matters related directly or indirectly to the work being done by the subcontractor. The quality of the assessment and the adherence to the processes described therein needs to be checked.

MANAGEMENT AND SUPERVISION OF UNIVERSITY TEST FACILITIES

The management and operational structures of powertrain test laboratories within universities differ from those of industrial facilities, as does the relevant training and experience of the facility user group. Housekeeping seems, to the author, to be a particular problem in many academic automotive test cells where, because of lack of storage space, it is not uncommon to find work spaces cluttered by stored equipment. In some cases this clutter causes blockage to human access or escape and adds to the facility fireload.

Housekeeping is a matter of primary safety, while some forms of physical guarding, which is often given greater management attention, can be of secondary concern.

To qualify for access to the test facility every student and staff member should have been taken through an appropriate formal facility safety briefing.

A recent "danger" for some academic test cell users is that the composite management structure will become too risk averse to allow all but very conservative engine running. Replacing such valuable practical experience with computer-based simulation should not be the intention of risk assessments or teaching staff. The rigorous enforcement and use by the senior manager of the use of test cell log book (see next section) will help overcome the inherent dangers of the sometimes tortuous communication paths in academic organizations and the frequent changing of the student body; it is strongly recommended.

The author has observed that in both university and government organizations there is too frequently an organizational fracture, in the worst cases verging on open hostility, between laboratory user groups and their internal facility maintenance group (estates department). Such situations and the time, effort, and fund wasting they cause has, from time to time, been a source of frustration and amazement to many contractors involved in affected facility construction and modification projects. It has been observed by the author that attention to instrument calibration routines in some university test laboratories is lax and ill prepares students for the rigors of industrial test work.

USE AND MAINTENANCE OF TEST CELL LOG BOOKS

In the experience of the author nothing is more helpful to the safe, efficient, and profitable operation of test cells than the discipline of keeping a proper log book that contains a summary of the maintenance and operational history of both the cell and the tests run therein.

The whole idea of a written log book may be considered by many to be old fashioned in an age in which the computer has taken over the world. However, test cell computer software is not usually designed to report those subjective recordings of a trained technician, nor are the records and data they hold always available to those who require them, either because the system is shut down or they are barred by a security system from its use, or data are hidden within the computer or network.

Where do the technicians, who worked until 1 a.m. fixing a problem, or what is more important not fixing a problem, leave a record of the cell's status to be read by the morning shift operator, if not in the log book? Do they use a post-it note stuck on a monitor, or perhaps an SMS text message to a colleague, who may or may not read it?

The log book is also a vital record of all sorts of peripheral information on such matters as safety, maintenance, suspected faults in equipment or data recording and, last but most important, as an immediate record of "hunches" and intuitions arising from a consideration of perhaps trivial anomalies and unexpected features of performance.

The facility accountant will benefit from a well-kept log book, since it should provide an audit trail of the material and time consumed by a particular contract and record the delays imposed on the work by "nonbookable" interruptions.

It must be obvious that to be of real benefit the log book must be read and valued by the facility management.

TEST EXECUTION, ANALYSIS, AND REPORTING

The execution, analysis, and reporting of a program of tests and experiments are difficult arts and involve a number of stages:

- First, the experimental engineer must understand the questions that his experiment is intended to answer and the requirements of the "customer" who has asked them.
- There must be an adequate understanding of the relevant theory.
- The necessary apparatus and instrumentation must be assembled and, if necessary, designed and constructed.
- The experimental program must itself be designed with due regard to the levels of accuracy required and with an awareness of possible pitfalls, misleading results, and undetected sources of error.
- The test program is executed, the engineer keeping a close watch on progress.
- The test data are reduced and presented in a suitable form to the "customer" and to the level of accuracy required.
- The findings are summarized and related to the questions the program was intended to answer.

Finally, the records of the test program must be put together and stored in a coherent form so that, in a year's time, when everyone concerned has forgotten the details, it will still be possible to find the data and to understand exactly what was done. Test programs are very expensive and often throw up information the significance of which is not immediately apparent, but which can prove to be of great value at a later date.

Formal reports may follow a similar logical sequence:

1. Objective of experimental program
2. Essential theoretical background
3. Description of equipment, instrumentation, and experimental method
4. Calculations and results
5. Discussion, conclusions, and recommendations.

In writing the report, the profile of the customer must be kept in mind.

A customer who is a client from another company will require rather different treatment from one who is within the same organization. There will be common characteristics. The customer:

- will be a busy person who requires a clear answer to specific questions;
- will probably not require a detailed account of the equipment—but will need a clear and accurate account of the instrumentation used and the experimental methods adopted;
- will be concerned with the accuracy and reliability of the results;
- must be convinced by an intelligent presentation that the problem has been understood and the correct answers given.

Although English is commonly used worldwide for papers and reports on powertrain development, native writers of English should be aware of the problems caused to readers for whom English is a second or third language; they should avoid idiomatic phrases and, most importantly, give the full meaning of acronyms.

DETERMINATION OF CAUSE AND EFFECT

Test engineers spend much of their working life determining the difference between cause and effect. Both in isolating the value of design changes observed through test results or trying to find the cause of a system malfunction, test and commissioning staff have to develop an intelligently applied skepticism. All instruments tend to be liars but even when the data is true, the cause of an effect observed within complex systems such as those discussed in this book can be difficult to determine, even counter-intuitive.

The Latin "tag" that should be in every test engineer's notebook is "*Post hoc, ergo propter hoc*", meaning "*after this, therefore because of this*". It has probably been used in the teaching of logic for millennia and is a very tempting logical fallacy much practiced by engineers today. It is an example of *correlation not causation*. The author, during his years of engine and test facility fault-finding, has also found useful the medical aphorism "When you hear hoofbeats, think horses not zebras",[3] now replaced by the cruder "KISS"—Keep It Simple, Stupid.

---

3. Attributed to Dr Theodore Woodward, University of Maryland.

KEY LIFE TESTING

Unless testing directly relates to stresses and wear of the powertrain's experiences in real life it will not discover how to correctly optimize the UUT to meet the customers' requirements, nor will it reveal inherent faults or modes of failure before they are discovered by those customers.

The problem is that the real life of a powertrain system, on average, lasts for several years while the pressure on test engineers is to reduce development test times from months to weeks.

Once a powertrain system has been calibrated to run within legislative limits and fit within a vehicle shell, the key operational variable becomes the end user and his or her perception of drivability. The design and marketing of vehicles goes some way in dealing with this variability by targeted model variants. Thus, a medium-sized car produced by a major manufacturer will have a range of models specifically calibrated for its target market and the archetypical driving characteristics of users within that target group will be used within the models used to calibrate the powertrain.

There is much work being done to ensure that testing, particularly durability test sequences, represent the envisaged real-life operating conditions and then to accelerate that set of conditions so that the UUT can fulfill its design life of 150,000 miles (240,000 km) in weeks or months rather than 10 years: this is the basis of key life testing.

Key life testing is carried out at component, subassembly, and transmission test rigs probably to a greater extent than with engine test facilities. At the lower, subassembly test level the techniques are called highly accelerated life testing (HALT) and might involve repeated thermal cycling or load cycling; special rigs for component HALT testing are to be seen at Tier 1 and Tier 2 production facilities.

While key life test sequences are a significant advance in the drive to reduce testing costs in the development process, test management must audit and have confidence in the veracity of models used in expensive endurance sequences, before test commencement. To quote an expert correspondent: "The greater the degree of test acceleration, the further the test will tend to depart from realistic duty conditions."

Once developed to production prototype the emphasis of testing becomes type approval, homologation and, when into full production, conformation of production (COP).

VEHICLE AND VEHICLE SYSTEMS TYPE APPROVAL, HOMOLOGATION, AND CONFIRMATION OF PRODUCTION

This is a complex area of international and national legislation that has been subject to change within the time taken to write this volume; references listed later in this chapter will lead readers to websites that should allow the latest status to be discovered.

In essence, in order to be sold around the world vehicles have to meet the many and varied environmental and safety standards contained in regulations in each country in which they are sold. These standards not only define exhaust emission limits of the powertrain systems, but also the performance and construction standards of vehicle components and the whole vehicle (see Figure 2.3).

After leading the way with the introduction of exhaust emission standards, the USA and Canada are currently (2011) trailing Europe in defining harmonized tests and standards for the full range of whole vehicles.[4] Thus, the following comments center on the most widely used source of type approval standards that are contained in the directives and regulations issued by the European Community (EC) and the United Nations Economic Community for Europe (ENECE). Type approval requires certification by an external approval authority; in the UK it is the Vehicle Certification Agency (VCA).

For example, electric vehicle type approval is currently covered by Regulation No. 100 of the UNECE "Uniform provisions concerning the approval of battery electric vehicles with regard to specific requirements for the construction, functional safety and hydrogen emission".

Type approval is gained by both the homologation of the production version and by the approval of a quality plan that ensures that subsequent production models continue to meet the type approved standard.

For the engine or vehicle manufacturer, homologation is a complex and expensive area of activity since all major versions of the vehicle must meet the formal requirements that are in force in each country in which the vehicle is to be sold. Whereas much of the test methodology for homologation is laid down by regulation covering emission and fuel consumption drive cycles, then "drive-by" noise tests, the COP tests are decided upon during the formulation of an externally approved COP quality plan. COP testing of engines seems, correctly, to have become the task of quality assurance test beds and in some cases displaced other tasks such as component quality investigations. Whatever the constraints on test capacity, COP testing is, in the author's opinion, a core OEM task unsuitable for contracting to third-party facilities.

CELL-TO-CELL CORRELATION

It is quite normal for the management of a test department to wish to be reassured that all the test stands in the test department "give the same answer". Therefore, it is not unusual for an attempt to be made to answer this question of cell correlation by the apparently logical procedure of testing the same engine on all the beds. The outcome of such tests, based on detailed comparison of

---

4. European whole vehicle type approvals exist for: passenger cars, motorcycles, bus and coaches, goods vehicles, ambulances, motor caravans, trailers, agricultural and forestry vehicles.

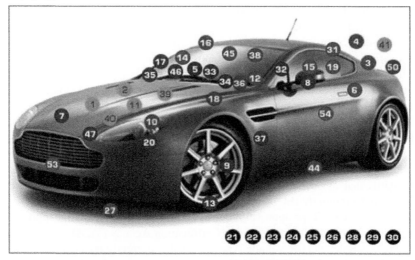

Environment

01. Sound Levels EC 2007/34
02. Emissions EC 2003/76
11. Diesel Smoke EC 2005/21
39. Fuel Consumption EC 2004/3
40. Engine Power EC1999/99
41. Diesel Emissions EC 2008/74

Active Safety

05. Steering Equipment EC 1999/7
07. Audible Warning EC 70/388
35. Wash / Wipe.EC 94/68
13. Antitheft EC 95/56
32. Forward Vision EC 90/630
17. Speedometer and Reverse Gear EC 97/39
08. Rear Visibility EC 2005/27
46. Tyres EC 2005/11
34. Defrost / Demist EC 78/317
09. Braking EC 2002/78
20. Lighting Installation EC 2008/89
33. Identification of Controls EC 94/53
37. Wheel Guards EC 94/78

Passive Safety

19. Safety Belt Anchorage EC 2005/41
16. Exterior Projections EC 2007/15
15. Seat Strength EC 2005/39
14. Protective Steering EC 91/662
03. Fuel Tank EC 2006/20
12. Interior Fittings EC 2000/4
31. Safety Belts EC 2005/40
06. Door Latches and hinges EC 2001/31
38. Head restraints EC 78/932
45. Safety glazing EC 2001/92
53. Frontal impact EC 1999/98
54. Side impact EC 96/27

Lighting Equipment

21. Reflex Reflectors EC 97/29
22. Side, Rear and Stop lamps EC 97/30
23. Direction indicator lamps EC 1999/15
24. Rear registration plate lamp EC 97/31
25. Headlamps (including bulbs) EC 1999/17
26. Front fog lamps EC 1999/18
28. Rear fog lamps EC 1999/14
29. Reversing Lamps EC 97/32
30. Parking Lamps EC 1999/16

Other Directives

27. Towing Hooks EC 96/64
10. Radio Interference Suppression EC 2009/19
04. Rear Registration Plate EC 70/222
18. Statutory Plates EC 78/507
36. Heating systems 2004/78
44. Masses and Dimensions EC 95/48
50. Mechanical Couplings EC 94/20

FIGURE 2.3 Passenger car image showing the European Community (EC) regulation references. *(Taken from VCA documentation.)*

results, is invariably a disappointment. Over many years, such tests have led to expensive disputes between the test facility management and the suppliers of the test equipment.

The need to run COP tests in test facilities distant and different from the homologation test cells has again raised the problem of variability of engine test results.

It cannot be too strongly emphasized that an IC engine, however sophisticated its management system, is not suitable for duty as a standard source of torque.

As discussed in Chapters 6 and 14, very substantial changes in engine performance can arise from changes in atmospheric conditions. In addition, engine power output is highly sensitive to variations in fuel, lubricating oil and cooling water temperature, and it is necessary to equalize these very carefully, over the whole period of the tests, if meaningful comparisons are to be hoped for.

Finally, it is unlikely that a set of test cells will be totally identical: apparently small differences in such factors as the layout of the ventilation air louvres and in the exhaust system can have a significant effect on engine performance.

Over the last 25 years there have been several correlation exercises carried out by different companies that involved sending a sealed "golden engine" to run identical power curve sequences in test cells around the world. Very few, if any, of the detailed results have ever been released into the public domain, simply because they were judged unacceptable.

As has been demonstrated by the test procedures of fuel and lube certification, to gain acceptable correlation results the "round robin"[5] experiment has to be carried out using a pallet mounted engine that is fully rigged with all its fluid, air, and immediate exhaust systems. The engines need to be fully fitted with the test transducers and each test cell should have combustion air supplied at a standard temperature. Using a reference fuel and making allowances for cell air ambient air pressure, correlation on the power curve points of better than $\pm1.5\%$ should be possible.

A fairly good indication of the impossibility of using an engine as a standard in this way is contained in the Standard BS 5514 [3]. This Standard lists the "permissible deviation" in engine torque as measured repeatedly during a single test run on a single test bed as 2%. This apparently wide tolerance is no doubt based on experience and, by implication, invalidates the use of an engine to correlate dynamometer performance.

A more recently observed reason for apparent differences in cell-to-cell data was caused by the poor management and loosely controlled use of data

---

5. This term is now used to describe any (testing) activity in which a group of facilities are involved singularly and in a circular order.

correction factors embedded in either the data acquisition system or, in the case of AC dynamometer systems, the drive control system.

There is no substitute for the careful and regular calibration of all the machines, and while cell-to-cell correlation testing is widely practiced, the results have to be judged on the basis of the degree of control over all critical variables external to the engine combustion chambers and by staff with practical experience in this type of exercise.

END-OF-LIFE VEHICLES (ELV) DIRECTIVE

The ELV Directive aims to reduce the amount of waste produced from vehicles when they are scrapped. This will not directly affect the working lives of most powertrain test engineers but all should be aware of the general idea and the International Material Data System (IMDS), which is a centralized data collection effort supported by the majority of the world's largest automakers and many smaller ones. It is a free and accessible online resource (http://www.mdsystem.com/) for manual data entry and accessible for electronic uploads for a fee.

POWER TEST CODES AND CORRECTION FACTORS

Several complex sets of rules are in general use for specifying the procedure for measuring the performance of an engine and for correcting this to standard conditions. Figure 2.3 shows diagrammatically the regulation references to the various light vehicle parts.

The most significant regulations covering powertrain test procedures in Europe are covered by ISO 15550 and ISO 3046 [4]; it is possible to purchase the full documentation directly from the ISO website (http://www.iso.org/iso/home.html). Quoting from that site:

ISO 15550 establishes the framework for ISO engine power measurement standards. It specifies standard reference conditions and methods of declaring the power, fuel consumption, lubricating oil consumption and test methods for internal combustion engines in commercial production using liquid or gaseous fuels. It applies to reciprocating internal combustion engines (spark-ignition or compression-ignition engines) but excludes free piston engines and rotary piston engines. These engines may be naturally aspirated or pressure-charged either using a mechanical pressure-charger or turbocharger.

ISO 3046-1:2002 specifies the requirements for the declaration of power, fuel consumption, lubricating oil consumption and the test method in addition to the basic requirements defined in ISO 15550.

So perhaps the best starting point for the test engineer who finds himself involved in type testing or elaborate acceptance tests is a study of these Standards. There are six parts, briefly summarized below.

Part 1. Standard Reference Conditions, Declarations of Power, Fuel and Lubricating Oil Consumptions, and Test Methods

In this Part standard atmospheric conditions are specified as follows:

1. Atmospheric pressure 1 bar ($=$ 750 mmHg)
2. Temperature 25 °C (298 K)
3. Relative humidity 30%.

The American SAE standards specify the same conditions (for marine engines standard conditions are specified as 45 °C (113 °F) and 60% relative humidity).

Specific fuel consumptions should be related to a lower calorific value (LCV) of 42,700 kJ/kg; alternatively, the actual LCV should be quoted.

The various procedures for correcting power and fuel consumption to these conditions are laid down and a number of examples are given. These procedures are extremely complicated and not easy to use: they involve 45 different symbols, 17 equations, and 10 look-up tables. The procedures in the SAE standards listed above are only marginally less complex. Rigorous application of these rules involves much work and for everyday test work the "correction factors" applied to measurements of combustion air consumption (the prime determinant of maximum engine power output) described in Chapter 14 should be perfectly adequate.

Definitions are given for a number of different kinds of rated power: continuous, overload, service, ISO, etc., and a long list of auxiliaries that may or may not be driven by the engine is provided. It is specified that in any declaration of brake power the auxiliaries operating during the test should be listed and, in some cases, the power absorbed by the auxiliary should be given.

The section on test methods gives much detailed advice and instruction, including tables listing measurements to be made, functional checks and tests for various special purposes. Finally, Part 1 gives a useful checklist of information to be supplied by the customer and by the manufacturer, including the contents of the test report.

Part 2. Test Measurements

This Part discusses "accuracy" in general terms but consists largely of a tabulation of all kinds of measurement associated with engine testing with, for each, a statement of the "permissible deviation". The definition of this term is extremely limited: it defines the range of values over which successive measurements made during a particular test are allowed to vary for the test to be valid. These limits are not at all tight and are now outside the limits required of most R&D test facilities: e.g. $\pm3\%$ for power, $\pm3\%$ for specific fuel consumption; they thus have little relevance to the general subject of accuracy.

Part 3. Speed Governing

This Part, which has been considerably elaborated in the 1997 issue, specified four levels of governing accuracy, M1–M4, and gives detailed instructions for carrying out the various tests. ISO 3046-4:2009 is the latest issue.

Part 4. Torsional Vibrations

This Part is mainly concerned with defining the division of responsibility between the engine manufacturer, the customer, and the supplier of the "set" or machinery to be driven by the engine, e.g. a generator, compressor, or ship propulsion system. In general the supplier of the set is regarded as responsible for calculations and tests.

Part 5. Specification of Overspeed Protection

This Part defines the various parameters associated with an overspeed protection system. The requirements are to be agreed between engine manufacturer and customer, and it is recommended that reset after overspeed should be manual.

Part 6. Codes for Engine Power

This Part defines various letter codes, e.g. ICN for ISO Standard Power, in English, French, Russian, and German.

It will be clear that this Standard gives much valuable guidance regarding many aspects of engine testing.

STATISTICAL DESIGN OF EXPERIMENTS

Throughout this book it is the aim to give sufficient information concerning often complex subjects to enable the reader to deal with the more straightforward applications without reference to other sources. In the case of the present topic this is scarcely possible; all that can be done is to explain its significance and indicate where detailed guidance may be found.

The traditional rule of thumb in engineering development work has been change one thing at a time. It was thought that the best way to assess the effect of a design change was to keep everything else fixed. There are disadvantages:

- It is very slow.
- The information gained refers only to the one factor.
- Even if we have hit on the optimum value it may change when other factors are changed.

While many modern design of experiment (DoE) techniques and algorithms capable of optimizing multivariate problems are embedded in modern software suites that are designed to assist in engine mapping and calibration (see Chapter 13), it is considered worthwhile describing one established method of approach.

Consider a typical multivariate development problem: optimizing the combustion system of a direct injection diesel engine (Figure 2.4). The volume of the piston cavity is predetermined by the compression ratio but there is room for maneuver in:

overall diameter	D
radius	r
depth	d
angle of central cone	θ
angle of fuel jets	ϕ
height of injector	h
injection pressure	$p,$

not to mention other variables such as nozzle hole diameter and length and air swirl rate.

Clearly any attempt to optimize this design by changing one factor at a time is likely to overrun both the time and cost allocations, and the question must be asked whether there is some better way.

In fact, techniques for dealing with multivariate problems have been in existence for a good many years. They were developed mainly in the field of agriculture, where the close control of experimental conditions is not possible and where a single experiment can take a year. Published material from statisticians and biologists naturally tended to deal with examples from plant and animal breeding and did not appear to be immediately relevant to other disciplines.

The situation began to change in the 1980s, partly as a result of the work of Taguchi, who applied these methods in the field of quality control. These

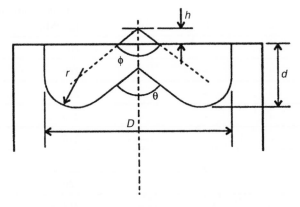

FIGURE 2.4 Diesel engine combustion system.

ideas were promoted by the American Supplier Institute, and the term "Taguchi methods" is often applied loosely to any industrial experiment having a statistical basis. A clear exposition of the technique as applied to the sort of work with which the engine developer is concerned is given by Grove and Davis [5], who applied these methods in the Ford Motor Company.

Returning to the case of the diesel combustion chamber, suppose that we are taking a first look at the influence of the injector characteristics:

included angle of fuel jets ϕ
height of injector h
injection pressure p.

For our initial tests we choose two values, denoted by + and −, for each factor, the values spanning our best guess at the optimum value. We then run a series of tests in accordance with Table 2.1. This is known as an orthogonal array, the feature of which is that if we write +1 and −1 for each entry the sum of the products of any two columns, taken row by row, is zero.

The final column shows the result of the tests in terms of the dependent variable, the specific fuel consumption. The chosen values were:

$$\phi_- = 110°$$
$$\phi_+ = 130°$$
$$h_- = 2 \text{ mm}$$
$$h_+ = 8 \text{ mm}$$
$$p_- = 800 \text{ bar}$$
$$p_+ = 1200 \text{ bar}.$$

TABLE 2.1 Test Program for Three Parameters

Run	ϕ	h	p	Specific Consumption (g/kWh)
1	−	−	+	207
2	+	−	+	208
3	−	+	+	210
4	+	+	+	205
5	−	−	−	218
6	+	−	−	216
7	−	+	−	220
8	+	+	−	212

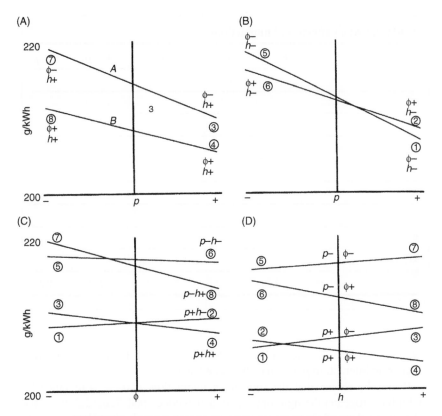

FIGURE 2.5 Relationship between parameters showing main effects.

One way of presenting these results is shown in Figure 2.5, in which each of the four pairs of results is plotted against each factor in turn. We can draw certain conclusions from these plots. It is clear that injection pressure is the major influence on fuel consumption and that there is some interaction between the three factors. There are various statistical procedures that can extract more information and guidance as to how the experimental program should continue.

The first step is to calculate the main effect of each factor. Line B of Figure 2.5A shows that in this case the effect of changing the injection pressure from p_- to p_+ is to reduce the specific consumption from 212 to 205 g/kWh. The main effect is defined as one-half this change or -3.5. The average main effect for the four pairs of tests is shown in Table 2.2.

A further characteristic of importance concerns the degree of interaction between the various factors: the interaction between ϕ and p is known as the $\phi \times p$ interaction. Referring to Figure 2.5A, it is defined as one-half the difference between the effect of p on the specific consumption for ϕ_+ and the effect for ϕ_-, or

$$1/2[(205 - 202) - (210 - 220)] = +1.5$$

TABLE 2.2 Main Effects and Interactions

Run	ϕ	h	p	$\phi \times h$	$\phi \times p$	$h \times p$
1	−	−	+	+	−	−
2	+	−	+	−	+	−
3	−	+	+	−	−	+
4	+	+	+	+	+	+
5	−	−	−	+	+	+
6	+	−	−	−	−	+
7	−	+	−	−	+	−
8	+	+	−	+	−	−
Main effect				Interaction		
	−1.75	−0.25	−4.5	−1.5	+0.75	+0.25

It would be zero if the lines in Figure 2.5A were parallel. It may be shown that the inverse interaction $p \times \phi$ has the same value.

It may be shown that the sign of an interaction + or − is obtained by multiplying together the signs of the two factors concerned. Table 2.2 shows the calculated values of the three interactions in our example. It will be observed that the strongest interaction, a negative one, is between h and ϕ. This makes sense, since an increase in h coupled with a decrease in ϕ will tend to direct the fuel jet to the same point in the toroidal cavity of the combustion chamber.

This same technique for planning a series of tests may be applied to any number of factors. Thus an orthogonal table may be constructed for all seven factors in Figure 2.5, and main effects and interactions calculated. The results will identify the significant factors and indicate the direction in which to move for an optimum result. Tests run on these lines tend to yield far more information than simple "one variable at a time" experiments. More advanced statistical analysis, beyond the scope of the present work, identifies which effects are genuine and which the result of random variation.

A more elaborate version of the same method uses factors at three levels +, 0, and −. This is a particularly valuable technique for such tasks as the mapping of engine characteristics, since it permits the derivation of the coefficients of quadratic equations that describe the surface profile of the characteristic. Figure 2.6 shows a part of an engine map relating speed, torque, and fuel consumption. The technique involves the choice of three values, equally spaced, of each factor. These are indicated in the figure.

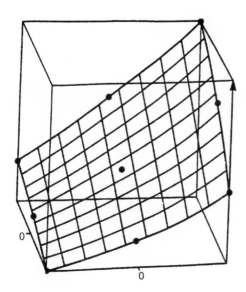

FIGURE 2.6 Engine map, torque, speed, and fuel consumption.

It is hoped that the student reader will gain some impression of the value of these methods and gain some insight into the three-dimensional graphs that frequently appear in papers relating to engine testing, from the above brief treatment.

REFERENCE DOCUMENTS

[1] BS EN 62061. Safety of Machinery. Functional Safety of Safety-Related Electrical, Electronic and Programmable Electronic Control Systems (2005).
[2] ISO/TR 23849. Guidance on the Application of ISO 13849-1 and IEC 62061 in the Design of Safety-Related Control Systems for Machinery (2010).
[3] BS 5514 Parts 1 to 6. Reciprocating Internal Combustion Engines: Performance.
[4] ISO 3046–1. Reciprocating Internal Combustion Engines—Performance. Part 1: Declarations of Power, Fuel and Lubricating Oil Consumptions, and Test Methods (2002).
[5] D.M. Grove, T.P. Davis, Engineering Quality and Experimental Design, Longmans, London, 1992.

USEFUL WEBSITES AND REGULATIONS

BS AU 141a. Specification for the Performance of Diesel Engines for Road Vehicles.
International Organization for Standardization: www.iso.org/iso/home.html
Rulebook, Chapter 8. Lloyd's Register of Shipping, London.
SAE J1995 Standard. Engine Power Test Code—Engine Power and Torque Certification (2007).

FURTHER READING

T.B. Barker, Engineering Quality by Design: Interpreting the Taguchi Approach, CRC Press, 1990. ISBN-13: 978-0824782467.

The Test Cell as a Thermodynamic System

The energy of the world is constant; the entropy strives towards a maximum.[1]

Rudolph Clausius (1822–1888)

INTRODUCTION

The closed engine test cell system makes a suitable case for students to study an example of the flow of heat and change in entropy. In almost all engine test cells the vast majority of the energy comes into the system as highly concentrated chemical energy entering the cell by way of the smallest penetration in the cell wall, the fuel line. It leaves the cell as lower grade heat energy through the largest penetrations: the ventilation ducts, engine exhaust pipes, and the cooling water pipes. In the case of cells fitted with electrically regenerative dyna-mometers, almost one-third of the energy supplied by fuel will leave the cell as electrical energy able to "slow down" the electrical energy supply meter.

Many problems are experienced in test cells worldwide when the thermo-dynamics of the cell have not been correctly catered for in the design of cooling systems. The most common problem is high air temperature within the test cell, either generally or in critical areas. The practical effects of such problems will

---

1. Der Energie der Welt ist konstant; die Entropy der Welt strebt einem Maximum zu.

Engine Testing. DOI: 10.1016/B978-0-08-096949-7.00003-0

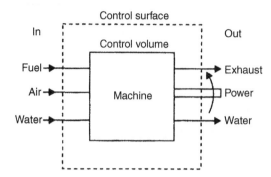

FIGURE 3.1 An open thermo-dynamic system.

be covered in detail in Chapter 6, but it is vital for the cell designer to have a general appreciation of the contribution of the various heat sources and the strategies for their control.

In the development of the theory of thermodynamics, much use is made of the concept of the open system. This is a powerful tool and can be very helpful in considering the total behavior of a test cell. It is linked to the idea of the control volume, a space enclosing the system and surrounded by an imaginary surface, the control surface (Figure 3.1).

The great advantage of this concept is that once one has identified all the mass and energy flows into and out of the system, it is not necessary to know exactly what is going on inside the system to draw up a "balance sheet" of inflows and outflows.

The various inflows[2] and outflows to and from a test cell are given in Table 3.1.

Balance sheets may be drawn up for fuel, air, water, and electricity, but by far the most important is the *energy balance*, since every one of these quantities has associated with it a certain quantity of energy. The same idea may be applied to the engine within the cell. This may be pictured as surrounded by its own control surface, through which the flows given in Table 3.2 take place.

Measurement of thermal losses from the engine is dealt with in Chapter 21, where the value of the method in the analysis of engine performance is made clear.

THE ENERGY BALANCE OF THE ENGINE

Table 3.3 shows a possible energy balance sheet for a cell in which a gasoline engine is developing a steady power output of 100 kW.

---

2. Compressed air may be considered as an energy input in facility design; however, for individual cell energy flow purposes its use tends to be intermittent and makes an insignificant contribution.

TABLE 3.1 Inflows and Outflows To and From a Test Cell

In	Out
Fuel	Ventilation air
Ventilation air (some may be used by the engine as combustion air)	Exhaust (includes air used by engine)
Combustion air (treated)	Engine cooling water
Charge air (when separately supplied)	Dynamometer cooling water or air
Cooling water	Electricity from dynamometer
Electricity for services	Losses through walls and ceiling

Note that where fluids (air, water, exhaust) are concerned the energy content is referred to an arbitrary zero, the choice of which is unimportant: we are only interested in the difference between the various energy flows into and out of the cell.

Given sufficient detailed information on a fixed engine/cell system, it is possible to carry out a very detailed energy balance calculation (see Chapter 6 for a more detailed treatment). Alternatively, there are some commonly used "rule-of thumb" calculations available to the cell designer; the most common of these relates to the energy balance of the engine, which is known as the "30–30–30–10 rule". This refers to the energy balance given in Table 3.4.

The key lesson to be learnt by the nonspecialist reader is that: any engine test cell has to be designed to deal with energy flows that are at least three times greater than the "headline" engine rating. To many, this will sound obvious but a common fixation on engine power and a casual familiarity with, but lack of appreciation of, the energy density of petroleum fuels still lead people to significantly underrate cell cooling systems.

TABLE 3.2 Inflows and Outflows To and From an Engine

In	Out
Fuel	Power
Air used by the engine	Exhaust
Cooling water	Cooling water
Cooling air	Cooling air
	Convection and radiation

TABLE 3.3 Simplified Energy Flows for a Test Cell Fitted with a Hydraulic Dynamometer and 100 kW Gasoline Engine

Energy Balance			
In		**Out**	
Fuel	300 kW	Exhaust gas	60 kW
Ventilating fan power	5 kW	Engine cooling water	90 kW
		Dynamometer cooling water	95 kW
		Ventilation air	70 kW
Electricity for cell services	25 kW	Heat loss, walls and ceiling	15 kW
	330 kW		330 kW

The energy balance for the engine is as follows:

In		**Out**	
Fuel	300 kW	Power	100 kW
		Exhaust gas	90 kW
		Engine cooling water	90 kW
		Convection and radiation	20 kW
	300 kW		300 kW

Like any rule of thumb this is crude, but it does provide a starting point for the calculation of a full energy balance and a datum from which we can evaluate significant differences in balance caused by the engine itself and its mounting within the cell.

TABLE 3.4 Example of the 30–30–30–10 Rule

In Via	Out Via
Fuel 300 kW	Dynamometer 30% (90+ kW)
	Exhaust system 30% (90 kW)
	Engine fluids 30% (90 kW)
	Convection and radiation 10% (30 kW)

First, there are differences inherent in the engine design. Diesels will tend to transfer less energy into the cell than petrol engines of equal physical size. For example, testers of rebuilt bus engines, which have both vertical and horizontal configurations, often notice that different models of diesels with the same nominal power output will show quite different distribution of heat into the test cell air and cooling water.

Second, there are differences in engine rigging in the cell that will vary the temperature and surface area of engine ancillaries such as exhaust pipes.

Finally, there is the amount and type of equipment within the test cell, all of which makes a contribution to the convection and radiation heat load to be handled by the ventilation system.

Specialist designers have developed their own versions of a test cell software model, based both on empirical data and theoretical calculation, all of which is used within this book. The version developed by colleagues of the author produces the type of energy balance shown in Figure 3.2. Table 3.5 lists just a selection, from an actual project, of the known data and calculated energy flows that such programs have to contain in order to produce Figure 3.2.

Such tools are extremely useful but cannot be used uncritically as the final basis of design when a range of engines need to be tested or the design has to cover two or more cells in a facility where fluid services are shared; in those cases the energy diversity factor has to be considered.

DIVERSITY FACTOR AND THE FINAL SPECIFICATION OF A FACILITY ENERGY BALANCE

To design a multi-cell test laboratory able to control and dissipate the maximum theoretical power of all its prime movers on the hottest day of the year will lead to an oversized system and possibly poor temperature control at low heat outputs. The amount by which the thermal rating of a facility is reduced from that theoretical maximum is the *diversity factor*. In Germany it is called the *Gleichzeitigkeits Faktor* and is calculated from zero heat output upwards, rather than 100% heat output downwards, but the results should be the same, providing the same assumptions are made.

The diversity factor often lies between 60% and 85% of maximum rating but individual systems will vary from endurance beds with high rating down to anechoic beds with very low rating.

In calculating, or more correctly estimating, the diversity factor it is essential that the creators of the operational and functional specifications use realistic values of actual engine powers, rather than extrapolations based on possible future developments.

A key consideration is the number of cells included within the system. Clearly one cell may at some time run at its maximum rating, but it may be considered much less likely that four cells will all run at maximum rating at the same time: the possible effect of this diminution from a theoretical maximum is

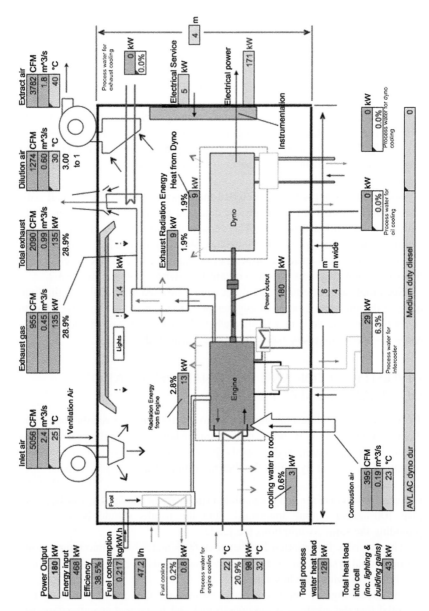

FIGURE 3.2 Output diagram from a test cell thermal analysis and energy flow program.

shown in Figure 3.3. There is a degree of bravery and confidence, based on relevant experience, required to significantly reduce the theoretical maximum to a contractual specification, but very significant savings in running costs may be possible if it is done correctly.

TABLE 3.5 A Selection of the Known, Estimated, and Calculated Data, Based on a Known Engine, Required to Produce a Complete Energy and Fluid Flow Diagram as Shown in Figure 3.2

Engine and Fuel Data			
Power output	180 kW	Engine max. power	180 kW
Fuel	Diesel	Primary energy from fuel	468 kW
Calorific value of fuel	43,000 kJ/kg	**Electricity Output**	
Density of fuel	0.830 kg/liter	From AC dyno.	171 kW
Combustion Air		**Cooling Water Loads**	
Temp. intake	23 °C	Lube oil HX (N/A)	0 kW
Temp. after compressor	185 °C	Engine jacket HX	98 kW
Temp. after intercooler	55 °C	Intercooler HX	29 kW
Combustion air temp.	70 °C	**Chilled Water**	
Exhaust Gas		Fuel cooling	1 kW
Manifold temp.	650 °C	**Cell Heat Loads** (radiation from)	
Temp. after turbine	434 °C	Engine block	13 kW
Temp. in cell system	400 °C	Exhaust in cell	9 kW
Dilution air temp.	30 °C	Cooling water	3 kW
Temp. at cell exit	100 °C	Dynamometer	9 kW
Dilution air ratio	3	**Exhaust System**	
Plant Cooling Water		Exhaust gas out of cell	135 kW
Glycol content (%)	50%		
Temperature in	22 °C		
Temperature out	32 °C		
Specific heat capacity	3.18 kJ/l K		
Density	1.06 kg/m^3		
Peak Fuel Consumption		**Exhaust Dilution**	
Mass rate	39.1 kg/h	Ratio (kg air/kg exh.)	3
Volume rate	47.2 liter/h	Mass flow at intake (air)	2519 kg/h
	0.79 liter/min	Density at intake (-50 Pa)	1.16 kg/m^3
Specific fuel consumption	0.217 kg/kWh	Volume rate at intake	2164 m^3/h

TABLE 3.5 A Selection of the Known, Estimated, and Calculated Data, Based on a Known Engine, Required to Produce a Complete Energy and Fluid Flow Diagram as Shown in Figure 3.2—cont'd

Combustion Air			0.60 m³/s
Mass rate	800 kg/h	Total mass flow (air + exh.)	3358 kg/h
Density at 1 bar abs.	1.19 kg/m³	Mixture temp. after intake	123 °C
Volume rate	672 m³/h	Mixture density out of cell	0.95 kg/m³
	0.19 m³/s	Mixture flow out of cell	3551 m³/h
Exhaust Gas			0.99 m³/s
Mass rate (air + fuel)	840 kg/h		
Density, after turbine	0.739 kg/m³		
Volume rate after turbine	1136 m³/h		
Density, out of cell	0.517 kg/m³		
Volume rate out	1622 m³/h		
of closed system	0.45 m³/s		

These types of models are a type of "expert system", the result of refinement based on experience of many man-years.

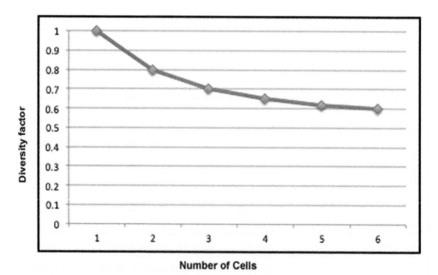

FIGURE 3.3 Diversity factor of thermal rating of facility services, plotted against number of test cells, based on empirical data from typical automotive test facilities.

"Future proofing" may be better designed into the facility by planning for possible incremental addition of plant rather than oversizing at the beginning.

COMMON OR INDIVIDUAL SERVICES IN MULTI-CELL LABORATORIES?

When considering the thermal loads and the diversity factor of a facility containing several test cells, it is necessary to decide on the strategy to be adopted in the design of the various services. The choice has to be based on the operation requirements rather than just the economies of buying and running the service modules.

Services such as cooling water (raw water) are always common and treated through a central cooling tower system. However, services such as cell ventilation and engine exhaust gas extraction may either serve individual cells or be shared. In these cases sharing may show cost savings and simplify the building design by reducing penetrations; it is prudent to build in some standby or redundancy to prevent total facility shutdown in the event of, for example, the failure of an induced-draught fan of a common exhaust duct.

A problem that must be avoided in the design of common services is "crosstalk" between cells where the action in one cell or in other industrial plant disturbs the control achieved in another. This is a particular danger when a service, for example chilled water, has to serve a wide range of thermal loads. In this case a central plant may be designed to circulate glycol/water mix at 6 °C through two or more cells in which the coolant is used by devices ranging from large intercoolers to small fuel conditioners; any sudden increase in demand may significantly increase the system return temperature and cause an unacceptable disturbance in the control temperatures. In such systems there needs to be individual control loops per instrument or a very high thermal inertia gained through the installation of a sufficiently large cold buffer tank.

SUMMARY

The "energy balance" approach outlined in this chapter will be found helpful in analyzing the performance of an engine and in the design of test cell services (Chapters 5–7).

It is recommended that at an early stage in the design of a new test cell, diagrams such as Figure 3.2 should be drawn up and labeled with flow and energy quantities appropriate to the capacity of the engines to be tested.

The large quantities of ventilation air, cooling water, electricity, and heat that are involved will often come as a surprise. Early recognition can help to avoid expensive wasted design work by ensuring that:

1. The general proportions of cell and services do not depart too far from accepted practice (any large departure is a warning sign)
2. The cell is made large enough to cope with the energy flows involved

3. Sufficient space is allowed for such features as water supply pipes and drains, air inlet grilles, collecting hoods, and exhaust systems—note that space is not only required within the test cell, but also in any service spaces above or below the cell and the penetrations within the building envelope.

FURTHER READING

T.D. Eastop, A. McConkey, Applied Thermodynamics for Engineering Technologists, Longmans, London, 1993. ISBN-13: 978-0582091931.

J.B. Heywood, Internal Combustion Engine Fundamentals, McGraw-Hill, Maidenhead, 1988. ISBN-13: 978-0071004992.

Chapter 4

Powertrain Test Facility Design and Construction

Engine Testing. DOI: 10.1016/B978-0-08-096949-7.00004-2

PART 1. CELL TYPES, SIZES, AND LAYOUT

As is stated elsewhere in this volume, the engine and powertrain test cell structure has always been considered as, and designed to be, a hazard containment box. Any test cell built today may be considered as "machinery" under the broad definition contained in the Machinery Directive 2006/42/EC. In this sense the cell doors are essentially now the guard for the machine. Any internal guards, such as those for the drive shaft, are essentially supplementary, i.e. additional protection to personnel when controlled access to the test cell is required and the controller is in a "reduced speed mode".

This role of the test cell as a complex programmable machine has to be kept in mind when considering all aspects of test cell design and its location within its immediate environment.

One of the early considerations in planning a new test facility will be the floor space required on one or more levels.

The interconnected areas to be considered are:

- The engine or powertrain test cell (the hazard containment space)
- The control room (whose role may differ widely between types of users and must be defined)
- The space required for cell services and support equipment
- The support workshop, UUT rig and de-rig areas
- The storage area required for rig items and consumables
- Office space for staff directly involved with the test cell(s).

Cell Sizing

A cramped cell in which there is no room to move around in safety and comfort is a permanent source of danger and inconvenience. The smaller the volume of the cell, the more difficult it is to control the ventilation system under conditions of varying load. However, unnecessarily large cells will lead to equipment, such as AC dynamometer drive cabinets, being installed in them that are not ideally suited to the highly variable conditions. Such cells also get used to house tool cabinets, leading to them being used as workshop spaces and to store occasionally used equipment, with the commensurate problems of housekeeping and the potential for damage.

As a rule of thumb, in automotive cells there should be an unobstructed walkway 1 m wide all round the rigged UUT and its connected dynamometer, but in most cases the exhaust tubes will form a barrier to a full 360° of access. This means that the access walkway is more often horseshoe shaped rather than an oval.

It is now often necessary, when testing automotive engines, to accommodate most of the exhaust system as used on the vehicle. The variation in engine and exhaust configurations to be used within a single cell may be the most significant factor in determining its size and internal configuration. Pre-planning activities need to include mock exhaust system layouts and should pay particular attention to the possible need to run some exhausts under or very close to the dynamometer plinth.

It must be remembered that much of the plant in a test cell requires regular calibration so there must be adequate access for the calibration engineer, his instruments and, in some cases, such as a fuel-weigher, a ladder.

The calibration of the dynamometer requires accommodation of a torque arm and calibration weights, usually on both sides of the machine. A classic layout error is to install electrical boxes, trunking or service pipes that clash with the hanging position of the dynamometer calibration weights.

Cell height may be determined by the provision or not of a crane beam in the structure; in practice, most modern automotive cells are between 4 and 4.5 m internal height.

A Note Concerning Lifting Beams in Test Cells

While an in-cell crane is essential when working on large machines such as medium-speed diesels, and were often in automotive cells built before 1990, they come at a construction and insurance cost, particularly in containerized or panel-built cells (see below). This added cost comes about because the whole building frame has to be strengthened to allow it to support the crane built to the allowed bending limits required by the insurer's overload capacity. Deep crane beams also lower effective cell height and can make airflow and lighting more difficult to optimize. Besides being one, albeit rather slow, method of engine mounting and de-mounting, a cell crane is certainly useful during initial installation of the cell. Subsequently crane beams can be used for, hopefully rare, major maintenance and tend to be used as a "sky-hook" for supporting various parts of the rigging looms and instrumentation. Where automotive engines are rigged on a removable pallet or trolley system, the cost/benefit calculation does not often justify a fixed crane installation.

Cells that are larger than the engine-transmission unit listed in Table 4.1 may be required for cells testing multiconfiguration transmission systems in which four-wheel dynamometers and one engine-simulating dynamometer have to be housed in various layouts. In the experience of many transmission test engineers their cells "are never big enough" and even the headroom can

TABLE 4.1 Examples of Actual Test Cell Dimensions Found in UK Industry

Dimensions	Cell Category
6.5 m long × 4 m wide × 4 m high	QA test cell for small automotive diesels fitted with eddy-current dynamometer
7.8 m long × 6 m wide × 4.5 m high	ECU development cell rated for 250 kW engines, containing workbench and some emission equipment
6.7 m long × 6.4 m wide × 4.7 m high	Gasoline engine development cell with AC dynamometer, special coolant, and inter-cooling conditioning
9.0 m long × 6 m wide × 4.2 m high (to suspended ceiling)	Engine and transmission development bed with two dynamometers in "tee" configuration. Control room runs along 9 m wall

become critical when a gear-change robot has to be rigged on a tall commercial vehicle gearbox.

Frequency of Change of Unit Under Test and Handling Systems

An important consideration in determining the size and layout of a test cell and its immediate environment is the frequency with which the UUT has to be changed.

At one extreme in the automotive sizes are "fuel and lube" cells, where the engine is virtually a permanent fixture, and at the other extreme are production hot test cells, where the engine's test duration is a few minutes and the process of rig and de-rig is fully automated.

The system adopted for transporting, installing, and removing the UUT has to be considered together with the layout and content of the support workshop, the joining corridors, and the position of the control room. All operational specifications for new or modified test facilities should include the intended frequency of change of the UUT in their first draft.

Seeing and Hearing the Unit Under Test

Except in the case of large marine diesel engines, portable dynamometer stands, and some production test beds, it is almost universal practice to have a physical barrier to separate the control space from the UUT.

Thanks to modern instrumentation and closed circuit television, a window between control room and test cell is rarely absolutely necessary, although many users will continue to specify one. The cell window causes a number of design problems; it may compromise the fire rating and sound attenuation of the cell, and uses valuable wall space in the control room.

The choice to have a window or not is often made on quite subjective grounds. Cells in the motor-sport industry all tend to have large windows because they are visited by sponsors and press, whereas original equipment manufacturer (OEM) multi-cell research facilities increasingly rely on remote monitoring and CCTV. The use of multiscreen data displays means that many cell windows have become partially blanked out by these arrays.

The importance or otherwise of engine visibility from a window is linked to a fundamental question: Which way round are the key units in the cell to be arranged? A number of variants are covered later in this chapter.

The experienced operator will be concentrating attention on the indications of instruments and display screen, and will tend to use peripheral vision to watch events in the cell by whatever means provided. It is important to avoid unnecessary visual distractions, such as dangling labels or identity tags that can flutter in the ventilation wind.

Hearing has always been important to the experienced test engineer, who can often detect an incipient failure by ear before it manifests itself through an alarm. Unfortunately modern test cells, with their generally excellent sound insulation, cut off this source of information and many cells need to have provision for in-cell microphones connected to external loudspeakers or earphones.

Containerized Test Cells

The nomenclature can be confusing since "containerized" cells have been built in several different forms, ranging from those using a modified ISO container, to a custom-designed suite of units constructed inside an existing building from prefabricated sections. Their advantage over conventional buildings is that of ease and speed of site construction rather than material cost. A further advantage, when the unit is constructed inside an already working factory, is that "wet trades"[1] are not required in their construction. The possibility of future relocation is often quoted as an advantage, but in reality it requires the unit to be specifically designed and expensively constructed for ease of later demounting.

The common feature of all true containerized cells is that they can be installed on a flat concrete base without any excavation for seismic blocks or service trenches. However, this flat-floor design feature requires a step or high sill between the floor of the cell and the concrete pad on which it sits, meaning

---

1. The wet trades referred to are those building tasks such as concrete pouring, block-laying, plastering, etc. None of this type of work is to be welcomed in a modern, working engineering production or test area.

that loading the UUT, such as engines mounted on pallets, has to be carried out by a custom-built conveyor system, a fork-lift truck, or customized lift system. Since internal space is often at a premium in these cells, service and subfloor areas tend to be very constricted, rather akin to marine engineering layouts, thus making any modification or even maintenance more difficult than in conventional buildings.

In most cases the weather-tight containerized test cells, installed outdoors, form a self-sufficient facility that requires only connections to cooling water (flow and return) and electrical power. Fuel supplies may be either part of the cell complex or plugged into an existing system. An advantage of this type of pre-packaged test cell is that the complete facility may be tested before delivery, thus giving the shortest possible installation and commissioning time at the customer's site.

The common types of cell built under the heading of "containerized" are:

1. *ISO shipping container* based, using either the standard 20-foot or 40-foot unit. These base ISO units have an unlined internal height and width of only 2.352 m, and therefore are limited in the size of test bed that can be fitted. As they are externally durable and highly portable they have been used for military test facilities, typically made up of three containers: the test cell, the control module, and the services module. These are plugged together so as to be quickly taken apart, transported and reassembled. However, this type of highly specialized and portable system has to be highly engine-range specific and is very expensive in terms of cost per volume of cell.
2. Standard "high" ($40' \times 8' \times 8'\,6''$) ISO Lloyd's A1 specification *cold store container* that has a stainless steel interior and is rated down to $-25\ °C$ may be used, either as a cost-effective form of climatic cell or as a "cold soak" facility in support of an exhaust emission test facility.

There is a range of non-ISO container-based cell types that differ from conventional concrete-built cells only by their construction materials but are still referred to as containerized cells, and they include the following variations:

3. All-weather, custom-built, panel-constructed cell and control room units are suitable for one-off or very occasional re-siting and are built for outside use. Such a construction can offer the advantage of quickly extending an existing manufacture or test facility without disruption to the main building and the need for substantial excavations. In some locations the "temporary" nature of such a facility may offer advantages in fast-tracking planning and building permits. Experience has shown that the long-term external durability of such cells has been very variable and requires careful design and choice of enclosure materials when installed in corrosive environments (tropical, saline, etc.).
4. Custom-built, panel-constructed cell units (similar to type 3) that are constructed inside an existing shell building (Figure 4.1).

FIGURE 4.1 A "building within a building" test cell and control room, constructed on existing factory floor using acoustic panels within a steel frame. Note the raised door sill. All the services, including air blast cooler, are roof mounted. *(Photo courtesy of Envirosound Ltd.)*

Both these "building within a building" type cells are made viable by the availability of industrial construction panels made of sound absorbent material sandwiched between metal sheets, of which the inner (cell) side is usually perforated. These 100-mm-thick panels are used with structural steel frames and, while not offering the same level of sound attenuation as dense block walls, give a quick and clean method of construction with a pleasing finished appearance.

It must be remembered that, when heavy equipment such as fuel conditioning units are to be fixed to the walls of such cells, precisely located "hard-mounting" points must be built into the panel structure.

5. There have been successful variations of the "building within a building" test cell construction where complete cell units have been built and pre-commissioned then craned and maneuvered into pre-prepared building compartments where pre-aligned service connections are made. However, such projects have to include multiples of identical units so that economies of scale can apply.

The Basic Minimum Engine Test Bed

There are some situations, such as in truck and bus fleet overhaul workshops, and occasionally in the specialist car after-market, in which there is a requirement to run engines under load, but so infrequently that there is no economic justification for building a permanent test cell.

To provide controlled load in such cases, a "bolt-on" dynamometer can be used (see Chapter 10 for details). These require no independent foundation and are bolted to the engine bell housing, using an adaptor plate, with a splined shaft connection engaging the clutch plate. Occasionally the dynamometer may be installed without removal of the engine from a truck chassis by dropping the propeller shaft and mounting the dynamometer on a hanger frame bolted to the chassis. This technique is useful for testing whole vehicle cooling systems without using a chassis dynamometer.

To support the needs of this sort of occasional test work all that is required is a suitable area provided with:

- Water supplies and drains with adequate flow capacity for the absorbed power (see Chapter 10)
- A portable fuel tank and supply pipe of the type sold for large marine outboard motor installations
- An adequately ventilated space, possibly out of doors
- Arrangements to take engine exhaust to exterior, if within a building
- Minimum necessary sound insulation or provision of personnel protection equipment (PPE)
- Adequate portable fire suppression and safety equipment.

Figure 4.2 shows a typical installation, consisting of the following elements:

- Portable test stand for engine
- Dynamometer mounted directly to the engine flywheel housing using a multi-model fixing frame
- Control console
- Flexible water pipes and control cable.

For small installations and short-duration tests the dynamometer cooling water may be simply run from a mains supply, providing it has the capacity and

FIGURE 4.2 Portable dynamometer installed on a truck diesel mounted on a simple wheeled support frame fitted with a vehicle-type radiator. *(Photo courtesy of Piper Test & Measurement Ltd.)*

constancy of pressure, and the warm outflow run to waste. Much more suitable is a pumped system, including a cooling system, operating from a sump into which the dynamometer drains (see Figure 7.1).

The engine cooling water temperature and pressure can be maintained by a mobile cooling column of the type shown in Figure 7.2, or by a vehicle radiator and fan as shown in Figure 4.2.

The control console requires, as a minimum, indicators for dynamometer torque and speed. The adjustment of the dynamometer flow control valve and engine throttle may be by cable linkage or simple electrical actuators. The console should also house the engine stop control and an oil pressure gauge. If required, a simple manually operated fuel consumption gauge of the "burette" type is adequate for this type of installation.

Common Variations of Multi-Cell Layouts

Test cells are often built in side-by-side multiples, but it is the chosen method and route of loading engines into the cells that can most influence the detailed layout.

In the case of the layout shown in Figure 4.3, the cells have corridors running at both ends; the "front" of the cells is taken up by a common,

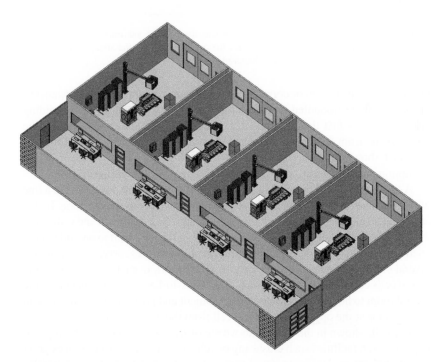

FIGURE 4.3 Cells arranged with a common control corridor at the front and with engine access at their rear.

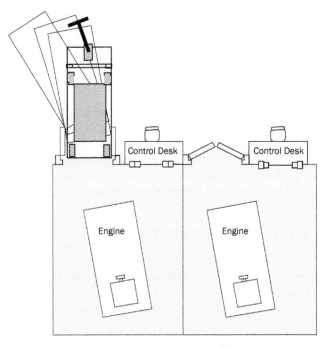

FIGURE 4.4 Side-by-side cell arrangement where the engines are transported via trolleys through the control corridor, reducing space for the control desk over the arrangement shown in Figures 4.2 and 4.4, but using a smaller footprint.

undivided, control room or corridor while the rear corridor is used for transport of rigged engines mounted on pallets or trolleys. This arrangement of building also determines that the position of the engine in the test cell will be at the opposite end from the control room and its window. It requires a larger footprint than that shown in Figure 4.4, but it keeps the control corridor clear of the disruption caused by periodic movement of engines through the space.

A third variant is shown in Figure 4.5, where the cells are positioned in a line with a common control room that is shared the by two cells on either side. In this arrangement engines enter the cell by way of a large door in the cell end wall while the operator may enter the cell by way of a door in the long wall to one side of the control desk. This arrangement finds some favor in facilities carrying out work for several different clients as it is easier to keep the work in each control room and pair of cells confidential. A variation of this "back-to-back" arrangement uses the central room to house shared emission equipment and positions the control room in a corridor at the opposite end of the cell from the engine entry.

The cells shown in the layouts above are of similar size and general type; in large research facilities such clustering of cells having similar roles makes good operational sense. However, some types of work need to be kept physically remote from others to whom they may be poor neighbors.

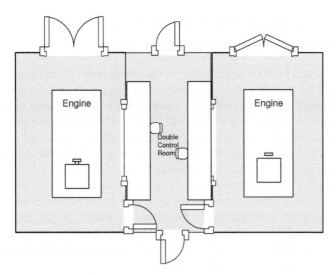

FIGURE 4.5 "Back-to-back" control room arrangement between two cells. Engines and operator access can be kept quite separate.

The prevailing wind direction should be marked on the ground plan of any engine test facility because it is important that exhaust emission facilities are kept out of any vapor plume from other exhaust systems or industrial processes emitting chemical pollutants, paint shops being a common source. Equally, NVH cells need to be as remote as possible from internal and external roadways or any source of ground transmitted vibration. Component tests using "shaker-tables" are particularly damaging to any rotating equipment having precision rotating element bearings that may suffer from brinelling. Ground-borne vibration from such facilities should be well isolated and they should be located remote from most other test cells.

Automotive R&D facilities built before the year 2000 may, a decade or more later, have had to find accommodation space for the following features, additional to their original contents, items such as:

1. Exhaust gas analysis equipment (see Chapter 16).
2. Dedicated combustion air treatment plant (see Chapter 6).
3. Ability to run variants of "in-vehicle" exhaust systems, some of which may require the dynamometer to be mounted on a raised frame in order to run exhaust lines under the shaft line to the rear of the cell.
4. High dynamic four-quadrant dynamometers, their drive cabinets, and supply systems (see Chapter 8).

The repositioning and addition of instrumentation and service plant in a suitable environment within constricted building spaces has been a feature of the lives of many test engineers over the last 10 years.

PART 2. CELL CONTENTS AND FITTINGS

Emergency Exits, Safety Signs, and Alarms

The number and position of emergency exits within both cell and control room spaces will be determined by national safety legislation and local building codes that will vary from country to country, but will be planned to give the shortest route to the building exterior. All doors and emergency exit routes have to be designated with the legislatively required battery-illuminated signs. All emergency doors must open outwards from the source of danger and their use must not be inhibited by stored or discarded objects blocking the escape route.

In the UK it is not uncommon to have only one door into the test cell as shown in Figure 4.4 but, as stated elsewhere, such facilities restrict entry to the cell when engines are running. In the USA one wall of engine test cells is normally formed by an external wall of the shell building, into which an emergency exit door is fitted with a "burst bar" opening device. Under some US building codes and in gas-fueled cells elsewhere, the cell has to be fitted with an explosion panel opening into an unoccupied external zone; the external rear cell door may also serve this purpose. Control rooms invariably have to be fitted with two entry/exit doors, as a minimum, including one designated and signed as the emergency exit.

Many test cells and control rooms are visually and audibly confusing spaces, particularly to visitors; good design and clear signage is required to minimize this confusion and thus improve the inherent safety of the facility.

The confusing use of multiple signs has become so common, particularly during facility construction, as to become a safety hazard in its own right, and is now warned against in British Standard 5944.

All workers in the facility must be trained and practiced in the required actions in the case of specified emergencies and all visitors should be briefed in the use of emergency exits in the case of any emergency identified by their host.

Transducer Boxes and Booms

The signals from each of the transducers with which the UUT is fitted (Chapter 12) flow through individual cables to a nearby marshaling box. In modern systems most signal conditioning and analog-to-digital conversion (ADC) is carried out by electronics within these transducer[2] boxes, which then send the data to the control computer system via one or more digital cables under the control of a communication bus such as the IEEE 1394 interface (Firewire). These boxes may be either mounted on an adjustable boom

---

2. The only transducers (energy converting and measuring devices) actually installed within the box are those measuring pressure, fed by Teflon lines from the UUT. Other measurements come into the dedicated conditioning cards inside the box by way of cables connected directly to transducers mounted on the UUT.

cantilevered and hinged from the cell wall, or on a pillar mounted from the bedplate but, since they may be housing very expensive electronics, due care needs taken in the support and location so that damage is not caused by gross vibration and overheating. Boom-mounted boxes that may be close to and immediately above the engine need to be force-ventilated and temperature monitored. Well-designed booms are made of hollow aluminum extrusions, of the type used in sailing-boat masts, and have a fan at the wall end blowing air through the whole structure.

Boom-mounted transducer boxes can be of considerable size and weight; because of the convenience of their location the contents and attachments tend to increase during the process of cell development.

Because of this high, cantilevered load, the ability of the wall to take the stresses imposed must be carefully checked; frequently it is found necessary to build into the wall, or attach to it, a special support pillar.

Fuel booms are often used to take fuel flow and return lines between the engine on test and wall-mounted conditioning units. These are of light construction and must not be used to attach other instrumentation or cables.

Test Cell Flooring and Subfloor Construction

The floor, or seismic block when fitted (see Chapter 7), must be provided with arrangements for bolting down the engine and dynamometer. A low-cost solution is to precisely level, and cast in position, two or more cast-iron T-slotted rails. The machined surfaces of these rails form the datum for all subsequent alignments and they must be set and leveled with great care and fixed without distortion. The use of fabricated steel-box beams has proven to be a false economy.

Sometimes complete cast-iron floor slabs containing multiple T-slots are incorporated in the concrete cell floor, but this configuration tends to trap liquids and debris unless effective drainage is provided. All cast-iron floor slabs can become very slippery so good housekeeping practices and equipment are essential.

The air-sprung bedplate discussed in detail in Chapter 9 is the modern standard design. However, the building construction plan must make arrangements for the maneuvering into place of these large and heavy objects before suitable access is closed by building work.

Concrete floors should have a surface finish that does not become unduly slippery when fluids have been spilt; special "nonslip", marine, deck paints are recommended over the more cosmetically pleasing smooth finishes. High-gloss, self-leveling floor surfaces have found wide favor in new factories but when fuels, lubricants, or cooling fluids are spilt on them they become dangerously slippery. Whatever the finish used, it must be able to resist chemical attack from the fluids, including fuels, used in the cell.

Even without the presence of a seismic block, which provides the opportunity, it is good practice to provide floor channels on each side of the bed, as they are particularly useful for running fluid services and drain pipes in a safe, uncluttered manner. However, regulations in most countries call for spaces below floor level to be scavenged by the ventilating system to avoid any possibility of the build-up of explosive vapor, thus slightly complicating the cell design (see Chapter 6).

All floor channels should be covered with well-fitting plates capable of supporting the loads that are planned to run over them. To enable easy removal for maintenance each plate should not weigh more than about 20 kg, and be provided with lifting holes. The plates can be cut as necessary to accommodate service connections.

Fuel service pipes, once commonly run underground and in floor trenches, should nowadays follow modern environmental and safety best practice, and remain in view above ground in cells and be easily accessible.

Facility and Cell Doors

Modern test facilities have to house many large electrical cabinets, all of which may be up to 2.2 m high when installed but over 2.4 m high during transportation through the building due to the additional height of pallets, skates, etc. The "simple" task of maneuvering heavy cabinets to their final installation position is frequently turned into an expensive problem due to intermediate doors of inadequate height or intermediate flooring of inadequate strength; foresight is required.

Cell doors that meet the requirements of noise attenuation and fire containment are inevitably heavy and require more than normal effort to move them; this is a safety consideration to be kept in mind when designing the cell. Forced or induced ventilation fans can cause pressure differences across doors, making it dangerous (door flies open when unlatched) or impossible to open a large door. The recommended cell depression for ventilation control is 50 Pa (see Chapter 6).

Test cell doors where operators are permitted to enter when tests are running must be either on slides or be outward opening. The double-walled and double-doored cell shown in Figure 4.6 is interlocked to prevent entry because the inner door opens inwards; the unusual double sections are required to contain the very high noise levels of F1 and other race engines.

There are designs of both sliding and hinged doors that are suspended and drop to seal in the closed position. Sliding doors have the disadvantage of creating "dead" wall space when open. Doors opening into normally occupied work spaces should be provided with small observation windows and may be subject to regulations regarding the provision of Exit signs.

Interlocking the cell doors with the control system to prevent human access during chosen operating conditions is a common safety strategy and is certainly

FIGURE 4.6 Double pairs of cell doors built into a high-noise, acoustic-panel-contructed engine test cell. *(Photo courtesy of Enviosound Ltd.)*

advised in educational and production facilities. The usual operation of such an interlock is to force the engine into a "no load/idle" state when a door begins to open.

Cell Windows

The degree to which a well-constructed test cell can form a hazard containment box may depend critically on the number and type of windows fitted into the cell walls and doors.

The different glass types suitable for cell windows are sold either as "bulletproof" (BS EN 1063:2000) or "bandit proof" (BS 5544). There are also quite separate fire-resistant types of glass such as Pilkington Pyrodur®. The decision as to whether it is more likely for a large lump of engine (bandit) or a high-speed fragment of turbocharger (bullet) to hit the window has the potential to prolong many an H&S meeting and will not be resolved in this volume. However, good practice would suggest that the glass (and its frame) on the cell side should have the highest fire resistance and the control room side glass should be the most impact resistant.

To achieve good sound attenuation, two sheets of glass with an air gap of some 80 mm is necessary. Many motor-sport or aero-engine cells have three panes of glass with at least one set at an angle to the vertical to minimize internal reflections.

Cell Walls and Roof

Test cell walls are required to meet certain special demands in addition to those normally associated with an industrial building. They, or the frame within which they are built, must support the load imposed by any crane installed in the cell, plus the weight of any equipment mounted on, or suspended below, the roof. They must be of sufficient strength and suitable construction to support wall-mounted instrumentation cabinets, fuel systems, and any equipment carried on booms cantilevered out from the walls. They should provide the necessary degree of sound attenuation and must comply with requirements regarding fire retention (usually a minimum of one hour containment).

High-density building blocks provide good sound insulation that may be enhanced by filling the voids with dried casting sand after being laid and before the roof is fitted; however, this leads to problems when creating wall penetrations after the original construction because of sand and dust leakage from the void above the penetration. Walls of whatever construction usually require some form of internal acoustic treatment, such as 50-mm-thick sound-absorbent panels, to reduce the level of reverberation in the cell. Such panels can be effective on walls and ceilings, even if some areas are left uncovered for the mounting of equipment. The alternative, easier to clean, option of having cast concrete walls that are ceramic tiled are seen in facilities remote from office spaces and where test operators are not allowed in the cell during engine running.

The key feature of *test cell roofs*, much disliked by structural engineers, is the number, position, shape, and size of penetrations required by the ventilation ducts and various services. It is vital that the major penetrations are identified early in the facility planning as they may affect the choice of best construction method. In the author's experience penetrations, additional to those planned and provided during the early build stage, are invariably required by one or more subcontractors even when their requirements have been discussed in detail at the planning stage.

The roof of a test cell often has to support the services housed above, which may include large and heavy electrical cabinets. Modern construction techniques, such as the use of "rib-decking" (Figure 4.7), which consists of

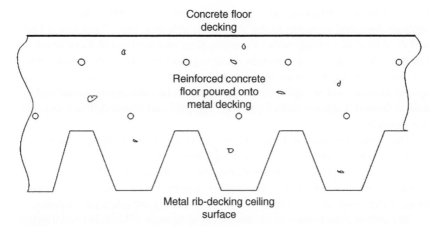

FIGURE 4.7   Section through metal and concrete "rib-decking" construction of a cell roof.

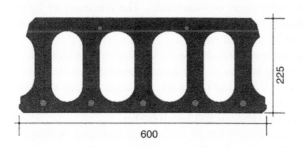

FIGURE   4.8 Section through a concrete plank, which gives a quick and high-strength cell roof construction but can be difficult to modify post-installation.

a corrugated metal ceiling that provides the base of a reinforced concrete roof that is poured in situ, are commonly used.

An alternative is hollow-core concrete planking (Figure 4.8), but the internal voids in this material mean that a substantial topping of concrete screed is required to obtain good sound insulation. The comments above concerning large penetrations are particularly true if concrete planking is used because major post-installation modification is extremely difficult.

Suspended ceilings made from fire-retardant materials that are hung from hangers fixed into the roof can be fitted in cells if the "industrial" look of concrete or corrugated metal is unacceptable, but this has to be shaped around every roof-penetrating service duct and pipe, and is usually only financially justified in high-profile research or motor-sport sites.

Lighting

The typical test cell ceiling may be cluttered with fire-sprinkler systems, exhaust outlets, ventilation ducting, and a lifting beam. The position of lights is often a late consideration, but is of vital importance. Lighting units must be securely mounted so as not to move in the ventilation "wind" and give a high and even level of lighting without causing glare into the control room window. Unless special and unusual conditions or regulations apply, cell lighting does not need to be explosion proof; however, units may be working in an atmosphere of soot- and oil-laden fumes and need to be sufficiently robust and be easily cleaned. Lighting units fitted with complex and flimsy diffuser units are not suitable.

The detailed design of a lighting system is a matter for the specialist.

The "lumen" method of lighting design gives the average level of illumination on a working plane for a particular number of "luminaries" (light sources) of specified power arranged in a symmetrical pattern. Factors such as the proportions of the room and the albedo of walls and ceiling are taken into account.

The unit of illumination in the International System of Units is known as the *lux*, in turn defined as a radiant power of one lumen per square meter. The unit of luminous intensity of a (point) source is the candela, defined as a source that emits one lumen per unit solid angle or "steradian". The efficiency of light sources in terms of candela per watt varies widely, depending on the type of source and the spectrum of light that it emits.

The IES Code lays down recommended levels of illumination in lux for different visual tasks. A level of 500 lux in a horizontal plane 500 mm above the cell floor should be satisfactory for most cell work, but areas of deep shadow must be avoided. In special cells where in-situ inspections take place, the lighting levels may be variable between 500 and 1000 lux.

Emergency lighting with a battery life of at least one hour and an illumination level in the range 30–80 lux should be provided in both test cell and control room.

It is sometimes very useful to be able to turn off the cell lights from the control desk, whether to watch for sparks or red-hot surfaces or simply to hide the interior of the cell from unauthorized eyes.

Cell Support Service Spaces

The design criteria of the individual services are covered in other chapters, but at the planning stage suitable spaces have to be reserved for the following systems listed in order of space required:

- Ventilation plant (including fans, inlet, and outlet louvres) and ducting (including sound attenuation section) (see Chapter 6)
- Engine exhaust system (see Chapters 9 and 16)
- Electrical power distribution cabinets, including large drive cabinets in the case of AC dynamometers (see Chapter 5)
- Fluid services, including cooling water, chilled water, fuel, and compressed air (see Chapter 7).

In addition to these standard services, space in some facilities needs to include:

- Combustion air treatment unit (see Chapter 6)
- Exhaust gas emissions equipment (see Chapter 16).

It is very common to mount these services above the cell on the roof slab and it is often desirable, although rarely possible, for the services of individual cells to be contained within the footprint of the cell below.

Control Room Design

The role and profile of the staff using the control room is of fundamental importance to the design of that work space and to the operation of the test department. There is no "correct" solution; the profile has to be optimized for the particular roles and needs of each site.

While there is no ideal control room layout that would suit all sites, there are some general features that should be avoided.

The multiple screens used in modern control rooms present the control system designer with a major problem in seeing "the wood for the trees" and organizing a coherent display. In most cases the suppliers of control and data acquisition systems have passed the problem to the user by allowing them freedom to produce their own customized display screens. The problem, while transferred, has not been solved and the answers will differ between cells having different tasks. Some form of optimization and discipline of data display is required within the user organization if serious misunderstandings or operational errors are to be avoided; such lack of discipline is not only inefficient but unsafe.

It is advisable to use articulated support booms for screens that are needed only during particular phases of tests. The displays and controls for secondary equipment such as cell services are usually housed in a 19″ racked cabinet positioned at the side of the control desk, while instruments used only inter-mittently, such as fuel or smoke meters, can be installed towards the top and operated if necessary from a standing position.

An analog dial display of speed, torque and, in the case of engine testing, oil pressure is considered by most experts to be vital for manual control panels in order for the operator to judge rate of change during initial start-up and running.

As an example of a poor layout and subsequent development, Figure 4.9 shows one of two control desk areas, in close proximity. These were incre-mentally modified from the mid 1980s to the early 2000s and show a number of design and consequential housekeeping problems that create problems for the operators:

- Operators' work area is confined and cramped with controls and displays in positions dictated by available floor or desk space rather than ergonomic considerations.
- The window has been rendered useless and unnecessary by display units.

FIGURE 4.9 Part of a control room area that has a number of shortcomings arising from poorly integrated and ad hoc development.

- Instrumentation that has been added in a piecemeal manner has meant that the cable routes are less than optimum and that signals are vulnerable to electrical noise and therefore data corruption (see Chapter 5).
- Room lighting has been designed for an empty space; it is now creating shadows and reflections and is not illuminating the work areas in a manner that is sympathetic to the tasks being carried out.
- Interconnecting wiring is exposed and vulnerable to the pedestrian door into the cell and being dislodged by anyone using the door.
- Cables joining shared equipment do not run through dedicated trunking but go across the floor covered by a loose floor mat.

While the above test facility could only have two individuals working in each cell control space safely, that shown in Figure 4.10, which is a control room of a modern Formula 1 powertrain test facility, has to house its two operators and a floating population of specialist development engineers but also, on occasions, VIP visitors. There are several key features to support its roles:

- A large window and free area behind the desk allows visitors, such as team sponsors, to visit and observe work in progress without disturbing the operators.
- The 12 screen displays are designed to display data in the required format best suited to the subsystem they serve and are placed in the optimum position and lighting to serve that purpose.
- Above the window are placed color CCTV screens displaying pictures from steerable and zoomable in-cell cameras; facility status screens are to the right of the operator's desk.

FIGURE 4.10 A general view of a modern powertrain test cell control room. Note that the screens above the cell window are dedicated to CCTV cameras inside the cell. The window is triple glazed. *(Photo courtesy of Mercedes GP Petronas Formula 1 Team.)*

- In motor-sport test facilities there are invariably desk positions for two operators, one acting as "driver" (although any highly dynamic test sequences will be computer controlled) and the other as "engineer", concentrating on the subject of the particular test. The screens have to be positioned to present information relevant to the two or more tasks.
- The only cables not in designated segmented cable-ducts are for ancillary systems such as 13-amp power supplies, phone, and computer network lines.
- The several keyboards and mice are cable attached to the devices they control rather than wireless-enabled to avoid potentially dangerous confusion caused by changing of positions.

As a general rule the operator(s), who during automatic test running may be physically inactive for lengths of time, must remain alert; they may therefore spend a great deal of time in the control room and its layout should be supportive of the role as defined within the facility.

It has been observed by the author that the cost of not providing a side trolley for operators' personal equipment has been counted as the cost of damage caused as a result of cans of sugar-rich drinks being spilt into keyboards or expensive under-desk instrumentation.

The operators' desk should be designed so that liquids spilled on the working surface cannot run down into electronic equipment mounted beneath, because in spite of sensible rules forbidding drinks and sample fluids to be put on the control desk, it will happen.

The floors of the control area should have nonslip and antistatic surfaces.

Some equipment often installed in control rooms, such as emissions instrumentation, can have a quite large power consumption and be a significant noise source. This must be taken into account when specifying the ventilation system that, for temperature, should conform to local office standards, which in Europe are within the range:

- Recommended—19 to 23 °C
- Minimum—16 °C
- Maximum—Currently there is no legal maximum workplace temperature, although employers are under a legal obligation to ensure temperatures in the workplace are reasonable; some trade unions use the figure of 30 °C.

Control and regulations concerning noise levels for all test facility areas are covered in Chapter 9.

In-Cell Control of the UUT

"Repeater", engine, and dynamometer controls, installed in the test cell as wall- or pedestal-mounted boxes, used to be a common feature of automotive test cells, but H&S practices have tended to inhibit their use in many facilities. However, they are very useful when carrying out some calibration and setting up functions (such as the throttle actuator stroke limits or fuel pump stops), and, provided sensible operational practices are enforced, their use can be recommended. Built into a control system that has two manual control stations has to be a robust safety system that inhibits dual control or control conflicts; therefore, the main control desk has to be fitted with a key switch that allows choice of either "desk control and auto" and "remote and manual". Control follows the single key that can only be used in the operational station.

A modern variation of the in-cell control box is the type of single, digital, control pod shown in Figure 4.11, which can be unplugged from the desk station

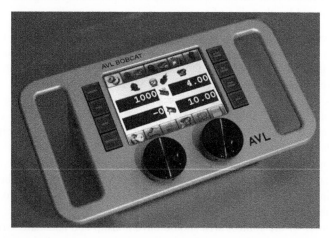

FIGURE 4.11 An example of a portable control pod that can be used both at the main control desk or at an in-cell station during calibration and commissioning. *(© AVL NA.)*

and plugged into the system at a convenient work point in the test cell; in this case control authority travels with the single unit, rather than being switched.

Manual Control

In all engine and powertrain test facilities it will be possible to carry out some of the primary control functions under manual control. So, by definition, the primary purpose of the control desk is either to permit the operator to regulate the engine or drive motor and/or the dynamometer directly. The system must also allow him to monitor the functions of a control system while it runs automated test sequences.

Central to any layout should be the controls that govern the torque produced or applied to the UUT and its rotational speed, under various modes of control (Chapter 10). Equally, the display of speed and torque should be prominent and in analog form in order for the operator to judge rate of change and "drive" the system.

Secondary displays and warning signals covering such features as coolant temperature, lubricating oil pressure, exhaust temperature, etc. are now displayed on computer-driven screens, the layout of which is vital if all channels are not to be of equal priority.

Manual control of the cell services should be enabled through a dedicated control panel that shows the running status of each major circuit and should be visible from the operator's normal work position.

PART 3. TEST CELL ROLES AND THEIR SPECIAL FEATURES

Hybrid and Electric Vehicle Powertrain Testing

Locomotive and marine engineers are quite familiar with electric propulsion motors and the diesel–electric powertrain, a configuration now referred to, in automotive engineering, as a series hybrid. Electric and hybrid cars have been designed from the first decade of the twentieth century but now, 100 years later, the environmental pressures applied by government through taxation regimes and the necessary reduction in vehicle use of hydrocarbon fuels are forcing the pace of development of hybrid and electrically propelled vehicles and are requiring facilities capable of testing them. As a reminder, the three major hybrid configurations are:

- *Series*, in which the vehicle is propelled only by one or more electrical motors and the internal combustion engine acts as a generator to supply power to those motors and to charge the vehicle batteries.
- *Parallel*, in which both an internal combustion engine and an electric traction motor drive the vehicle in a proportion depending on driver demand and vehicle status (battery charge level, etc.).

- *Combined series/parallel*, in which either the internal combustion engine or the electric traction motor can be decoupled from propelling the vehicle, again depending on driver demand and vehicle status.

The testing of the non-automotive diesel generator, using a load-bank and propulsion motors using electrical dynamometers, is a mature technology both as individual units and as systems. However, in their automotive form, hybrid vehicles provide a number of new challenges that are largely concerned with testing the sum of the parts, directly or by simulation through HiL.

Batteries and Battery Simulators

The most significant development work that lies at the heart of the future acceptance of hybrid vehicles is that of the battery technology required in most automotive hybrid configurations. This development work is outside the remit of this book and the lives of most automotive test engineers, but for hybrid powertrain testing it is vital to have an electrical power source that simulates the dynamic performance of different battery types and sizes. The voltage range of hybrid vehicle batteries currently varies from 280 to 440 VDC. The development and increased use of "super-capacitors", whose ability to charge much faster than batteries makes them particularly suitable for energy recovery through regenerative braking, also require simulation in individual and combined battery–capacitor configurations.

Investment in the latest generation of fully programmable, multiconfiguration, powerful battery–capacitor simulators is increasingly becoming essential for any test facility involved in hybrid powertrain work.

These devices currently take the form of large electrical cabinets having sizes typically of 2200 mm high, 800 mm deep, and between 800 and 3000 mm wide. The typical weight, when fully connected, will be up to 1000 kg; therefore, adding these devices into an existing cell or service space requires careful planning and an operational risk assessment.

Most hybrid cars use three-phase AC electric motors rated at up to 380 VAC controlled by IBGT control technology similar to that used in AC dynamometer control (Chapter 10).

The similarities may be instructive since both systems can be vulnerable to problems of torsional vibration in the mechanical connection of their engine and motor/generator, and both can be emitters of electromagnetic radiation in the RF bands.

For the automotive test facility there are a vast number of different test tasks and major design changes specific to hybrid configurations that have to be handled. Eight obvious examples are:

1. The development and refinement of model-based methods of testing electronic powertrain control systems.

2. Developing cell test-stand layouts to accommodate UUT that are of a significantly different shape than a single IC engine.
3. Detailed system calibration tasks such as optimizing performance, range, and emissions of the powertrain over the full climatic range.
4. Development and emission optimization of the downsized IC engines running operational cycles required to power and/or charge vehicle batteries over existing and developing urban and highway drive cycles.
5. Drivability is already a significant issue and hybrids provide new problems to solve, such as the "blending" of regenerative braking loads with that of vehicle brakes over the full range of driver force inputs and vehicle speeds.
6. Transmission testing is seeing changes in the type of units tested, which include the operation of the epicyclic units required in the power-splitting units of parallel hybrids.
7. Safety issues concerning the restraint of heavy battery packs in crashes, to the fire safety of high-energy storage systems; all the new designs will have to be examined and tested.
8. The minimization and suppression of electromagnetic emissions from the hybrid power systems provide EMC test facilities with a new generation of testing tasks.

Gearbox and Transmission Test Rigs

The majority of automotive gearbox designs, both manual and automatic, and final drive units have for the last 30 years been developments and refinements of existing units with really novel designs limited to motor-sport or off-road vehicles. Now, in addition to power-splitting hybrid transmissions, the control and actuation systems are undergoing considerable development. The wider use of dual-clutch transmissions (DCT), and the electrical actuation of clutch and gear selection, are requiring new rigs and test routines. Transmission test cells have had to evolve to support this change to highly integrated powertrains and are therefore having to use HiL techniques in order to test the drivability of the transducer and actuator functions.

Meanwhile there continue to be a range of purely gearbox and final drive test rigs that have to fulfill tasks in a number of rig configurations, including:

• Two-wheel drive (three motors) development and endurance rigs
• Four-wheel drive (five motors) development and endurance rigs
• Tilting lubrication rigs
• Gearbox NVH fully and semi-anechoic rig (Chapter 18)
• Durability rigs for both gear-form and gear-shift systems
• Component test rigs, synchronizer, etc.
• Clutch operation loading.

A common factor shared by the test facilities listed above is that they do not have to support IC engine running, so have lower thermal loads and less complex safety interlocking. However, many of the other comments in this book concerning the operators' work environment, cell layout, rigging, test control, and data acquisition remain the same. Transmission rigs that are designed to run IC engines are classified as full powertrain facilities and are discussed below.

Full Powertrain Test Rigs

Powertrain test rigs can now be built to be capable of using either the engine as the prime mover, or an electric motor simulating the engine. The evolution of such cells has been made possible by the comparatively recent development of permanent magnet motors (PMMs) and their associated controls in the automotive power ranges. These units are capable of engine simulation including that of most driveline dynamics and combustion pulses.

The same motor technology has produced dynamometers having low inertia yet capable of absorbing high torque at low speeds, thus providing road wheel load simulation that, with customized controllers, includes tire-stiffness and wheel-slip simulation.

An important logistical justification for such "all-electric" powertrain cells, besides not having to install and maintain all the cell services required by running an IC engine, is that the required engine may not be available at the time of the transmission test.

One cost-effective arrangement that suffers from similar logistical problems of unit availability but which overcomes several rig design problems is shown in Figure 4.12, where a complete (dummy or modified production) vehicle is mounted within either a two-wheel or "four-square" powertrain test rig.

It should be noted that, in spite of advances in motor and drive technology, flywheels still have a valuable part to play in transmission and powertrain testing and are often fitted to the free end of "wheel" dynamometers, a position that allows various flywheel masses to be fitted according to the demands of the test and UUT (see Chapter 11 for a discussion of flywheels).

Powertrain rigs have to be designed to be able to take up different configurations on a large bedplate, as required by the UUT layout. A large tee-slotted test floor, made up of sections of cast-iron bedplates bolted together and mounted on "air springs" (see Chapter 9), is the usual way to enable the various drive-motor or dynamometer units to be moved and aligned in typical powertrain configurations. However, a cheaper alternative for multiconfiguration transmission test rigs, which usually experience lower vibration levels than engine rigs, is shown in Figure 4.13, where steel slideways set into the concrete floor allow relative movement, albeit restricted, of the major dynamometer frames.

To allow fast transition times the various test units should be pallet mounted in a system that presents a common height and alignment to the cell interface points.

FIGURE 4.12 A "wheel dynamometer" system installed within a garage workshop space having a simple exhaust gas extract system and being used to carry OBD tests. A large cooling fan would be needed in front of the car radiator when running under power. *(Photo courtesy of Rototest Ltd.)*

FIGURE 4.13 The five-axis transmission test rig fitted with an electric motor as the engine simulator and a gear-change robot.

Inclined Engine Test Beds

For the simulation of special operating conditions experienced in "off-the-road" vehicles and race cars under high lateral "*g*" forces, there is a requirement for test beds capable of handling engines and gearboxes running with the crankshaft centerline inclined to the horizontal. These tests not only affect inclined oil levels in the sump but have transient effects on oil pump "pickup" and in turbochargers.

Electric motor-based dynamometers are easily adapted to inclined running, as are eddy-current dynamometers with closed-circuit cooling systems. However, such arrangements present problems with the high-powered hydraulic dynamometers required in aero-engine rigs having open water outlet connections. Some rigs have been built where the whole engine and dynamometer bedplate is mounted upon a system of hydraulic actuators allowing for dynamic movement in three planes while the engine is running. These high-value rigs present high operational complexity and require both a large footprint and vertical space.

For engines with a vertical crankshaft, e.g. outboard boat engines tested without dummy transmissions, the electrical dynamometer is the obvious choice and may generally be used without modification, although dry gap, eddy current, machines have also been used.

Special arrangements need to be made for torque calibration.

Automotive Engine Production Test Cells (Hot Test)

These cells are highly specialized installations forming part of an automation system lying outside the scope of this book. The objective is to check, in the minimum possible process time, that the engine is complete and runs. Typical "floor-to-floor" times for small automotive engines range between 5 and 8 minutes.

The whole procedure—engine handling, rigging, clamping, filling, starting, draining, and the actual test sequence—is highly automated, with interventions, if any, by the operator limited to dealing with fault identification. Leak detection may be difficult in the confines of a hot-test stand, so it is often carried out at a special (black-light) station following test while the engine is still warm.

The test cell is designed to read from identity codes on the engine, recognize variants, and to adjust the pass or fail criteria accordingly.

Amongst the measurements made during a production test, two vital build integrity checks are carried out by checking cranking torque and time taken for oil pressure to reach normal level.

Even in the, increasingly rare, cases of gasoline engines being subjected to an end-of-line (EOL) hot test, they are no longer loaded by any form of dynamometer.

During the EOL test of diesel engines, loaded test sequences are still the standard practice, although 100% cold test and some percentage of engines hot tested is becoming more common for small automotive diesels.

Automotive Engine Production Cold-Test Stations

Along with pressure and rotational testing "in process" of subassemblies, cold testing is increasingly being applied to (near) completely built engines. It has considerable cost and operational advantages over hot testing, which requires an engine to be fully built and dressed for running in a test cell with all supporting services. Cold-test areas also have the cost advantage of being built and run without any significant enclosure other than safety guarding and can form an integral section of the engine production line.

Cold-test sequences are of short duration and some potential faults are easier to spot than in hot testing when "the bang gets in the way".

The principal technique of cold testing is that the engine is spun at lower than normal running speed, typically 50–200 rpm, by an electric motor with an in-line torque transducer.

Early in the test sequence, the "torque to turn" and rate of oil pressure rise figures are used to check for gross assembly errors. The engine wiring loom is usually connected through the ECU connector to a slave unit that is programmed to check the presence, connection, and correct operation of the engine's transducers. Thereafter, vibration and noise patterns are recorded and compared to a developing standard model.

This is a highly automated process that uses advanced computer models and pattern recognition technology. The maximum long-term benefit of cold testing is derived by feeding back field service data to refine the "pass–fail" algorithms that pick out production faults at their incipient stage.

End-of-Line (EOL) Test Station Facility Layout

In the design phase of either type of production EOL test facility, a number of fundamental decisions have to be made, including:

- Layout, e.g. in-line, branch line, conveyor loop with workstations, carousel, etc.
- What remedial work, if any, is to be carried out on the test stand
- Processing of engines requiring minor rectification
- Processing or scrapping and parts recycling of engines requiring major rectification
- Engine handling system, e.g. bench height, conveyor and pallets, "J" hook conveyor, automated guided vehicle
- Engines rigged and de-rigged at test stand or remotely
- Storage and recycling of rigging items
- Test-stand maintenance facilities and system fault detection
- Measurements to be made, handling and storage of data.

Production testing, hot or cold, imposes heavy wear and tear on engine rigging components, which need constant monitoring and spares to be available. The use of vehicle standard plug or socket components on the rig side has proved to be totally unsatisfactory; these components have to be significantly toughened versions to survive in the EOL environment.

Automatic shaft docking systems may represent a particularly difficult design problem where multiple engine types are tested, and where faulty engines are cranked or run for periods leading to unusual torsional vibration and torque reversals.

Shaft docking splines need adequate tooth lubrication to be maintained and, like any automatic docking item, can become a maintenance liability if one damaged component is allowed to travel round the system, causing consequential damage to mating parts.

Modular construction and the policy of holding spares of key subassemblies will allow repairs to be carried out quickly by replacement of complete units, thus minimizing production downtime.

Large and Medium-Speed Diesel Engine Test Areas

As the physical size of engines increases, the logistics of handling them becomes more significant; therefore, the test area for medium-speed diesel engines is, more often than not, located within the production plant close to the final build area.

Above a certain size, engines are tested within an open shop in the position in which they have been finally assembled. The dynamometers designed to test engines in ranges about 20 MW and above are small in comparison to the prime mover; therefore, the test equipment is brought to the engine rather than the more usual arrangement where the engine is taken to a cell.

Cells for testing medium-speed diesels require access platforms along the sides of the engine to enable rigging of the engine and inspection of the top-mounted equipment, including turbochargers, during test. There is a design temptation to install services under these platforms but these spaces can be difficult and unpleasant to access; therefore, the maintenance items such as control valves should, where possible, be wall or boom mounted.

Rig items can be heavy and unwieldy; indeed, the rigging of engines of this size is a considerable design exercise in itself. The best technique is to pre-rig engines of differing configurations in such a way that they present a common interface when put in the cell. This allows the cell to be designed with permanently installed semi-automated or power-assisted devices to connect exhausts, intercooler, and engine coolant piping. The storage of rig adaptors will need careful layout in the rig/de-rig area.

Shaft connection is usually manual, with some form of assisted shaft lift and location system.

Special consideration should be given in these types of test areas to the draining, retention, and disposal of liquid spills or wash-down fluids.

Large engine testing is always of a duration exceeding the normal working day, therefore running at night or weekends is common and may lead to complaints of exhaust noise or smoke from residential areas nearby. Each cell or test area will have an exhaust system dedicated to a single engine; traditionally and successfully the silencers have been of massive construction built from the ground at the rear of the cells. Modern versions may be fitted with smoke dilution cowls and require well-maintained condensate and rain drains to prevent accelerated corrosion.

PART 4. FIRE SAFETY AND FIRE SUPPRESSION IN TEST FACILITIES: EUROPEAN ATEX CODES APPLIED TO ENGINE TEST CELLS

ATmospheric EXplosion regulations were introduced as part of the harmonization of European regulations for such industries as mines and paper mills, where explosive atmospheres occur; the engine and vehicle test industry was not explicitly identified within the wording, therefore the European automotive test industry has had to negotiate the conditions under which test cells may be excluded where conformity would make operation impossible. As with all matters relating to regulation, it is important for the local and industrial sector-specific interpretation to be checked. The regulation classification of zones is shown in Table 4.2. Areas classified into zones 0, 1, and 2 for gas–vapor–mist must be protected from effective sources of ignition.

As has been determined over many years, secondary explosion protection measures such as using EX-rated equipment (even if it was available and in many cases it is not) in the engine test cell makes little sense since the ignition source is invariably the engine itself. Therefore, it is necessary to use primary

TABLE 4.2 ATEX Designation of Zones in Which Gas, Fuel Vapor, and Oil Mist May Form an Explosive Mixture

USA Divisions	ATEX Zone Designation	Explosive Gas Atmosphere Exists	Remark
	Zone 0	Continuously, or for long periods	>1000 h/year
Division 1 = Zones 0 and 1	Zone 1	Occasionally	10–1000 h/year
Division 2 = Zone 2	Zone 2	For a short period only	<10 h/year

explosion prevention methods that prevent the space from ever containing an explosive atmosphere covered by ATEX.

These primary precautions, which cover both gasoline- and diesel-fueled beds without distinction, are certified in Europe by the relevant body TÜF and are:

- The cell space must be sufficiently ventilated both by strategy and volume flow to avoid an explosive atmosphere.
- There has to be continuous monitoring and alarming of hydrocarbon concentration (normally "Warning" at 20% of lower explosive mixture and "Shutdown" at 40%).
- Leak-proof fuel piping using fittings approved for use with the liquids contained.
- The maximum volume of fuel "available" in the cell in the case of an emergency or alarm condition is 10 liters.

With these conditions fulfilled, the only EX-rated electrical devices that need to be included in the cell design are the gas detection devices and the purge extraction fan.

In the USA the treatment and classification of "hazardous locations" such as engine and vehicle test cells are defined in the National Electrical Code (NFPA70). Like most regulations having relevance to our industry, parts of the code are subject to local interpretation by the "authority having jurisdiction" (AHJ), who may be a fire marshal or city planning officer.

The code often refers to the "adequate ventilation" and "sound engineering judgment" being required in classification of industrial spaces. Best and usual practice in the USA often uses changes-per-hour figures in classified zones. In engine test cells subfloor trenches and in areas up to 18″ above the floor "where volatile fuels are transferred" require specific ventilation flows. In general the resulting classifications give similar air flow requirements to European practice, but close working with the local AHJ is advised from the inception and initial planning of any test facility.

Refer to Chapter 6 for a discussion of detailed design strategies of ventilation systems compliant with ATEX regulations.

Fire Stopping of Cable Penetrations in Cell Structure

Where ventilation ducts, cables, or cable trunking break through the test cell walls, roof, or floor they must pass through a physical "fire block" to preserve the one-hour minimum fire containment capability. All wall penetrations carrying cables between control space and cell should be, as a minimum, sealed by using wall boxes having "letterbox" brush seals in steel wallplates fixed on either side of the central void, which is then stuffed with intumescent fire-stopping material. However, poor maintenance or incomplete closure of these firebreaks, commonly caused by frequent installation of temporary looms, can allow fire or the pressurized extinguishant, such as CO_2, to escape explosively into the control room.

A hole in the control room wall loosely stuffed with rag not only compromises the building's fire safety, but may also invalidate insurance.

Alternatively, the more robust Hawke Gland™ boxes may be used; these devices provide rigid clamping of the individual cables and although the fire block can be disassembled and extra cables added, it is not particularly easy to use after a year or more in service, so it is advisable to build in a number of spare cables at the time of the initial closure.

Intumescent materials are widely used to "fire-stop" test cell penetrations. Often based on a graphite mixture, such materials swell up and char in a fire, thus sealing their space and creating a barrier of poor heat conduction. Intumescent material is available in bags of various volumes that is ideal for stuffing between the ventilation ducting and the roof or wall, since it allows thermal movement without being dislodged. However, it should be held in place, top and bottom, by plates fixed to the concrete to prevent it being dislodged by any pressure pulses in the cell. Foams are available to use for sealing pipe and cable ducts, but in all cases "listing and approval for use and compliance" should be sought from local fire authorities and facility insurers before particular products are used.

Plastic "soil pipes" cast into the floor are a convenient way of carrying cables between test cell and control room. Several such pipes need to be dedicated to cables of the same type to avoid crosstalk and signal corruption. These ducts should be laid to fall slightly in the direction of the cell to prevent liquid flow into the control room and should have a raised lip of 20 mm or more to prevent drainage of liquid into them but positioned so as not to create a trip hazard. Spare cables should be laid during installation. These pipes can be "capped" by foam or filled with dried casting sand to create a noise, fire, and vapor barrier.

Large power cables may enter a cell through cast concrete trenches cast under the wall and filled with dense dry sand below a floor plate; this method gives both good sound insulation and a fire barrier, and it is relatively easy to add cables later.

Fire and Gas Detection and Alarm Systems for Test Facilities

There are three separate system design subjects:

- *The prevention of explosion and fire* through the detection of flammable, explosive, or dangerous gases in the cell and the associated remedial actions and alarms.
- *The suppression of fire* and the associated systematic actions and alarms.
- *Detection of gases* injurious to health.

An engine test cell's gas detection system, supplied by a specialist subcontractor, will be fitted to transducers designed to detect various levels of hydrocarbon vapor (see ATEX regulations). In facilities testing fuel cells hydrogen detectors should be fitted in the roof spaces.

Depending on the requirements of the risk analysis valid for the facility, some areas within and outside the cell may be fitted with carbon monoxide sensors. This is particularly important if pressurized and undiluted engine exhaust ducting is routed through a building space, a design feature to be avoided if possible.

Gas detection systems must be linked to the cell control system and, where it exists, the building management system (BMS) in conformity to local and national regulations.

The gas hazard alarm and fire extinguishing systems are always entirely independent of, and hierarchal to, the test cell's emergency stop circuit system.

The operation of both systems, which can vary significantly, should be specified within the control and safety interlock matrix covered in Chapter 5.

There is much legislation relevant to industrial fire precautions and a number of British Standards. In the UK the Health and Safety Executive, acting through the Factory Inspectorate, is responsible for regulating such matters as fuel storage arrangements, and should be consulted, as should the local fire authority.

The choice of audible alarms and their use requires careful thought when designing the facility safety matrix.

Audible alarms, separate from those visual alarm displays that are a function of the test controller, are usually reserved to warn of fire. The fire alarms usually use electronic solid-state sounders with multi-tone output, normally in the range of 800–1000 Hz, or can be small sirens operating in the range of 1200–1700 Hz. Regulations require that they output a sound level 5 decibels above ambient; since ambient even in the control area may be 80 decibels, most fire alarms fitted into engine test cell areas tend to be painfully loud and thus successful in quickly driving humans out of their immediate area.

However, it is not uncommon for hot, sometimes incandescent, engine parts to trigger false fire alarms through in-cell detectors, so it is most important that the operator is trained and able to identify and kill such false alarm states and avoid consequential automatic shutdown of unassociated plant, or inappropriate release of a fire suppression system.

Fire Extinguishing Systems

Whatever the type of fire suppression systems fitted in test cells, it is absolutely vital that all responsible staff are formally trained and certified in its correct operation and that this training is kept up to date with changes in best practice and the nature of the units under test.

In the period between the second and third editions of this book, some gas-based fire suppressant systems were banned for use in new building systems since they were judged injurious to the environment.

Table 4.3 lists the common fire suppression technologies and summarizes their characteristics, which are covered in more detail in the following paragraphs.

Microfog Water Systems

Microfog or high-pressure mist systems, unlike other water-based fire extinguishing systems, have the great advantage that they remove heat from the fire source and its surroundings and thus reduce the risk of reignition when it is switched off. The system is physically the smallest available and therefore makes its integration within the crowded service space much easier than the gas-based systems.

Microfog systems use very small quantities of water and discharge it as a very fine spray. They are particularly efficient in large cells, such as vehicle anechoic chambers, where they can be targeted at the fire source, which is likely to be of small dimensions relative to the size of the cell.

Other advantages of these high-pressure mist systems is that they tend to entrain the black smoke particles that are a feature of engine cell fires and prevent, or considerably reduce, the need for a major cleanup of ceiling and walls. Such systems are used in facilities containing computers and high-powered electrical drive cabinets with proven minimum damage after activation and enabling a prompt restart.

Carbon Dioxide (CO_2)

CO_2 was used extensively in the industry until the environmental impact was widely understood. While it can be used against flammable liquid fires it is hazardous to life in confined spaces; breathing difficulties become apparent above a concentration of 4% and a concentration of 10% can lead to unconsciousness or, after prolonged exposure, to death. Therefore, a warning alarm period must be given before activation to ensure pre-evacuation of the cell.

CO_2 is about 1.5 times denser than air and it will tend to settle at ground level in enclosed spaces. The discharge of a CO_2 flood system is likely to be violent and frightening to those in the region and the pressure pulse will blow out any incompetent blockage of holes made in the cell walls for the transit of cables, etc. The sudden drop of cell temperature causes dense misting of the atmosphere to take place, obscuring any remaining vision through cell windows.

Dry Powder

Powders are discharged from hand-held devices and are designed for high-speed extinguishing of highly flammable liquids such as petrol, oils, paints, and alcohol; they can also be used on electrical or engine fires. It must be remembered that dry chemical powder does not cool nor does it have a lasting smothering effect and therefore care must be taken against reignition.

TABLE 4.3 Characteristics of Major Fire Suppression Systems

	Water Sprinkler	Inert Gas (CO_2)	Chemical Gases	High-Pressure Water Mist
Cooling effect on fire source	Some	None	None	Considerable
Effect on personnel in cell	Wetting	Hazardous/fatal	Minor health hazards	None
Effect on environment	High volume of polluted water	Greenhouse and ozone layer depleting	Greenhouse and ozone layer depleting	None
Damage by extinguishing agent	Water damage	None	Possible corrosive/hazardous by-products	Negligible
Warning alarm time before activation	None required	Essential	Essential	None required
Effect on electrical equipment	Extensive	Small	Possible corrosive by-products	Small
Oxygen displacement	None	In entire cell space	In entire cell space	At fire source

Halons

Following the Montreal Protocol in 1987, halons (halogenated hydrocarbons) including Halons 1211, 1301, and 2402 were subject to a slow phase-out down to zero by 2010. These contain chlorine or bromine, thought to be damaging to the ozone layer, and their production has been banned. Existing stocks of Halons 1211 and 1301, both hitherto used for total flooding and in portable extinguishers for dealing with flammable liquid fires, should, by the publication date of this book, have been used up and replacement systems installed.

Inergen

Inergen is the trade name for the extinguishing gas mixture of composition 52% nitrogen, 40% argon, 8% carbon dioxide. It works by replacing the air in the space into which it is discharged and taking the oxygen level down to <15% when combustion is not sustained.

There are other alternative fire suppressants of the same type as Inergen, including pure argon, many of which may be used in automatic mode even when the compartment is occupied, provided the oxygen concentration does not fall below 10% and the space can be quickly evacuated.

With all gaseous systems, precautions should be taken to ensure that accidental or malicious activation is not possible. In particular, with carbon dioxide systems, automatic mode should only be used when the space is unoccupied and in conjunction with alarms that precede discharge.

Total flood systems, of whatever kind, are usable only after the area has been sealed. They must be interlocked with the doors and special warning signs must be provided.

Foam

Foam extinguishers could be used on engine fires but they are more suited to flammable liquid spill fires or fires in containers of flammable liquid. If foam is applied to the surface or subsurface of a flammable liquid it will form a protective layer. Some powders can also be used to provide rapid knock-down of the flame. Care must be taken, however, since some powders and foams are incompatible.

USEFUL WEBSITES

http://live.planningportal.gov.uk/buildingregulations/: UK planning law and good practice guides.
http://www.firesafe.org.uk/specadvc.htm: A useful general reference site for free advice on fire regulations and for useful links.
http://www.hse.gov.uk/fireandexplosion/atex.htm: Contains no information on ATEX regulation that is specific to Engine Test facilities but contains useful links.
http://www.pilkington.com: Lists special glass for use in hazardous building environments.

Chapter 5

Electrical Design Requirements of Test Facilities

INTRODUCTION

This chapter, as part of its content, deals with the subject of electromagnetic compatibility (EMC), but only in the context of the automotive test facility, both as a source and a victim of electromagnetic noise. The subject of testing the EMC of the whole vehicle, powertrain units, and their components is treated in Chapter 18.

The electrical system of a test facility provides the power, nerves, and operating logic that control the test piece, the test instrumentation, and building services. The power distribution to, and integration of, these many parts falls significantly within the remit of the electrical engineer, whose drawings will be

Engine Testing. DOI: 10.1016/B978-0-08-096949-7.00005-4

used as the primary documentation in system commissioning and for any subsequent fault-finding tasks.

The theme of system integration is nowhere more pertinent than within the role of the electrical designer, nor is any area of engineering more constrained by differing national regulations worldwide, making integration of international projects fraught with traps for the unwary. On either side of the Atlantic electrical engineers can use similar words that have dissimilar meanings or technical implications. The integration of test cell equipment in Europe with equipment that is built to American standards can create detailed compatibility problems and vice versa. Such problems may not be appreciated by specialists at either end of the project unless they have experience in dealing with each others' practices.

Note: In this text and in context with electrical engineering the UK English terms "*earth*" and "*earthing*" have generally been used in preference to the US English terms "*ground*" and "*grounding*" with which they are, in the context of this book, interchangeable.

THE ELECTRICAL ENGINEER'S DESIGN ROLE

The electrical engineer's prime guidances are the operational and functional specifications (see Chapter 1) from which he or she will develop the final detailed functional specification, system schematics, power requirement, distribution, and alarm logic matrices.

Before the electrical engineer can start to calculate the electrical power required in order to produce an electrical distribution scheme, all of the major test and service plant and their electrical loads have to be identified. Before transformers, power distribution boards, control cabinets, and interconnecting cable ways can be located, all of which may have implications for the architect and structural engineer, the general building layout has to be known. The electrical engineer is therefore a key design team member from the start of the project and throughout the iterative design and system integration processes.

Engine test facilities are, by the nature of their component parts and the low power of their measurement signals, particularly vulnerable to signal distortion caused by various forms of electrical noise. This vulnerability has changed significantly during the lifetime of this publication, with the developments in high-power pulse-width modulation (PWM) speed controllers and wireless communications causing a consequential change in the nature of the electromagnetic noise within the test cell environment.

To avoid the possibility of signal corruption and cell downtime due to instrumentation errors, special and detailed attention must be given to the standard of electrical installation within a test facility. The electrical designer and installation supervisor must be able to take a holistic approach and be aware of the need to design and install an electrically integrated and mutually compatible system. Laboratory facilities that are built and developed on an "ad hoc" basis often fall foul of unforeseen signal interference or control logic errors within interacting subsystems.

GENERAL CHARACTERISTICS OF THE ELECTRICAL INSTALLATION

Perhaps more than any other aspect of test cell design and construction, the electrical installation is subject to regulations, most of which have statutory force. National differences in the details of regulations can lead to quite significant differences in the morphology of power distribution and control cabinets; for this reason it is always good practice to have such units as motor control panels built in the country of operation, even if schematically designed outside of that country. However, over and above compliance with regulations is the need for all electrical installation within an entire test facility to be of the highest possible standard. The integrity of data produced by the facility relies on the correct separation of undistorted cables and the high quality of their connection; such work should not be carried out by inexperienced or unsupervised staff. The author has experience of a significant delay in commissioning test plant that was caused by an electrical contractor employing an individual subcontractor, experienced in domestic power electrical installation, but suffering from undetected "red-end" color blindness. This individual was contracted to install a system containing 40-core, individual-screened, and color-coded cable. The lesson is clear: attention to every detail and relevant experience is vital.

It is essential that any engineer responsible for the design or construction of a test cell in the UK should be familiar with BS 7671, "Requirements for Electrical Installations". There are also many British Standards specifying individual features of an electrical system; other countries will have their own national standards. In the USA the relevant electrical standards may be accessed through the ANSI and NEMA websites (http://www.nema.org/stds/ and www.ansi.org).

The rate of change of regulations, particularly European regulation, will outpace that of any general textbook, but in 2011 the regulations listed in Table 5.1 were appropriate in Europe.

BS EN 60204 is of particular relevance to test cell design since it includes general rules concerning safety interlocking and the shutting down of rotating plant as in the section in Chapter 9 covering shutting down of engine/dynamometer combinations.

While these regulations cover most aspects of the electrical installation in test cells, there are several particular features that are a consequence of the special conditions associated with the test cell environment that are not explicitly covered. These will be included within this chapter, as will the special electrical design features required if four-quadrant electrical dynamometers (see Chapter 10) are being included within the facility.

TABLE 5.1

IEC 60204—1/97 EN 60204—1/97	Safety of machinery—Electrical equipment of machines
IEC 1010—1 EN 61010—1/A2	Safety requirement for electrical equipment for measurement, control, and laboratory use
EN 50178/98	Electronic equipment of power installation
IEC 61800—3/96 EN 61800—3/96	Adjustable speed electrical power drive systems EMC Product Standard
EN 61000-6-4	Electromagnetic influence—Emission
EN 61000-6-2	Electromagnetic influence—Immunity
EN 61010-1 IEC 61010-1	Safety requirements of laboratory (test cell) equipment designed for measurement and control
IEC 61326	Electrical equipment for measurement, control, and laboratory use—EMC requirements

PHYSICAL ENVIRONMENT

The physical environment inside a test cell can vary between the extremes used to test the engine or vehicle. The inside of subsystem control cabinets or transducer boxes can become overheated unless they are protected from radiant heat sources and well ventilated. Engine cells, particularly those running diesel engines, may have particulates entrained in the airflow that, over time, block filters fitted to control cabinets. It is not advisable to locate the power cabinets for AC dynamometers inside the engine test cell or in dusty or damp service spaces, since they can be particularly vulnerable to ingested dirt and moisture.

The guidelines for acceptable ambient conditions for most electrical plant are as follows:

- The ambient temperature of the control cabinet should be between +5 and +35 °C.
- The ambient temperature for a control room printer is restricted to +35 °C and for the PC to +38 °C.
- The designed ambient temperature of a nonclimatic test cell should be between +5 and +45 °C.
- The air must be free of abnormal amounts of dust, acids, corroding gases, and salt (use filter).
- The relative humidity at +40 °C must not be more than 50% and 90% at +20 °C (use heating).

If air-conditioning of control room or service spaces is installed in atmospheres of high humidity, the electrical designer must consider protection against condensation within control cabinets and dynamometer drive cabinets. In such conditions, both in operation and during installation, anticondensation heaters should be installed and switched on (see "Electrical Cabinet Ventilation" section later in this chapter).

The Ingress Protection (IP) rating of electrical devices such as motors and enclosures must be correctly defined in the design stage once their location is known.

In some vehicle and engine cells, particularly large diesel and engine rebuild facilities, it is common practice for a high-pressure heated water washer to be used, in which case the test-bed enclosures should have a rating of at least IPx5 (see Table 5.2).

TABLE 5.2 Ingress Protection Rating Details

IP54 = IP Letter Code _____ IP
First Digit _____5
Second Digit _____ 4

First Digit		Protection From Solid Objects	Second Digit		Protection From Moisture
0		Nonprotected	0		Nonprotected
1		Protected against solid objects greater than 50 mm	1		Protected against dripping water
2		Protected against solid objects greater than 12 mm	2	15°	Protected against dripping water when tilted up to 15°
3		Protected against solid objects greater than 2.5 mm diameter	3	60°	Protected against spraying water
4		Protected against solid objects greater than 1.0 mm diameter	4		Protected against splashing water
5		Dust protected	5		Protected against water jets

(Continued)

TABLE 5.2 Ingress Protection Rating Details—cont'd

6		Dust tight	6		Protected against heavy seas
			7		Protected against the effects of immersion
			8		Protected against submersion

ELECTRICAL SIGNAL AND MEASUREMENT INTERFERENCE

The protection methodology against signal interference in engine test cells has changed significantly over the period since the mid 1990s because of the arrival of electromagnetic interference in the radio frequencies (RF) as the dominating source of signal corruption.

The pulse-width-modulating drive technology based on fast switching insulated bipolar gate transistor (IGBT) devices used with AC dynamometers has reduced the total harmonic distortion (THD) experienced in power supplies from that approaching the 30% that was produced by DC thyristor-controlled drive, to <5% THD:

$$\text{THD } (\%) = \sqrt{\sum_{2}^{k} \left(\frac{H_x}{H_f}\right)^2}$$

where H_x is the amplitude of any harmonic order and H_f is the amplitude of the fundamental harmonic (first order). Most sensitive devices should be unaffected by a THD of less than 8%.

The harmonic distortion produced by thyristor drive-associated DC machines is load dependent and causes a voltage drop at the power supply (see Figure 5.1).

While IGBT technology, associated with AC dynamometers, has reduced THD that caused problems in facilities fitted with DC dynamometers, it has joined other digital devices in introducing disturbance in the frequency range of 150 kHz to 30 MHz. IBGT drive systems produce a common-mode, load-independent disturbance that causes unpredictable flow through the facility earth system. This has meant that previous standard practices concerning signal cable protection and provision of "clean earth" connections, separate from "protective

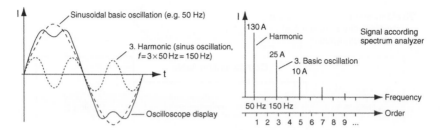

FIGURE 5.1 Harmonic distortion.

earths", are tending to change in order to fight the new enemy: electromagnetic interference (EMI) in the radiofrequency range. The connection of shielded cables between devices shown in Figure 5.2 *was* a standard method aimed at preventing earth loop distortion of signals in the shielded cable.

Earth loops are caused by different earth potential values occurring across a measuring or signal circuit that induces compensating currents.

The single end connection of the cable shield shown in Figure 5.2 offers no protection from EMI, which requires connection at both ends (at both devices); this requires that in installations containing IBGT drives, the ground loops are defeated by different measures. At the level of shielded cable connection the countermeasure is to run a compensation lead of high surface area in parallel with the shield, as shown in Figure 5.3. This method also reduces the vulnerability of the connection to external magnetic flux fields.

EARTHING SYSTEM DESIGN

The earthing systems used in many industrial installations are primarily designed as a human protection measure. However, electromagnetic

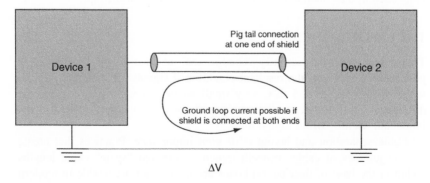

FIGURE 5.2 Recommended method of connecting shielded cable between devices to prevent ground loop current distortion of signals.

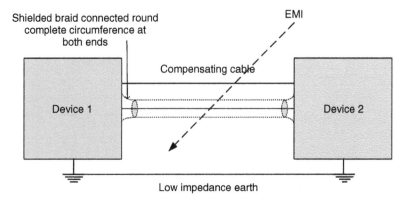

FIGURE 5.3 Recommended method of connecting shielded cable between devices, in test facilities containing high-powered AC dynamometers with IBGT drives, to prevent EMI and ground loop corruption of signals.

compatibility (EMC) requirements increasingly require high-frequency equipotential bonding, achieved by continuous linking of all ground potentials. This modern practice of earthing devices has significantly changed layouts in which there was once a single protective earth (PE) connection plus, when deemed necessary, a "clean earth" physically distant for the PE.

A key feature of modern EMC practice is the provision of multiple earth connections to a common earthed grid of the lowest possible impedance. The physical details of such connection practice are shown in Figure 5.4.

To ensure the most satisfactory functioning of these electromagnetic immunity systems, the earthing system needs to be incorporated into a new building design and the specification of the electrical installation. Ideally, the building should be constructed with a ground mat made of welded steel embedded in the concrete floor and a circumferential earth-strap.

THE LAYOUT OF CABLING

Transducer signals are usually "conditioned" as near to the transducer as possible; nevertheless, the resultant conditioned signals are commonly in the range 0–10 VDC or 0–5 mA, very small when compared with the voltage differences and current flows that may be present in power cables in the immediate vicinity of the signal lines.

Cable separation and layout is of vital importance. Practices that create chaotic jumbles of cables beneath trench covers and "spare" cable lengths coiled in the base of distribution boxes are simply not acceptable in modern laboratories. Figure 5.5 shows an example of a very poor test facility installation.

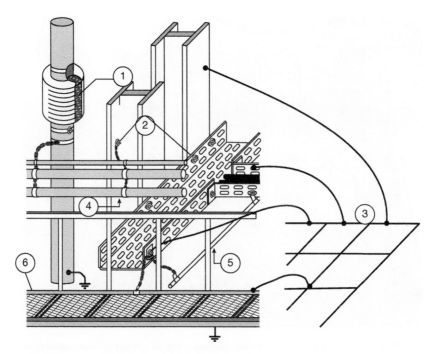

FIGURE 5.4 Continuous linked network of all ground potentials. (1) Braided conducting strip bridging pipe joints. (2) Bolts welded into the metal building frame that is connected to the reinforcement mat. (3) Foundation earthing mat installed in new buildings and/or run-around grounding strap in existing buildings providing an EMC reference potential. (4) Braided strap linking conductive clamps to service pipes. (5) Protective earth wire. (6) Cable trays, steelwork grid, etc.

Coiling of overlong power cables is a common error and it causes inductive interference in control and transducer cables; the more coils, the stronger the effect.

Most engine test facilities will contain the following types of wiring, which have distinctly different roles and which must be prevented from interfering with each other:

- Power cables, mains supply ranging from high-power wiring for dynamometers through three-phase and single-phase distribution for services and instruments to low-power supplies for special transducers, also high current DC supply for starter systems.
- Control cables for inductive loads, relays, etc.
- Signal cables:
  - Digital control with resistive load
  - Ethernet, RS232, RS422, IEEE1394
  - Bus systems such as CAN.

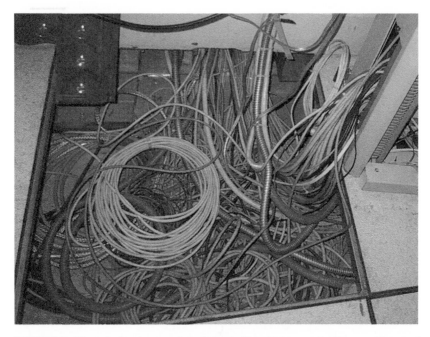

FIGURE 5.5   The underfloor horrors of a bad electrical installation revealed! Power cables mixed with signal cables in close parallel proximity, overlong cables coiled and producing magnetic interference, and the creation of a fault-finding nightmare.

- Measuring cables associated with transducers and instrumentation transmitting analog signals.
- 24 VDC supplies.

In the following paragraphs the common causes of signal interference are identified and practices recommended to avoid the problems described.

Inductive interference is caused by the magnetic flux generated by electrical currents inducing voltages in nearby conductors. Counter measures include:

- Do not run power cables close to control or signal cables (also see capacitive interference below).

  Use either segmented trunking or different cable tray sections, as shown in Figure 5.3.
- Use twisted pair cables for connection of devices requiring supply and return connection (30 twists per meter will reduce interference voltage by a factor of 25).
- Use shielded signal cables and connect the shield to earth at both ends.

Figure 5.6 shows a suggested layout of cables running in an open cable tray. The same relationship and metal division between cable types may be achieved

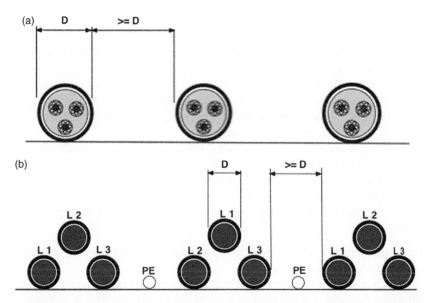

FIGURE 5.6 (a) Recommended spacing of multicore power cables in a trench or tray. (b) Recommended spacing orientation and layout of individual phase power cables with protective earth (PE) cables between each bunch (only two shown).

using segmented trunking, trays, or ladders. It is important that the segments of such metallic support systems are connected together as part of the earth bonding system; it is not acceptable to rely on metal-to-metal contact of the segments.

Note that cables of different types crossing through metal segments at 90° to the main run do not normally cause problems.

Capacitive interference can occur if signal cables with different voltage levels are run closely together. It can also be caused by power cables running close to signal lines. Counter measures include:

- Separate signal cables with differing voltage levels.
- Do not run signal cables close to power cables.
- If possible use shielded cable for venerable signal lines.

Electromagnetic interference can induce both currents and voltages in signal cables. It may be caused by a number of "noise" transmitters ranging from spark plugs to mobile communication devices.

RF noise produced by motor drives based on IBGT devices is inherent in the technology with inverter frequencies of between 3 and 4 kHz, and is due to the steep leading and failing edge of the, typically 500 V, pulses that have a voltage rise of around 2 kV/µs. There may be some scope in varying the pattern of frequency disturbances produced at a particular drive/site combination by adjustment of the pulsing frequency; the supplier would need to be consulted.

Counter measures for AC drive-induced noise include:

- Choice and layout of both motor/drive cabinet and drive cabinet to supply cables is very important:
  - Motor cable should have three multicore power conductors in a symmetrical arrangement within a common braided and foil screen. The bonding of the screen at the termination points must contact 360° of the braid and be of low impedance.
  - The power cables should contain symmetrically arranged conductors of low inductivity within a concentric PE conductor.
  - Single-core per-phase cables should be laid as shown in Figure 5.7.
- Signal cables should be screened.
- Signal cables should be encased in metal trunking or laid within metal cable tray.

The layout of power cables of both multicore and single core is shown below as following recommended practice.

Conductive coupling interference may occur when there is a supply voltage difference between a number of control or measuring devices. It is usually caused by long supply line length or inappropriate distribution layout. Counter measures include:

- Keep supply cables short and of sufficient conductor size to minimize voltage drop.
- Avoid common return lines for different control or measurement devices by running separate supply lines for each device.

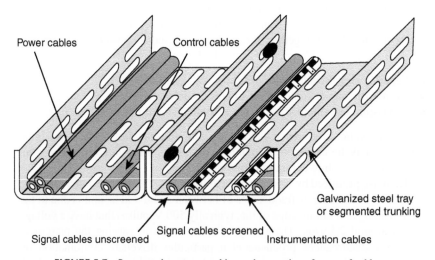

FIGURE 5.7   Segmented trays or trunking and separation of types of cable.

The same problem may occur when two devices are fed from different power supplies.

To minimize interference in analog signal cables, an equipotential bonding cable or strap should connect the two devices running as close as possible to the signal cable. It is recommended that the bonding cable resistance be less than one-tenth of the cable screen resistance.

INTEGRATION OF AC DYNAMOMETER SYSTEMS

Careful consideration should be made when integrating an AC drive system into an engine test facility. Where possible it is most advisable to provide a dedicated electrical supply to an AC drive, or number of AC drives.

There needs to be a clear understanding of the status of the existing electrical supply network, and the work involved in providing a new supply for an AC dynamometer system. It is important to calculate the correct rating for a new supply transformer. This is a highly specialized subject and the details may change depending on the design of the supply system and AC devices, but the general rule is that the "mains short-circuit apparent power (S_{SC})" needs to be at least 20 times greater than the "nominal apparent power of one dynamometer (S_N)". Hence: $S_{SC}/S_N > 20$.

In a sample calculation based on a single 220 kW dynamometer, the following sizing of the supply transformer may be found:

$P_N = 220$ kW nominal power, cos $\Pi = 1$, overload 25%
$S_N = P_N \times \cos \Pi = 220$ kVA
$S_{SC} = S_N \times 20 = 4.400$ kVA
Standard transformer (ST) ~ $U_{SC} = 6\%$
ST $= S_{SC} \times 0.06 = 264$ kVA, R transformer $= 315$ kVA
ST $= S_N \times 1.25 = 275$ kVA, R transformer $= 315$ kVA.

In the case of multiple dynamometer installation, some manufacturers in some conditions will allow a reduction in the S_{SC}/S_N ratio; this can reduce the cost of the primary supply.

Shown below is the basic calculation of power supply transformer in the case of 3 × 220 kW dynamometers:

$P_N = 220$ kW nominal power, cos $\Pi = 1$, overload 25%, diversity 1, 0.8
$S_N = 3 \times P_N \times \cos \Pi = 660$ kVA
$S_{SC1} = S_N \times 20 = 13{,}200$ kVA in the case of no reduction for multiple machines
$S_{SC2} = S_N \times 10 = 660$ kVA in the case of ratio reduction to minimum
Standard transformer ~ $U_{SC} = 6\%$
ST $= S_{SC1} \times 0.06 = 792$ kVA, R transformer $= 800$ kVA
ST $= S_{SC2} \times 0.06 = 396$ kVA, R transformer $= 400$ kVA
ST $= S_N \times 1.25 \times 0.8 = 660$ kVA, R transformer $= 800$ kVA.

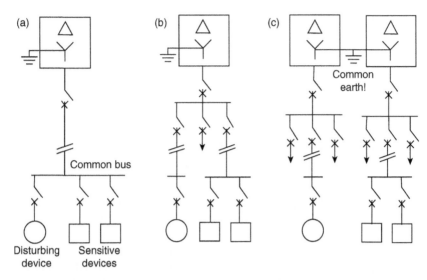

FIGURE 5.8   Different power connection layouts, ranging from poor (a) to recommended (c).

SUPPLY INTERCONNECTION OF DISTURBING AND SENSITIVE DEVICES

An infinite variation of transformer and connection systems is possible, but Figure 5.8 shows the range from poor (a) to a recommended layout (c). The worst connection scheme is shown in Figure 5.8a, where both the sensitive and a disturbing device, such as a pulse-width modulation (PWM) drive, are closely connected on a common local bus. Figure 5.8b shows an improved version of Figure 5.8a, where the common bus is local to the transformer rather than the devices. The best layout is based on Figure 5.8c, where the devices are fed from separate transformers that share a common earth connection.

Power utility representatives drawn into projects involving AC dynamometer or chassis dynamometer installations may not at first appreciate that, for the majority of their working life, these machines are not operating as motors in the usual sense but are working as engine/vehicle power absorbers; in electrical terms they are exporting, rather than importing, electrical power. It may be necessary to use the expertise of the equipment suppliers at the planning stage if upgraded power systems are being provided.

POWER CABLE MATERIAL AND BEND RADII

The wide use of AC motors and dynamometers in modern test cells has increased the need for the careful choice of power cables, all of which must be highly resistant to gasoline, oils, and hydraulic fluids but normally need not be gasoline or oil "proof" or "fast". The conductor copper must be stranded rather than massive and the bending radius should not exceed the following figures:

Fixed cables:	15 times the cable outside diameter
Moving or movable cables:	10 times the cable outside diameter.

ELECTRICAL POWER SUPPLY SPECIFICATION

Most suppliers of major plant will include the required details of power supply conditions their equipment will accept; in the case of AC dynamometers this may require a dedicated isolating transformer to provide both isolation and the required supply voltage.

The EEC directive 89/336/EEC gives the acceptable limits of high-frequency distortion on mains supply; the categories EN_55011_QP_A2* and A2 are applicable to supplies for AC automotive dynamometers.

A typical mains power specification in the UK is shown in Table 5.3.

ELECTRICAL CABINET VENTILATION

Instrument errors caused by heat are particularly difficult to trace, as the instrument will probably be calibrated when cold, so the possibility of such damage should be eliminated in the design and installation phases. Many

TABLE 5.3 Typical UK Electrical Supply Specifications

Voltage	230/240 VAC ± 10%
Frequency	0.99–1.01 of nominal frequency 50 or 60 Hz continuously, 0.98–1.02 of nominal frequency 50 or 60 Hz short time
Harmonics distortion	Harmonics distortion is not to exceed 10% of total r.m.s. voltage between the live conductors for the sum of the second through to fifth harmonics, or 12% max of total r.m.s. voltage between the live conductors for the sum of the sixth through to the 30th harmonics
Voltage interruption	Supply must not be interrupted or at zero voltage for more than 3 ms at any time in the supply cycle and there should be more than 1 second between successive interruptions
Voltage dips	Voltage dips must not exceed 20% of the peak voltage or the supply for more than one cycle and there should be more than 1 second between successive dips

instrumentation packages produce quite appreciable quantities of heat and if mounted low down within the confined space of a standard 19-inch rack cabinet they may raise the temperature of the apparatus mounted above them over the generally specified maximum of 40 °C.

Control and instrument cabinets should be well ventilated, and it may be necessary to supplement individual ventilation fans by extraction fans high in the cabinet. Cabinet ventilation systems should have filtered intakes that are regularly changed or cleaned.

Special attention should be given to heat insulation and ventilation when instruments are carried on an overhead boom. When signal conditioning and complex instruments are within a boom-mounted box that may be positioned above the engine under test, then forced ventilation must be provided and particular attention must be paid to the choice of equipment installed; some signal conditioning modules are specially designed to run in such situations and at temperatures of up to 60 °C. It is recommended that thermal "tell-tale" strips are installed in boom boxes and in cabinets as a good maintenance device and aid to fault-finding.

EUROPEAN SAFETY STANDARDS AND CE MARKING

The European Community has, since 1985, been developing regulations to achieve technical harmonization and standards to permit free movement of goods within the Community. There are currently four directives of particular interest to the builders and operators of engine test facilities (see Table 5.4).

The use of the "CE mark" (abbreviation for "Conformité Europeen") implies that the manufacturer has complied with all directives issued by the EEC that are applicable to the product to which the mark is attached.

There may be some confusion as to the difference between CE marking and the e-mark, the latter being part of the vehicle homologation requirements for type-approved vehicles and vehicle components sold in the EU. "E" and "e"

TABLE 5.4 EC Directives for CE Marking Relevant to Engine Test Facilities

Directive	Reference	Optional	Mandatory
Electromagnetic compatibility	89/336/EEC	Jan 1, 1996	
Machinery	98/37EC	Jan 1, 1993	Jan 1, 1995
Low voltage	73/23/EWG plus 93/68/EEC	Jan 1, 1995	Jan 1, 1997
Pressure equipment		Nov 1999	May 2002

type approval marks are required on devices related to all safety-relevant functionality of the vehicle. Such devices will have to be e-mark certified by a certifying authority; unlike most CE marking, manufacturers cannot self-declare and affix an e-mark (see "Type Approval 'E' and 'e' Marking" section in Chapter 18).

An engine test cell must be considered as the sum of many parts. Some of these parts will be items under test that may not meet the requirements of the relevant directives. Some parts will be standard electrical products that are able to carry their individual CE marks, while other equipment may range from unique electronic modules to assemblies of products from various manufacturers. The situation is further complicated by the way in which electronic devices may be interconnected.

If standard and tested looms join units belonging to a "family" of products, then the sum of the parts may comply with the relevant directive. If the interconnecting loom is unique to the particular plant, the sum of the CE marked parts may not meet the strict requirements of the directive.

It is therefore not sensible for a specification for an engine test facility to include an unqualified global requirement that the facility "be CE marked". Some products are specifically excluded from the regulations while others are covered by their own rules; for example, the directive 72/245/EEC covers radio interference from spark-ignition vehicle engines. Experimental and prototype engines may well fail to comply with this directive, just as HiL tests running unproven simulation models may not comply in every respect with some strict safety standards. These are possible examples of the impossibility of making any unqualified commitment to comply in all respects and at all times with bureaucratic requirements drawn up by legislators unfamiliar with our test industry.

There are three levels of CE marking compliance that can be considered:

- All individual control and measuring instruments should individually comply with the relevant directives and bear a CE mark.
- "Standardized" test-bed configurations that consist exclusively of compliant instrumentation, are configured in a documented configuration, installed to assembly instructions/codes, and have been subjected to a detailed and documented risk analysis may be CE marked.
  This requires a test-bed equipment supplier to define and document such a package, which would allow the whole "cell" to be CE marked.
- Project-specific test cells. As stated above, the CE marking of the complete hybrid cell at best will require a great deal of work in documentation and at worst may be impossible, particularly if required retrospectively for cells containing instrumentation of different generations and manufacturers.

The reader is advised to consult specific "Health and Safety" literature, or that produced by trade associations, if in doubt regarding the way in which these directives should be treated.

TABLE 5.5 An Example of the Recommended Layout of a Test Cell Safety Matrix with Events Listed in the Left Row, the Instrumentation Modules on the Top Row, and Their Status Within the Matrix

Test Cell Safety Matrix	Cell Control System	400 VAC Electrical Supply	Test Cell Power Sockets	Ventilation Fans	Ventilation System Dampers	Combustion Air System	AC Dynamometer	Engine Control	Test Cell Incoming Fuel Solenoids	Fuel Conditioning Unit	Compressed Air System
Controlled by:	CS	ESR	ESR	BMS	BMS	BMS	CS	ESR	ESR-DO	ESR	ESR-DO
Main building fire panel	NO reaction	Enabled	Enabled	BMS shuts down service after 4 min			Manual shutdown of test cell before evacuating the facility				
Test cell fire system (automatic)	Stop	Enabled	Disabled	Disabled	Closed	Disabled	Stop	Stop	Closed	Power off	Closed
Level 1 HC/CO gas alarm	Message display	Enabled	Enabled	Vent to purge	Open	Enabled	Power on	Enabled	Open	Power on	Open
Level 2 HC/CO gas alarm	Fast stop	Enabled	Enabled	Vent to purge	Open	Power off	Regen. stop	Stop	Closed	Power off	Open
Emergency stop (Cat 1)	Fast stop	Enabled	Disabled	Disabled	Closed	Disabled	Regen. stop	Stop	Isolated	Power off	Closed
Fast stop—via control desk button	Fast stop	Enabled	Enabled	Freeze vent	Open	Enabled	Regen. stop	Stop	Open	Power on	Open

TABLE 5.5 An Example of the Recommended Layout of a Test Cell Safety Matrix with Events Listed in the Left Row, the Instrumentation Modules on the Top Row, and Their Status Within the Matrix—cont'd

Test Cell Safety Matrix	Cell Control System	400 VAC Electrical Supply	Test Cell Power Sockets	Ventilation Fans	Ventilation System Dampers	Combustion Air System	AC Dynamo-meter	Engine Control	Test Cell Incoming Fuel Solenoids	Fuel Conditioning Unit	Compressed Air System
Test cell doors opened	Fast stop	Enabled	Enabled	Freeze vent	Open	Enabled	Regen. stop	Stop	Open	Power on	Open
Test sequence engine alarm	Message display	Enabled	Enabled	Enabled	Open	Enabled	Power on	Enabled	Open	Power on	Open
Engine stop (automatic)	Stop	Enabled	Enabled	Enabled	Open	Enabled	Power on	Stop	Open	Power on	Open

The actions listed may not be appropriate for readers; systems are shown only as an example.

Notes:

- "Freeze vent" means that the ventilation fans remain at the speeds set at alarm initiation.
- "Isolated" means that power to the device is latched off and requires resetting.
- "Fast stop" requires the dynamometer to apply torque at a pre-programmed ramp (regenerative stop) to bring rotation to a stop.
- BMS, building management system; CS, control system; DO, digital out (of control system); ESR, emergency stop relay.
- For categories of emergency stop according to BS EN 60204, see Chapter 12.

SAFETY INTERACTION MATRIX

A key document of any integrated system is some form of safety or alarm interaction matrix, and its first draft is usually the responsibility of the electrical engineer having system design authority. It is the documentary proof that a comprehensive risk analysis has been carried out (see Table 5.5 for an example of a recommended matrix layout). To be most effective the base document should be verified between the electrical design engineer, the system integrator, and user group. It is the latter that have to decide, within a framework of safety rules, upon the secondary reactions triggered in the facility by a primary event.

The control logic of the building management system (BMS) has to be integrated with that of the test control system; if the contractual responsibility for the two systems is split then the task of producing an integrated safety matrix needs to be allocated and sponsored.

Note on "As Built" Electrical Documentation (Drawing) Standards

As stated at the start of this chapter, the electrical schematics of a test facility will be used, more than other documentation, by technical staff having to modify and maintain a test facility. It is therefore vital that such drawings reflect the "as built" state of the facility at the time of acceptance, and that they are arranged in a logical manner. There are many national and company standards covering the layout of electrical schematics and the symbols used. Some are significantly different from others, which can cause problems for system integrators and maintenance staff. Documentation standards should be stated in the functional specification (Chapter 1) and the final contract payment should be conditional upon receipt of the "as built" drawings.

USEFUL TEXTS ON EMC SUITABLE FOR ELECTRICAL ENGINEERS INVOLVED WITH TEST FACILITY DESIGN

BS EN 60204-1:2006 + A1. Safety of Machinery—Electrical Equipment of Machines (2009).

BS 7671:2008. Requirements for Electrical Installations: IEE Wiring Regulations, seventeenth ed. ISBN: 978-0-86341-844-0.

J. Goedbloed, Electromagnetic Compatibility, Prentice Hall, New Jersey, 1992. ISBN: 0-13249293-8.

Institution of Engineering and Technology, Electromagnetic Compatibility and Functional Safety, Latest versions available through: www.theiet.org/factfiles.

J. Middleton, The Engineer's EMC Workbook, Marconi Instruments Ltd, Sterenage, UK, 1992. ISBN: 0-95049413-5.

T. Williams, EMC for Product Designers, Newnes, Oxford, 1992. ISBN: 0-75061264-9.

Chapter 6

Ventilation and Air-Conditioning in Powertrain Test Facilities

Engine Testing. DOI: 10.1016/B978-0-08-096949-7.00006-6

PART 1. TEST CELL VENTILATION STRATEGIES

The purpose of air-conditioning and ventilation is the maintenance of an acceptable or specified environment in an enclosed space. This is a comparatively simple matter where only human activity is taking place, but becomes progressively more difficult as the energy flows into and out of the space increase. An engine test cell represents perhaps the most demanding environment encountered in industry. Large amounts of power will be generated in a comparatively small space, surfaces at high temperature are unavoidable, and large flows of cooling water, air and electrical power have to be accommodated, together with rapid variations in thermal load.

An internal combustion engine of whatever breed is essentially an air engine. Whether the air used by the engine comes from the cell's ventilation air or from a special treatment unit outside the cell, the performance and power output of the engine is affected by the condition, temperature, pressure, and humidity of both ingested and surrounding air.

In this chapter the strategies of test cell ventilation are reviewed and the concept of the test cell as an open system is applied to the analysis of thermal loadings and ventilation requirements with a worked example.

Much of this chapter covers the most commonly used method of removing engine-generated heat from the test cell, by forced ventilation using ambient (outside) air that provides a balanced pressure in the closed-cell system. An alternative pressure-balanced system recirculates some or all of the cell air through a conditioning system and this will be described. Open-cell ventilation systems use fans that draw air through the cell that enters through ground-level grilles, usually in an external wall; these can be satisfactory in more limited test conditions than a balanced system and suffer from high noise breakout.

The final choice will be influenced by the range of ambient conditions at a geographic location, building space restraints, and the type of testing being carried out.

TABLE 6.1 Suggested Air Changes per Hour, "Rule of Thumb"

Banks	2−4	Laboratories	5−10
Cafés	10−12	Offices	5−7
Shower rooms	15−20	Toilets	6−10
Garages	6−8	Factories	8−10

Some authorities still use "room volume changes per hour" as a measure of ventilation requirements; in this book the measure is only used to deal with legislative or safety guidelines such as those in Table 6.1.

Ventilation systems for test cells not only remove heat but also prevent the build-up of dangerous levels of gases and vapors. Such requirements are dealt with by specifically designed vapor purge systems (see below) and by ensuring sufficient air flow through the cell even at times of low thermal load.

The correct control strategy for a balanced ventilation system should ensure that the cell space runs slightly below ambient pressure (-50 Pa), thus preventing engine exhaust and hydrocarbon fuel fumes from being pushed into the control and other work areas.

Purge Fans: Safety Requirements to Reduce Explosion Risk

In closed cells using volatile fuels the ventilation system will have to incorporate a purge fan, the purpose of which is to remove heavier-than-air and potentially explosive fumes from the lowest points of the cell complex. An open system fitted with (closable) ground-level air inlet grilles will not be immune from such vapor build-up, particularly if it has floor trenches, so the same purge requirement has to be met.

The purge fan plays an essential role in reducing the fire and explosion risk in a test cell (for relevance to ATEX regulations, see Chapter 4); it has to be of a "nonsparking" construction and certified to be ATEX compliant. The purge system should be integrated into the test cell control system in such a way as to ensure, on cell start-up, it has run for a minimum of 10 minutes, with no hydrocarbon sensors showing an alarm state, before engine ignition and therefore fuel inlet valves are allowed to be energized. The rating of the purge fan may be covered by local legislation, but as a minimum should provide for 30 air changes of the enclosed cell space per hour. The purge suction duct should extend to the lowest point of the cell, which may include any services trench system. In some designs, usually multi-cell installations, the purge fan system is also used for providing exhaust dilution.

Purge fan extraction flow is additive to the outgoing flow of the main ventilation extract fan, which has to be balanced by the inflow volume in balanced systems. The source of the replenishment may be the main inlet fan alone or the combined flow of the combustion air system and the main inlet air fan. Whatever strategy is used, the cell should run, when the doors are closed, at a negative pressure of around 30–50 Pa compared with ambient; any greater than this will make it difficult to initiate door opening. If doors of the cell open for more than a set number of seconds, then the cell can be considered as operating in "workshop mode" with ventilation fans dropping down to low speed and the fuel system turned off.

One method of providing workshop mode conditioning is to design a system whereby treated combustion air flow, warmed or cooled, is balanced with that of the purge fan to provide a space heating system with the main fans switched off. In cells with no combustion air system or circulation through an air handling unit (see below), then a separate "comfort" mode air heating and cooling system may have to be provided for the cell.

Air Handling Units (AHUs)

Recirculation of the majority, or variable proportion, of test cell air through a temperature-conditioning AHU system can have operational advantages over a forced "in-and-out" ambient air system, particularly in cold climates. They may reduce the problem of engine noise breakout through the external air ducting and conveniently provide workshop mode air-conditioning. Designs are usually based on packaged units, supplied with building services, such as chilled and medium-pressure hot water (MPHW) supply. The units used must be designed to deal with air contaminated with the type of particulates produced during some types of engine testing. They are normally mounted above the cell in the services room and aligned on the long axis of the cell to minimize ducting. A rarer alternative arrangement, internal within high-roofed cells such as chassis dynamometer facilities, uses ceiling-mounted units that are based on direct evaporative cooling and electrical heating units.

AHUs, particularly the direct evaporative type, can be energy efficient but a detailed and site-specific calculation of initial and running costs is needed to make a valid comparison. Since closed systems will recirculate entrained pollutants, a separate combustion air system is recommended and a fresh air make-up source will always be required to prevent build-up and to balance the loss through the low-level purge.

The Heat Capacity of Cooling Air

By definition, the test cell environment is mainly controlled by regulating the quantity, temperature, and in some cases the humidity of the air passing through it. Air is not the ideal heat transfer medium: it has low density and low specific heat; it is transparent to radiant heat, while its ability to cool hot surfaces is much inferior to that of liquids.

The main properties of air of significance in air-conditioning may be summarized as follows:

The gas equation:

$$p_a \times 10^5 = \rho R(t_a + 273) \tag{6.1}$$

where p_a = atmospheric pressure (bar), ρ = air density (kg/m^3), R = gas constant for air = 287 J/kg·K, and t_a = air temperature (°C).

Under conditions typical of test cell operation, with $t_a = 25$ °C (77 °F), and standard atmospheric pressure (see Units and Conversion Factors), the density of air, from equation (6.1), is:

$$\rho = \frac{1.01325 \times 10^5}{287 \times 298} = 1185 \text{ kg/m}^3$$

or about 1/850th that of water.

The specific heat at constant pressure of air at normal atmospheric conditions is approximately:

$$C_p = 1.01 \text{ kJ/kg·K}$$

or less than one-quarter that of water.

The air flow necessary to carry away 1 kW of power with a temperature rise of 10 °C is:

$$m = \frac{1}{1.01 \times 10} = 0099 \text{ kg/s} = 0084 \text{ m}^3/\text{s} = 29 \text{ ft}^3/\text{s}$$

This is a better basis for design than any rule of thumb regarding number of cell air changes per hour.

Heat Transfer From the Engine

It is useful to gain a feel for the relative significance of the elements that make up the total of heat transferred from a running engine to its surroundings by considering rates of heat transfer from bodies of simplified form under test cell conditions.

Consider a body of the shape sketched in Figure 6.1. This might be regarded as roughly equivalent, in terms of projected surface areas in the horizontal and vertical directions, to a gasoline engine of perhaps 100 kW maximum power output, although the total surface area of the engine could be much greater. Let us assume the surface temperature of the body to be 80 °C and the temperature of the cell air and cell walls as 30 °C.

Heat loss occurs as a result of two mechanisms: natural convection and radiation. The rate of heat loss by natural convection from a vertical surface in *still air* is given approximately by:

$$Q_v = 1.9(t_s - t_a)^{1.25} \text{ W/m}^2 \tag{6.2}$$

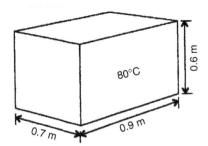

FIGURE 6.1 Simplified model of example 100 kW engine, for analysis of heat transfer to surroundings.

The total area of the vertical surfaces in Figure 6.1 = 1.9 m². The corresponding convective loss is therefore:

$$1.9 \times 1.9 \times (80 - 30)^{1.25} = 480 \text{ W}$$

The rate of heat loss from an upward-facing horizontal surface is approximately:

$$Q_h = 2.5(t_s - t_a)^{1.25} \text{ W/m}^2 \qquad (6.3)$$

giving in the present case a convective loss of:

$$0.63 \times 2.5(80 - 30)^{1.25} = 210 \text{ W}$$

The heat loss from a downward facing surface is about half that for the upward facing case, giving a loss = 110 W.

We thus arrive at a rough estimate of convective loss of 800 W. For a surface temperature of 100 °C, this would increase to about 1200 W.

However, this is the heat loss in still air: the air in an engine test cell is anything but still and very much greater rates of heat loss will occur.

This effect must never be forgotten in considering cooling problems in a test cell: an increase in air velocity greatly increases the rate of heat transfer to the air and may thus aggravate the problem of high air temperatures within a cell (see "Use of Spot Fans" below).

As a rough guide, doubling the velocity of air flow past a hot surface increases the heat loss by about 50%. The air velocity due to natural convection in our example is about 0.3 m/s.

An air velocity of 3 m/s would be moderate for a test cell with ventilating fans producing a vigorous circulation, and such a velocity past the body of Figure 6.1 would increase convective heat loss fourfold, to about 3.2 kW at 80 °C and 4.8 kW at 100 °C.

The rate of heat loss by radiation from a surface depends on the emissivity of the surface (the ratio of the energy emitted to that emitted by a so-called black body of the same dimensions and temperature) and on the temperature difference between the body and its surroundings. Air is essentially transparent to radiation, which thus serves mainly to heat up the surfaces of the surrounding cell; this heat must subsequently be transferred to the cooling air by convection, or conducted to the surroundings of the cell.

Heat transfer by radiation is described by the Stefan–Boltzmann equation, a form of which is:

$$Q_r = 5.77\varepsilon \left[\left(\frac{t_s + 273}{100} \right)^4 - \left(\frac{t_w + 273}{100} \right)^4 \right] \tag{6.4}$$

A typical value of emissivity (ε) for machinery surfaces would be 0.9, t_s = temperature of hot body (°C), and t_w = temperature of enclosing surface (°C).

In the present case:

$$Q_r = 5.77 \times 0.9 \left[\left(\frac{353}{100} \right)^4 - \left(\frac{303}{100} \right)^4 \right] = 370 \text{ W/m}^2$$

In the example, total surface area = 3.16 m^2, giving a radiation heat loss of 1170 W. For a surface temperature of 100 °C, this would increase to 1800 W.

Heat Transfer From the Exhaust System

The other main source of heat loss associated with the engine is the exhaust system. In the case of turbocharged engines, this can be particularly significant. Assume in the present example that exhaust manifold and exposed exhaust pipe are equivalent to a cylinder of 80 mm diameter $\times$ 1.2 m long at a temperature of 600 °C, surface area 0.3 m^2.

Heat loss at this high temperature will be predominantly by radiation and equal to:

$$0.3 \times 5.77 \times 0.9 \left[\left(\frac{873}{100} \right)^4 - \left(\frac{303}{100} \right)^4 \right] = 8900 \text{ W}$$

from equation (6.4).

Convective loss is:

$$0.3 \times 1.9 \times (600 - 30)^{1.25} = 1600 \text{ W}$$

from equation (6.2).

It is clear that this can heavily outweigh the losses from the engine, and points to the importance of reducing the run of unlagged exhaust pipe as much as possible.

The heat losses from the bodies sketched in Figure 6.1 and described above, representing an engine and exhaust system, in surroundings at 30 °C is summarized in Table 6.2.

Heat Transfer From Walls

Most of the heat radiated from the engine and exhaust system will be absorbed by the cell walls and ceiling, also by instrument cabinets and

TABLE 6.2 Heat Losses From the Bodies in Figure 6.1

	Convection	Radiation
Engine, jacket, and crankcase at 80 °C	3.2 kW	1.2 kW
Engine, jacket, and crankcase at 100 °C	4.8 kW	1.8 kW
Exhaust manifold and tailpipe at 600 °C	1.6 kW	8.9 kW

control boxes, and subsequently transferred to the ventilation air by convection.

Imagine the "engine" and "manifold" considered above to be installed in a test cell of the dimensions shown in Figure 6.2. The total wall area is 88 m². Assuming a wall temperature 10 °C higher than the mean air temperature in the cell and an air velocity of 3 m/s, the rate of heat transfer from wall to air is in the region of 100 W/m², or 8.8 kW for the whole wall surface, roughly 90% of the heat radiated from engine and exhaust system; the equilibrium wall temperature is perhaps 15 °C higher than that of the air.

While an attempt to make a detailed analysis on these lines, using exact values of surface areas and temperatures, would not be worthwhile, this simplified treatment may clarify the principles involved.

Sources of Heat in the Test Cell

The Engine

As introduced in Chapter 3, various estimates of the total heat release to the surroundings from a water-cooled engine and its exhaust system have been published. One authority quotes a maximum of 15% of the heat energy in the fuel, divided equally between convection and radiation. This would correspond to about 30% of the power output of a diesel engine and 40% in the case of a gasoline engine.

In the experience of the author, a figure of 40% (0.4 kW/kW engine output) represents a safe upper limit to be used as a basis for design for water-cooled

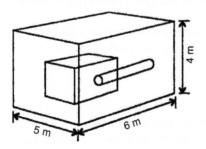

FIGURE 6.2 Simplified test cell for heat transfer calculation.

engines. This is divided roughly in the proportion 0.1 kW/kW engine to 0.3 kW/kW exhaust system. It is thus quite sensitive to exhaust layout and insulation.

In the case of an air-cooled engine the heat release from the engine will increase to about 0.7 kW/kW output in the case of a diesel engine and to about 0.9 kW/kW output for a gasoline engine. The proportion of the heat of combustion that passes to the cooling water in a water-cooled engine, in an air-cooled unit, must of course pass directly to the surroundings. High-powered air-cooled engines, such as aeronautical radial engines, tend to be tested on thrust and torque measuring cradles in "hanger" type buildings with free flow of air through open wall ends, precisely because of the need to provide sufficient cooling flows, cost-effectively.

The Dynamometer

A water-cooled dynamometer, whether hydraulic or eddy current, runs at a moderate temperature and heat losses to the cell are unlikely to exceed 5% of power input to the brake. Usually AC and DC machines are air cooled, and heat loss into the cell is around 6–10% of power input; the contribution to cell air heating of the (rarer) water-cooled type of AC dynamometer is usually negligible for these calculations.

Other Sources of Heat

AC drive cabinets generate heat that may prove significant if they are located in control spaces. Heat output can vary with the manufacturer and the work cycle but for general ventilation calculations a figure of 5% of the maximum rated power of the dynamometer the unit drives should be used. In some cases and dynamometer system models the heat load is more than this 5% of rated power figure, thus with AC drives and devices such as battery simulator cabinets, the manufacturer's specification should be obtained.

Effectively all the electrical power to lights, fans, and instrumentation in the test cell will eventually appear as heat transmitted to the ventilation air. The same applies to the power taken to drive the forced-draught inlet fans: this is dissipated as heat in the air handled by the fans.

The secondary heat exchangers used in coolant and oil temperature control will similarly make a contribution to the total heat load; here long lengths of unlagged pipework holding the primary fluid will not only add heat to the ventilation load, but will adversely affect control of the fluid temperature.

Heat Losses From the Cell

The temperature in an engine test cell is generally higher than usual for an industrial environment. There is thus in some cases appreciable transfer of heat through cell walls and ceiling, depending on the configuration of the site but,

TABLE 6.3 Heat Transfer to Ventilation Air

Heat Source	kW/kW Power Output
Engine, water cooled	0.1
Engine, air cooled	0.7–0.9
Exhaust system (manifold and silencer)	0.3
Hydraulic dynamometer	0.05
Eddy-current dynamometer	0.05
AC or DC dynamometer air blast cooled	0.15

except in the case of a test cell forming an isolated or containerized unit, these losses may probably be neglected.

Recommended values as a basis for the design of the ventilation system are given in Table 6.3. In all cases they refer to the maximum rated power output of the engines to be installed.

Calculation of Ventilation Load

The first step is to estimate the various contributions to the heat load from engine, exhaust system, dynamometer, lights, and services. This information should be summarized in a single flow diagram. Table 6.3 shows typical values.

Heat transfer to the ventilation air is to a degree self-regulating: the cell temperature will rise to a level at which there is an equilibrium between heat released and heat carried away. The amount of heat carried away by a given air flow is clearly a function of the temperature rise ΔT from inlet to outlet.

If the total heat load is H_L kW, then the required air flow rate is:[1]

$$Q_A = \frac{H_L}{101 \times 1.185 \Delta T} = 0.84 \frac{H_L}{\Delta T} \ \text{m}^3/\text{s} \qquad (6.5)$$

Formula (6.5) will be found very useful for general ventilation air flow calculations and is worth memorizing; of course, the main difficulty in its use is the realistic estimation of the value of the heat load H_L.

A temperature rise $\Delta T = 10\ °\text{C}$ is a reasonable basis for design. Clearly, the higher the value of ΔT, the smaller the corresponding air flow. However, a reduction in air flow has two influences on general cell temperature: the higher the outlet temperature, the higher the mean level in the cell, while a smaller air flow implies lower air velocities in the cell, calling for a greater

---

1. For air at standard atmospheric conditions, $\rho = 1.185$ kg/m^3. A correction may be applied when density departs from this value, but is probably not worthwhile.

temperature difference between cell surfaces and air for a given rate of heat transfer.

PART 2. DESIGN OF VENTILATION DUCTS AND DISTRIBUTION SYSTEMS

Pressure Losses

The layout of ventilation ducting in the services space above the cell is made more difficult if headroom is constricted and the ducts have to run at the same level as installed plant such as drive cabinets or combustion air treatment plant. The space layout needs careful advanced planning to obtain an optimum balance between the use and access of floor-mounted equipment and tortuous, energy-wasting, noisy ducting.

The velocity head or pressure associated with air flowing at velocity V is given by:

$$p_v = \frac{\rho V^2}{2} \text{ Pa}$$

This represents the pressure necessary to generate the velocity. A typical value for ρ, the density of air, is 1.2 kg/m^3, giving:

$$p_v = 0.6V^2 \text{ Pa}$$

The pressure loss per meter length of a straight duct is a fairly complex function of air velocity, duct cross-section, and surface roughness. Methods of derivation with charts are given in Ref. [1]. In general, the loss lies within the range 1–10 Pa/m, the larger values corresponding to smaller duct sizes. For test cells with individual ventilating systems, duct lengths are usually short and these losses are small compared with those due to bends and fittings such as fire dampers.

The choice of duct velocity is a compromise depending on considerations of size of ducting, power loss, and noise. If design air velocity is doubled by the size of the ducting being reduced, the pressure losses are increased roughly fourfold, while the noise level is greatly increased (by about 18 dB for a doubling of velocity). Maximum duct velocities recommended are given in Table 6.4.

TABLE 6.4 Maximum Recommended Duct Velocities

Volume of Flow (m^3/s)	Maximum Velocity (m/s)	Velocity Pressure (Pa)
<0.1	8–9	38–55
0.1–0.5	9–11	55–73
0.5–1.5	11–15	73–135
>1.5	15–20	135–240

It is general practice to aim for a flow rate of 12 m/s through noise atten-uators built within ducting.

The total pressure of an air flow p_t is the sum of the velocity pressure and the static pressure p_s (relative to atmosphere):

$$p_t = p_s + p_v = p_s + \frac{pV^2}{2}$$

The design process for a ventilating system includes the summation of the various pressure losses associated with the different components and the choice of a suitable fan to develop the total pressure required to drive the air through the system.

Ducting and Fittings

Various codes of practice have been produced covering the design of ventilation systems. The following brief notes are based on the *Design Notes* published by the Chartered Institution of Building Services.

Galvanized sheet steel is the most commonly used material, and ducting is readily available in a range of standard sizes in rectangular or circular sections. Rectangular section ducting has certain advantages: it can be fitted against flat surfaces and expensive round-to-square transition lengths for connection to components of rectangular section such as centrifugal fan discharge flanges, filters, and coolers are avoided.

Spiral wound galvanized tubing has been used satisfactorily and cost-effectively in workshop and garage space ventilation schemes, but has not proved suitable for test cell work and particularly not in systems carrying diluted engine exhaust, where it suffers from condensate accelerated corrosion.

Once the required air flow rate has been settled and the general run of the ducting decided, the next step is to calculate the pressure losses in the various elements in order to specify the pressure to be developed by the fan. In most cases a cell will require both a forced-draught fan for air supply and an induced-draught fan to extract the air. The two fans must be matched to maintain the cell pressure as near as possible to atmospheric. For control purposes cell pressure is usually set at 50Pa below atmospheric, which gives the safest set of condi-tions concerning door pressurization and fume leakage.

Figure 6.3 shows various components in diagrammatic form and indicates the loss in total pressure associated with each. This loss is given by:

$$\Delta p_t = K_e \frac{pV^2}{2}$$

Information on pressure losses in plant items such as filters, heaters, and coolers is generally provided by the manufacturer.

The various losses are added together to give the cumulative loss in total pressure (static pressure + velocity pressure), which determines the required fan performance.

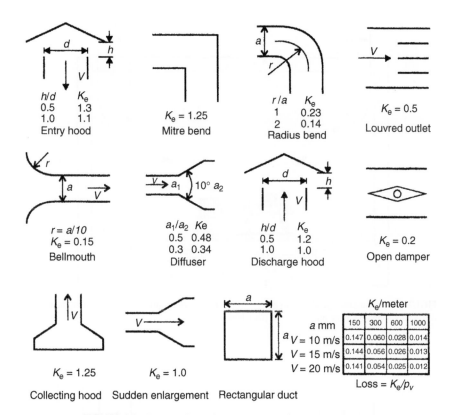

FIGURE 6.3 Pressure losses in components of ventilation systems.

Inlet and Outlet Ducting

The ducts also have to be designed so that noise created in the test cell does not break out into the surrounding environment. The problem has to be solved by the use of special straight attenuating sections of duct that are usually of a larger section and that should be designed in such a way as to give an air velocity around an optimum value of 12 m/s (see "Ventilation Duct Silencers" section below).

The arrangement of air inlets and outlets calls for careful consideration if short-circuiting and local areas of stagnant air within the cell are to be avoided.

There are many possible layouts based on combinations of high-level, low-level, and above-engine direction ducts; four commonly used layouts are shown in Figures 6.4–6.7.

Figure 6.4 shows a commonly used design in cells without large subfloor voids; designers have to ensure that the air flow is directed over the engine rather than taking a higher level short-cut between inlet and outlet. In this and the next design, the purge of floor-level gases would be carried out through the exhaust gas dilution duct (see comments concerning exhaust systems in Chapter 7).

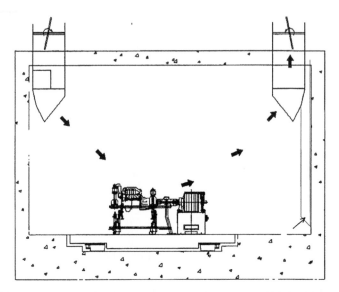

FIGURE 6.4 Variable-speed fans in a pressure-balanced system using high-level inlet and outlet ducts plus supplementary purge duct at floor level.

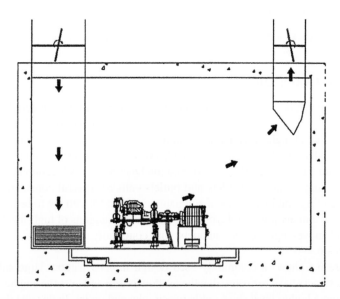

FIGURE 6.5 Low-level inlet and high-level outlet ducted system; the low-level duct can either be part of a balanced fan system as in Figure 6.4 or drawing outside air through an inlet silencer. Purge duct has been omitted but needs to be fitted as in Figure 6.4.

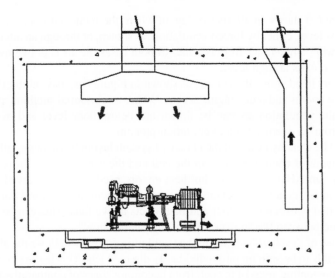

FIGURE 6.6   A balanced fan system as in Figure 6.4, but inlet ducted through nozzles from an over-engine plenum and extracted at low level.

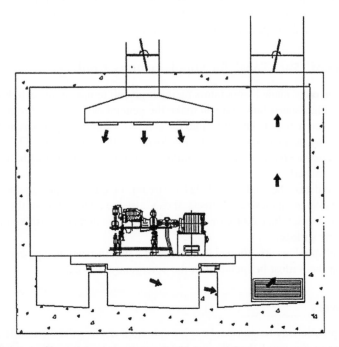

FIGURE 6.7   A balanced fan system in a cell with large subfloor services space, where ventilation air is drawn over the engine and extracted from low level.

Figure 6.5 shows an alternative layout where the intake air comes into the cell at low level, either by forced ventilation, as shown, or through an attenuated duct drawing ambient air from outside. In both cases the air is drawn over the engine and exits at high level.

An inlet hood above the engine, as shown in Figure 6.6, may inhibit the use of service booms and other engine access. Where a substantial subfloor space is present the ventilation air can be drawn out below floor level and over the engine from an overhead inlet ventilation plenum.

It will be clear by now that the choice of system layout has a major influence on the layout of equipment both in the cell and the service space above.

Where an AHU is not fitted but heat energy needs to be conserved in the ventilation air, then a recirculation duct, complete with flow control dampers, can be fitted between the outlet and inlet ducts above fans. This can be set to recirculate between 0% to over 80% of the total airflow.

The use of *"eyeball" nozzles* in the ventilation entry duct allows the air flow into the cells to be split up into individually directed jets. This feature certainly helps in preventing areas of the cell from becoming zones of comparatively still air, as may happen with single outlet entry ducts. However, once the optimized flow has been discovered and unless major movements of plant are made within the cell, the nozzles need to be left fixed, otherwise significantly varied air flows may result in subtle variations in engine performance that will affect repeatability of tests. Due to the distance between such nozzles and the UUT, the problems related to spot fans discussed below should not arise.

The Use of "Spot Fans" for Supplementary Cooling

A common error made by operators facing high cell temperatures is to bring in auxiliary, high-speed fans that can make the general cell temperature problem worse by increasing the heat flow into the cell. Well-directed spot fans and "air-movers" may be successful in reducing *localized* hotspots on the UUT, but it must be remembered that the cell ventilation system has to cope with the extra heat load created.

It should also be understood that a stream, or eddy, of hot air can have unpredictable and undesirable effects; it can raise the temperature of the engine inlet air and upset the calibration of force transducers that are sensitive to a ΔT across their surface.

So that spot fans are stopped in the case of a fire, the switched power outlets supplying them must be included in the emergency stop chain.

Air-Movers

A more precise way to spot-cool engine-mounted devices than a mobile fan is to use an air-mover of the type made by companies such as Brauer®. These devices use a supply of compressed air to induce and entrain surrounding air in

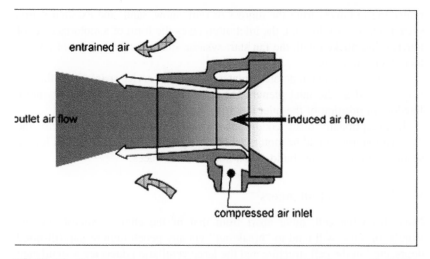

entrained air

outlet air flow

induced air flow

compressed air inlet

FIGURE 6.8 Diagrammatic representation of an "air-mover" or "air flow amplifier" using a compressed air feed for power. These units may be used for spot cooling with more directional precision than a spot fan.

an accelerated jet up to some 100 times the free volume of the compressed supply. Figure 6.8 shows in diagrammatic form these small and relatively quiet spot-cooling devices.

Fire Dampers

Designs of inlet and outlet ducting should obey the rule of hazard containment; therefore, fire dampers must be fitted in "ventilation" ducts at the cell boundary. These devices act to close off the duct and act as a fire barrier at the cell boundary. The two types commonly used are:

- *Normally closed, motorized open.* These are formed by stainless steel louvres within a special framework having their own motorized control gearing and closure mechanism. They are fitted with switches allowing the BMS or cell control system to check their positional status.
- *Normally open, closed through release of a thermal-link-retained steel curtain.* These only operate when the air flow temperature reaches their release temperature and have to be reset manually.

A major advantage of the motorized damper is that when the cell is not running or being used as a workshop space during engine rigging, the flow of outside (cold) air can be cut off.

External Ducting of Ventilation Systems

The external termination of ducts can take various forms, often determined by architectural considerations but always having to perform the function of protecting the ducts from the ingress of rain, snow, sand, and external debris such as leaves. To this end, the inlet often takes the form of a motorized set of louvres, interlocked with the fan start system, immediately in front of coarse, then finer, filter screens.

In some cases, in temperate climates, the whole service space above the cell may be used as the inlet plenum for an air-handling unit; this allows more freedom in position of the inlet louvres.

In exceptionally dusty conditions such as exist in arid desert areas, the external louvres should be motorized to close and be fitted with vertical louvre sections to prevent dust build-up on horizontal surfaces.

Ventilation Duct Silencers

Noise from the cell, quite apart from that of the engine exhaust and the ventilation fans, will tend to "break out" via any penetration or gap (ill-fitted doors, etc.) in the cell structure and the large ventilation ducts are a significant pathway. The classic method of minimizing noise disturbance in the area surrounding a facility is to direct noise vertically through roof-mounted stacks, but this is often a partial solution at best.

Both inlet and outlet air ducting may need to be fitted with attenuators or duct silencers. The effectiveness of such attenuating duct sections is roughly proportional to their length, a commonly made standard found in test cell systems being 2.4 m long. Their internals usually consist of baffles of glass cloth or wool encased in perforated galvanized steel sheet and work on the same principle as the absorption mufflers described in Chapter 7. While attenuation above 500 Hz is usually good, it reduces at the lower frequencies. The balance between building space, noise spectrum, and resistance to air flow is a complicated balance to achieve and requires expert site-specific design work.

Control of Ventilation Systems

The simplest practical ventilation control system is that based on a two-speed extraction fan in the roof of the cell drawing air though a low-level duct at one end of the cell (a variant of Figure 6.5). The fan runs at low speed during times when the cell is used as a workshop and at high speed during engine running. In this case the cell pressure is ambient unless the inlet filter is blocked.

Closed cells require both inlet and extract fans to be fitted with variable-speed drives and a control system to balance them. There are several alternatives, but under a commonly used control strategy operating with a nonrecirculation system, one fan operates under temperature control and the other under pressure. Thus, with both fans running and with a cell pressure of about 50 Pa below ambient, when the cell air temperature rises the extract fan increases in speed to increase air flow; this speed increase creates a drop in cell pressure, which is detected by the control system, so the inlet fan speed is increased until equilibrium is restored. The roles of the fans are somewhat interchangeable, the difference being that in the case described above the control transients tend to give a negative cell pressure and if the control roles are reversed it gives a, less desirable, positive cell pressure.

The fans controlled in the way described above have to be able to deal with the additional flows in and out of the cell by way of purge and combustion air flows where these systems are fitted. The control system also has to be sufficiently damped to prevent surge and major disturbance in the case of a cell door being opened.

Fans

Methods of testing fans and definitions of fan performance are given in British Standard 848-7:2003. This is not an entirely straightforward matter. A brief summary follows.

Again, the control volume technique will be found useful. Figure 6.9 shows a fan surrounded by a control surface.

The various flows into and out of the control volume are as follows:

In	air flow Q_F at velocity V_1 and pressure p_{s1}, power input P_A.
Out	air flow Q_F at velocity V_2 and pressure p_{s2}, where p_{s1} and p_{s2} are the static pressures at inlet and outlet respectively.

Total pressure at inlet:

$$p_{t1} = p_{s1} + \frac{\rho V_1^2}{2}$$

Total pressure at outlet:

$$p_{t2} = p_{s2} + \frac{\rho V_2^2}{2}$$

Fan total pressure:

$$p_{tF} = p_{t2} - p_{t1}$$

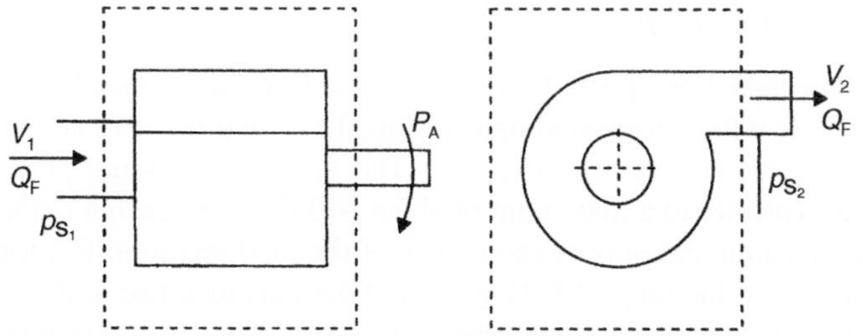

FIGURE 6.9 Centrifugal fans as an open system.

Air power (total):

$$P_{tF} = Q_F P_{tF}$$

Most fan manufacturers quote fan static pressure, which is defined as:

$$p_{sF} = p_{tF} - \frac{\rho V_2^2}{2}$$

This ignores the velocity pressure of the air leaving the fan and does not equal the pressure difference $p_{s2} - p_{s1}$ between the inlet and outlet static pressures. (It is worth noting that, in the case of an axial flow fan with free inlet and outlet, fan static pressure as defined above will be zero.)

The total air power P_{tF} is a measure of the power required to drive the fan in the absence of losses.

The air power (static) is given by:

$$P_{sF} = Q_f\, p_{sF}$$

Manufacturers quote either fan static efficiency η_{sA} or fan total η_{tA}.

The shaft power is given by:

$$P_A = \frac{P_{sF}}{\eta_{sA}} = \frac{P_{tF}}{\eta_{tA}}$$

Fan Noise

Taken in isolation, within a ventilation system the fan is usually the main source of system noise. The noise generated varies as the square of the fan pressure head, so that doubling the system resistance for a given flow rate will increase the fan sound power fourfold, or by about 6 dB.

As a general rule, to minimize noise, ventilation fans should operate as close to the design point as possible.

Reference [2] is recommended for serious students of the subject and gives examples of designs of fan inlet and discharge; a poor design leads to increased noise generation. Reference [3], Part 2, gives guidance on methods of noise testing fans.

Classification of Fans

1. *Axial flow fans.* For a given flow rate an axial flow fan is considerably more compact than the corresponding centrifugal fan, and fits very conveniently into a duct of circular cross-section. The fan static pressure per stage is limited, typically to a maximum of about 600 Pa at the design point, while the fan dynamic pressure is about 70% of the total pressure. Fan total efficiencies are in the range 65–75%. Axial flow fans mounted within a bifurcated duct, so that the motor is external to the gas flow, are a type commonly used in individual exhaust dilution systems. An axial flow fan is a good

choice as a spot fan, or for mounting in a cell wall without ducting, but should not be used in highly contaminated air flows or diluted exhaust systems. Multistage units are available, but tend to be fairly expensive.

2. *Centrifugal fans, flat blades, backward inclined.* This is probably the first choice in most cases where a reasonably high pressure is required, as the construction is cheap and efficiencies of up to 80% (static) and 83% (total) are attainable. A particular advantage is the immunity of the flat blade to dust collection. Maximum pressures are in the range 1–2 kPa.

3. *Centrifugal fans, backward curved.* These fans are more expensive to build than the flat-bladed type. Maximum attainable efficiencies are 2–3% higher, but the fan must run faster for a given pressure and dust tends to accumulate on the concave faces of the blades.

4. *Centrifugal fans, aerofoil blades.* These fans are expensive and sensitive to dust, but are capable of total efficiencies exceeding 90%. There is a possibility of discontinuities in the pressure curve due to stall at reduced flow. They should be considered in the larger sizes where the savings in power cost are significant.

5. *Centrifugal fans, forward curved blades.* These fans are capable of a delivery rate up to 2.5 times as great as that from a backward inclined fan of the same size, but at the cost of lower efficiency, unlikely to exceed 70% total. The power curve rises steeply if flow exceeds the design value.

The various advantages and disadvantages of each fan are summarized in Table 6.5.

Design of Ventilation System: Worked Example

By way of illustration, consider the case of the 250 kW turbocharged diesel engine for which an energy balance is given in Chapter 21, the engine to be coupled to a hydraulic dynamometer.

Ventilation Airflow

Assume convection and radiation losses as follows:

Engine, 10% of power output	25 kW
Exhaust manifold	15 kW
Exhaust tail pipe and silencer	15 kW
Dynamometer, 5% of power input	12 kW
Lights and services	20 kW
Forced-draught fan	5 kW
Subtotal	92 kW
Less losses from cell by conduction	5 kW
Total	87 kW

Assume air inlet temperature is 20 °C, ΔT of 11 °C, therefore an outlet temperature of 31 °C.

TABLE 6.5 Fans: Advantages and Disadvantages

Fan Type	Advantages	Disadvantages
Axial flow	Compact Convenient installation Useful as free-standing units	Moderate efficiency Limited pressure Fairly expensive
Centrifugal, flat blades, backward inclined	Cheap Capable of high pressure Good efficiency Insensitive to dust	
Centrifugal, backward curved	Higher efficiency	More expensive Higher speed for given pressure Sensitive to dust
Centrifugal, aerofoil	Very high efficiency	Expensive Sensitive to dust May stall
Centrifugal, forward curved	Small size for given duty	Low efficiency Possibility of overload

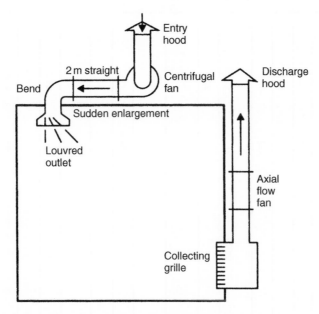

FIGURE 6.10 Simplified cell ventilation system layout use in example calculation. Purge air is taken out in a combined extract duct situated low in the cell.

Then air flow rate, from equation (6.5), is:

$$Q_A = \frac{0.84 \times 87}{11} = 6.6 \text{ m}^3/\text{s}$$

plus induction air, 0.3 m³/s; say 7 m/s in total (= 101 m³/hour per kW engine power output). Assuming cell dimensions of 8 m × 6 m × 4.5 m high, cell volume = 216 m³; this gives 117 air changes per hour.

Table 6.3 suggests a mean duct velocity in the range 15–20 m/s as appropriate, giving a cross-sectional area of 0.37–0.49 m². Heinsohn [2] gives a range of recommended standard duct dimensions of which the most suitable in the present circumstances is 600 mm × 600 mm, giving a duct velocity of 19.5 m/s and a velocity pressure of 228 Pa.

Figure 6.10 shows one possible layout for the ventilation system. The inlet or forced draught system uses a centrifugal fan and the duct velocity assumed above. For the extraction system, with its simpler layout and smaller pressure losses, an axial flow fan has been chosen. The pressure losses are calculated in Table 6.6 (this process lends itself readily to computer programming) and indicates fan duties as follows:

Forced inlet fan
    Flow rate   7 m³/s
    Static pressure   454 Pa
Extraction fan
    Flow rate   7 m³/s
    Static pressure   112 Pa.

TABLE 6.6 Calculation of System Pressure Losses

Item	Size (mm)	Area (m²)	Volume Flow Rate (m³/s)	Velocity (m/s)	Velocity Pressure (Pa)	Fitting Loss Factor K_e	Pressure Drop K_e/m	Length (m)	Pressure Loss (Pa)	Cumulative Loss (Pa)
Forced Draught										
Entry hood	800 dia	0.5	7.0	14	118	1.1			129	129
Straight	800 dia	0.5	7.0	14	118		0.02	2	5	134
Bend	800 dia	0.5	7.0	14	118	0.23			27	161
Fan										
Straight	600 × 600	0.36	7.0	19	217		0.025	2	11	172
Bend	600 × 600	0.36	7.0	19	217	0.23			50	222
Sudden enlargement	600 × 600	0.36	7.0	19	217	1.0			217	439
Louvred outlet	1000 × 1000	1.00	7.0	7	29	0.5			15	454
								Total		454
Extraction										
Collecting hood	1000 dia*	0.79	7.0	9	49	1.25			61	61
Straights	1000 dia	0.79	7.0	9	49		0.02	2	2	63
Fan										
Discharge hood	1000 dia	0.79	7.0	9	49	1.0			49	112
								Total		112

* Diameter of branch connection.

A manufacturer's catalog offers a fan for the forced-draught situation to the following specification:

Centrifugal, backward inclined
    Impeller diameter   900 mm
    Speed   850 rev/min
    Fan static efficiency   65%,

giving, from the equations earlier in the chapter:

$$\text{Shaft power} = \frac{7.0 \times 454}{0.65} = 4900 \text{ W}$$

$$\text{Motor power} = 5 \text{ kW}$$

For the extraction fan, an axial flow unit has the following specifications:

Diameter   1 m
Speed   960 rev/min
Velocity pressure   51 Pa
Fan total efficiency   80%.

$$\text{Shaft power} = \frac{7.0 \times (112 + 51)}{0.80} = 1400 \text{ W}$$

In this case the manufacturer recommends a motor rated at 2.2 kW.

Ventilation of the Control Room

This is in general a much less demanding exercise for the test installation designer. Heating loads are moderate, primarily associated with lights and heat generated by electronic apparatus located in the room. Regulations regarding air flow rate per occupant and general conditions of temperature and humidity are laid down in various codes of practice and should be equivalent to those considered appropriate for offices.

Air-Conditioning

Most people associate this topic with comfort levels under various conditions: sitting, office work, manual work, etc. Levels of air temperature and humidity, along with air change rates, are laid down by various national statutes (in the UK: BS5720:1979). Such regulations must be observed with regard to the control room. However, the conditions in an engine test cell are far removed from the normal, and justify special treatment, which follows.

    Two properties of the ventilating air entering the cell (and, more particularly, of the induction air entering the engine) are of importance: the temperature and the moisture content. Air-conditioning involves four main processes:

- Heating the air
- Cooling the air
- Reducing the moisture content (dehumidifying)
- Increasing the moisture content (humidifying).

Fundamentals of Psychometry

The study of the properties of moist air is known as psychometry. It is treated in many standard texts and only a very brief summary will be given here.

Air-conditioning processes are represented on the psychometric chart in Figure 6.11. This relates the following properties of moist air:

- *The moisture content or specific humidity*, ω kg moisture/kg dry air. Note that even under fairly extreme conditions (saturated air at 30 °C) the moisture content does not exceed 3% by weight.
- *The percentage saturation or relative humidity*, φ. This is the ratio of the mass of water vapor present to the mass that would be present if the air were saturated at the same conditions of temperature and pressure. The mass of vapor under saturated conditions is very sensitive to temperature. A consequence of this relationship is the possibility of drying air by cooling. As the temperature is lowered the percentage saturation increases until at the dew point temperature the air is fully saturated and any further cooling results in the deposition of moisture.

Temperature (°C)	10	15	20	25	30
Moisture content ω (kg/kg)	0.0076	0.0106	0.0147	0.0201	0.0273

- *The wet- and dry-bulb temperatures.* The simplest method of measuring relative humidity is by means of a wet- and dry-bulb thermometer. If unsaturated air flows past a thermometer having a wetted sleeve of cotton around the bulb the temperature registered will be less than the actual temperature of the air, as registered by the dry-bulb thermometer, owing to evaporation from the wetted sleeve. The difference between the wet- and dry-bulb temperatures is a measure of the relative humidity. Under saturated conditions the temperatures are identical, and the depression of the wet-bulb reading increases with increasing dryness. Wet- and dry-bulb temperatures are shown in a psychometric chart.
- *The specific enthalpy of the air*, relative to an arbitrary zero corresponding to dry air at 0 °C. We have seen earlier that on this basis the specific enthalpy of dry air,

$$h = C_p t_a = 1.01 t_a \text{ kJ/kg} \tag{6.6a}$$

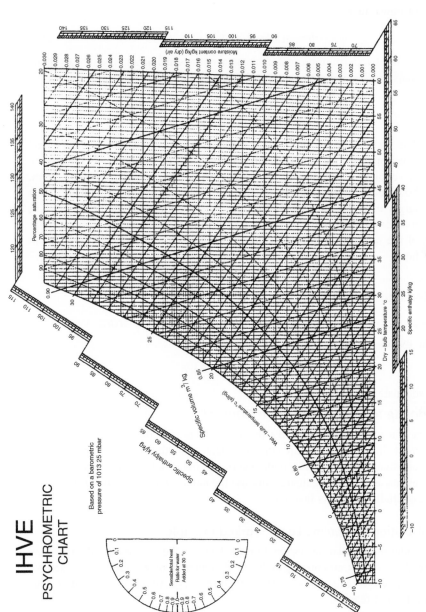

FIGURE 6.11 Psychometric chart.

The specific enthalpy of *moist* air must include both the sensible heat and the *latent heat of evaporation* of the moisture content. The specific enthalpy of moist air is:

$$h = 1.01t_a + \omega(1.86t_a + 2500) \tag{6.6b}$$

the last two terms representing the sum of the sensible and latent heats of the moisture. Taking the example of saturated air at 300 °C:

$$h = 1.01 \times 30 + 0.0273(1.86 \times 30 + 2500)$$
$$= 30.3 + 1.5 + 68.3 \text{ kJ/kg}$$

The first two terms represent the sensible heat of air plus moisture, and it is apparent that ignoring the sensible heat of the latter, as is usual in air cooling calculations, introduces no serious error. The third term, however, representing the latent heat of the moisture content, is much larger than the sensible heat terms. This accounts for the heavy cooling load associated with the process of drying air by cooling: condensation of the moisture in the air is accompanied by a massive release of latent heat.

Air-Conditioning Processes

Heating and Cooling Without Deposition of Moisture
The expression for dry air:

$$H = \rho Q_A C_p \Delta T \text{ kW} \tag{6.7}$$

is adequate.

Cooling to Reduce Moisture Content
This is a very energy-intensive process, and is best illustrated by a worked example. Increasing moisture content is achieved either by spraying water into the air stream (with a corresponding cooling effect) or by steam injection. With the appearance of Legionnaires' disease the latter method, involving steam that is essentially sterile, is favored.

Calculation of Cooling Load: Worked Example
Consider the ventilation system described above. We have assumed an air flow rate $Q_A = 7$ m³/s, air entering at 20 °C. If ambient temperature is 25 °C and we are required to reduce this to 20 °C, then from equation (6.7), assuming saturation is not reached,

$$\text{Cooling load} = 1.2 \times 7.0 \times 1.01 \times 5 = 42.4 \text{ kW}$$

Now let us assume that the ambient air is 85% saturated, $\varphi = 0.85$ and that we need to reduce the relative humidity to 50% at 20 °C.

The psychometric chart shows that the initial conditions correspond to a moisture content:

$$\omega_1 = 0.0171 \text{ kg/kg}$$

The final condition, after cooling and dehumidifying, corresponds to:

$$\omega_2 = 0.0074 \text{ kg/kg}$$

The chart shows that the moisture content ω_2 corresponds to saturation at a temperature of 9.5 °C.

From equation (6.6b) the corresponding specific enthalpies are:

$$h_1 = 1.01 \times 25 + 0.0171(1.86 \times 25 + 2500) = 68.80 \text{ kJ/kg}$$
$$h_2 = 1.01 \times 9.5 + 0.0074(1.86 \times 9.5 + 2500) = 28.22 \text{ kJ/kg}$$

The corresponding cooling load is:

$$L_C = 1.2 \times 7.0 \times (68.80 - 28.22) = 341 \text{ kW}$$

If it is required to warm the air up to the desired inlet temperature of 20 °C, the specific enthalpy is increased to:

$$H_3 = 1.01 \times 20 + 0.0074(1.86 \times 20 + 2500) = 39 \text{ kJ/kg}$$

The corresponding heating load is:

$$L_H = 1.2 \times 7.0 \times (39.00 - 28.22) = 90.5 \text{ kW}$$

To summarize:

Air flow of 7 m³/s at 25 °C, 85% saturated

Cooling load to reduce temperature to 20 °C	42.4 kW
Cooling load to reduce temperature to 9.5 °C with dehumidification	341 kW
Heating load to restore temperature to 20 °C 50% saturated	90.5 kW

Cooling of air is usually accomplished by heat exchangers fed with chilled water. See Ref. [4] for a description of liquid chilling packages.

This illustrates the fact that any attempt to reduce humidity by cooling as opposed to merely reducing the air temperature without reaching saturation conditions imposes very heavy cooling loads.

Calculation of Humidification Load: Worked Example

Let us assume initial conditions $Q_A = 7$ m³/s, temperature 20 °C, relative humidity $\varphi = 0.3$, and that we are required to increase this to $\varphi = 0.7$ for some experimental purpose.

Relative moisture contents are:

$$\omega_1 = 0.0044 \text{ kg/kg}$$
$$\omega_2 = 0.0104 \text{ kg/kg}$$

Then,

Rate of addition of moisture $= 7 \times 1.2 \times (0.0104 - 0.0044) = 0.050$ kg/s

Taking the latent heat of steam as 2500 kJ/kg,

$$\text{Heat input} = 2500 \times 0.050 = 125 \text{ kW}$$

This is again a very large load, calling for a boiler of at least this capacity.

Effects of Humidity: A Warning

Electronic equipment is extremely sensitive to moisture. Large temperature changes, when associated with high levels of humidity, can lead to the deposition of moisture on components such as circuit boards, with disastrous results.

This situation can easily arise in hot weather: the plant cools down overnight and dew is deposited on cold surfaces. Some protection may be afforded by continuous air-conditioning. The use of chemical driers is possible; the granular substances used are strongly hygroscopic and are capable of achieving very low relative humidities. However, their capacity for absorbing moisture is of course limited and the container should be removed regularly for regeneration by a hot air stream.

The operational answer, adopted by many facilities, is to leave vulnerable equipment switched on in a quiescent state while not in use, or to fit anti-condensation heaters at critical points in cabinets. In the period of facility construction, between delivery of equipment and commissioning, it is vital to fit an auxiliary power supply to run a rudimentary set of anticondensation heaters such as a chain of 60 W incandescent light bulbs.

Legionnaires' Disease

This disease is a severe form of pneumonia and infection is usually the result of inhaling water droplets carrying the causative bacteria (*Legionella pneumophila*). Factors favoring the organism in water systems are the presence of deposits such as rust, algae and sludge, a temperature between 20 and 45 °C, and the presence of light. Clearly all these conditions can be present in systems involving cooling towers.

Preventive measures include:

- Treatment of water with scale and corrosion inhibitors to prevent the build-up of possible nutrients for the organisms.
- Use of suitable water disinfectant such as chlorine (1–2 ppm) or ozone.
- Steam humidifiers are preferable to water spray units.

If infection is known to be present, flushing, cleaning, and hyperchlorination are necessary. If a system has been out of use for some time, heating to about 70–75 °C for 1 hour will destroy any organisms present.

This matter should be taken seriously. This is one of the few cases in which the operators of a test facility may be held criminally liable, with consequent ruinous claims for compensation.

PART 3. COMBUSTION AIR (CA) TREATMENT AND CLIMATIC CELLS

The influence of the condition of the combustion air (its pressure, temperature, humidity, and purity) on engine performance and variations in these factors can have a very substantial effect on performance. In an ideal world engines under test would all be supplied with air in "standard" conditions. In practice there is a trade-off between the advantages of such standardization and the cost of achieving it.

For routine (non-emissions) production testing, variations in the condition of the air supply are not particularly important, but the performance recorded on the test documents, requiring some degree of correlation with other test beds or facilities, should be corrected to standard conditions.

The simplest and most widely used method of supplying the combustion air is to allow the engine to take its supply directly from the test cell atmosphere. The great advantage of using cell air for vehicle engines in particular is that rigging of complex "in-vehicle" air filtration and ducting units is straightforward.

The major disadvantage of drawing the air from within the cell is the uncontrolled variability in temperature and quality arising from air currents and other disturbances in the cell. These can include contamination with exhaust and other fumes and may be aggravated by the use of spot fans.

For research and development testing, and particularly critical exhaust emissions work, it is necessary that, so far as is practicable, the combustion air should be supplied with the minimum of pollution and at constant conditions of temperature, pressure, and humidity. While the degree of pollution is a function of the choice of site, the other variables may be controlled in the following order of difficulty:

- Temperature only
- Temperature and humidity
- Temperature, humidity, and pressure
- Temperature and dynamic pressure control during transient engine conditions.

Centralized Combustion Air Supplies

Centralized combustion air-conditioning units can be designed to supply air at "standard" conditions to a number of cells, and this may be a cost-effective solution providing that all cells require the same standardized conditions of combustion air and if the air is used to condition nonrunning cell spaces. Feedback from several sites indicates that individual units offer the best operational solution for the majority of R&D test cells.

TABLE 6.7 Rating of Two Standard CA Treatment Units

Model	Flow Rate	Suitable for Engine of Rating Below
Unit 1	800 m³/h	Diesel 140 kW, gasoline 200 kW
Unit 2	1600 m³/h	Diesel 280 kW, gasoline 400 kW

Specification of Operational Envelope for Humidity and Temperature Control

To create a realistic operational specification for temperature and humidity control of combustion air requires that the creator has a clear understanding of a psychometric chart (Figure 6.11). The implications of requiring a large operational envelope plus a clear understanding of the temperature and humidity points required by the planned testing regimes are vital. It is recommended that cell users mark on a psychometric chart the operating points prescribed by known (emission homologation, etc.) tests or investigate the specification of available units before producing specifications that may impose unnecessarily high energy demands on the requested unit.

Dedicated Combustion Air Treatment Units

It should be noted that standard combustion air treatment units are sold by specialist companies worldwide. The ratings of these units match the common sizes of automotive engines, but due to the different air consumption of gasoline and diesel engines the power rating of the engines supported by one unit will differ for the two engine types as shown in a typical example in Table 6.7.

Temperature-Only Control, Flooded Inlet

The simplest form controls air temperature only and supplies an excess of air via a flexible duct terminating in a trumpet-shaped nozzle close to the engine's inlet; this is known as a "flooded" inlet. The unit should be designed to supply air at a constant volume, calculated by doubling the theoretical maximum air demand of the engine. Because the flow is constant, good temperature control can be achieved; the excess air is absorbed into the cell ventilation inlet flow and compensated for by the ventilation temperature and pressure control.

The normal design of such units contains the following modules:

- Inlet filter and fan
- Chiller coils

- Heater matrix
- Insulated duct, taken through a fire damper, terminating in flexible section and bell-mouth.

By holding the chiller coils' temperature at a fixed value of say 7 °C and heating the air to the required delivery temperature, it is possible to reduce the effect of any humidity change during the day of testing without installing full humidity control.

Switching combustion air units on and off frequently will cause the system control accuracy to decline and therefore it is sensible to use the delivered air in a space heating role when the engine is not running, allowing the main ventilation system to be switched off or down to a low extract level. An energy-efficient comfort ventilation regime can be designed by using the combustion air balanced with the purge fan extraction.

Humidity-Controlled Units

Systems for the supply of combustion air in which both temperature and humidity is controlled are expensive, the expense increasing with the range of conditions to be covered and the degree of precision required. They are also energy intensive, particularly when it is necessary to reduce the humidity of the atmospheric air by cooling and condensation. Before introducing such a system a careful analysis should be made to ensure that the operational envelope proposed is achievable from a standard unit and the services available to it and, if not, if it is really justified.

It is sensible to consider some operational restraints on the scheduling of tests having different operational levels of humidity in order to avoid condensation build-up in the system. Corrugated ducting should not be used in systems providing humid air, since they tend to collect condensate that has to dry out before good control can be restored.

To indicate more exactly what is involved in full conditioning of combustion air, let us suppose that it is necessary to supply air at standardized conditions to the 250 kW diesel engine for which the energy balance is given in Chapter 21, Table 21.3.

The calculated air consumption of the engine is 1312 kg/h, corresponding to a volumetric flow rate of about 0.3 m^3/s.

The necessary components of an air-conditioning unit for attachment to the engine air inlet are shown diagrammatically in Figure 6.12 and may comprise in succession:

1. Air inlet with screen and fan
2. Heater, either electrical or hot water
3. Humidifier, comprising steam or atomized water injector with associated generator
4. Cooling element, with chilled water supply
5. Secondary heating element.

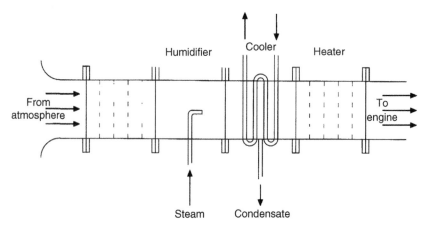

FIGURE 6.12 Schematic of typical combustion air system contents.

It may be asked why it is necessary to include two heater elements, one before and one after the humidifier. The first element is necessary when it is required to humidify cold dry air, since if steam or water spray is injected into cold air supersaturation will result and moisture will immediately be deposited. On the other hand, it is commonly necessary, in order to dry moist air to the desired degree, to cool it to a temperature lower than the desired final temperature and reheat must be supplied downstream of the cooler.

The internal dimensions of the duct in the present example would be approximately 0.3 m × 0.3 m. Assume that the air is to be supplied to the engine at the standard conditions specified in BS 5514 (ISO 3046):

Temperature 25 °C
Relative humidity (r.h.) 30%.

Let us consider two fairly extreme atmospheric conditions and determine the conditioning processes necessary:

Hot and humid: 35 °C and 80% r.h.
Cool to 7 °C, cooling load 34 kW (flow of condensate 0.5 l/min)
Reheat to 25 °C, heating load 6.5 kW.

It is necessary to cool to 7 °C in order to reduce the moisture content, originally 0.030 kg water/kg air, to the required value, 0.006 kg water/kg air; the latter corresponds to saturation at 7 °C:

Cold and dry: 0 °C and 50% r.h.
Heat to 25 °C, heating load 9 kW
Inject steam at rate of 0.1 l/min, heating load 4 kW.

A typical standard unit available for this type of duty has the following technical details:

Unit 2 with delivery of 1600 m³/h at temperature range 15–30 °C with option of humidity control 8–20 g/kg:

Total power consumption	64 kVA
C.W. supply at	4–8 °C
Pressure	2–4 bar
Flow rate	8 m³/h.

Note the above equates to about 68 kW, pressure across coils approx. 120 kPa.

Demineralized water for humidifier:

Temperature range	10–30 °C
Pressure	2–4 bar
Flow rate	65 dm³/h
Condensate drain	1/2″
Flow rate	80 dm³/h
Compressed air	6–8 bar
Ambient air temp. range	10–35 °C
(in order for unit to deliver air as per top line)	
Humidity	3–30 g/kg

Size of unit: length 2755 mm × width 1050 mm × height 2050 mm.

It will be observed that the energy requirements, particularly for the chiller either internal or from separate chilled water supply, are quite substantial. In the case of large engines and gas turbines, any kind of combustion air-conditioning is not really practicable and reliance must be placed on correction factors.

Any combustion air supply system must be integrated with the fire alarm and fire extinguishing system, and must include the same provisions for isolating individual cells as the main ventilation system.

Important System Integration Points Concerning CA Systems

Combustion air units of all types will require condensate drain lines with free gravity drainage to a building system; backup or overspill of this fluid can cause significant problems. Those fitted with steam generators will require both condensate and possibly steam "blow-down" lines piped to safe drainage.

If steam generators are fitted the condition of feed water will be critical and suitable water treatment plant needs to be installed, particularly where supply water is hard.

Pressure- and Temperature-Controlled Combustion Air Units

A close coupled air supply duct attached to the engine is necessary in special cases such as anechoic cells, where air intake noise must be eliminated, at sites running legislative tests that are situated at altitudes above those where standard conditions can be achieved, and where a precisely controlled conditioned supply is needed.

Since engine air consumption varies more or less directly with speed, a parameter can vary more rapidly than any air supply system can respond, and since it is essential not to impose pressure changes on the engine air, some form of excess air spill or engine bypass strategy has to be adopted to obtain pressure control with any degree of temperature and humidity stability.

For testing at steady-state conditions the pressure control system can operate with a slow and well-damped characteristic, typically having a stabilization time, following an engine speed set-point change, of 30 seconds or more. This is well suited to systems based on regulating the pressure of an intake plenum fitted with a pressure-controlling spill valve/flap.

The inlet duct must be of sufficient size to avoid an appreciable pressure drop during engine acceleration and it needs to be attached to the engine inlet in such a way as to provide a good seal without imposing forces on the engine.

It is sensible to encapsulate or connect to the normal engine air inlet filter, if it can sustain the imposed internal pressure imposed by the supply system. If the filter has to be encapsulated it often requires a bulky rig item that has to be suspended from a frame above the engine.

Various strategies have been adopted for fast pressure control, the best of which are able to stabilize and maintain pressure at set point within <2 seconds of an engine step change of 500 rpm.

Control of inlet pressure only does not simulate true altitude; this requires that both inlet and exhaust are pressurized at the same pressure, a condition that can be simulated using a dynamic pressure control system patented by AVL List GmbH, as shown in Figure 6.13. Such systems as those shown in Figure 6.13 find use in motor sport and other specialist applications, but where the "ram effect" of the vehicle's forward motion is required, as is the case in F1 and motor cycle engine testing, then considerable energy is required to move the air and the test cell begins to resemble a wind tunnel.

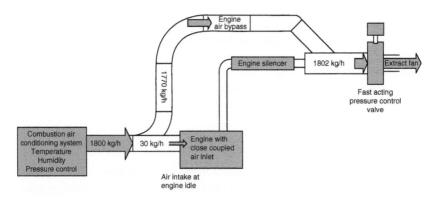

FIGURE 6.13 AVL dynamic combustion air pressure control system—Example flows shown at engine idle condition.

Climatic Testing of Engine and Powertrains

Cold chambers built exclusively for powertrain testing are quite rare, since much of the research work in cold starting and operation in extreme environments tends to be carried out in whole vehicle systems (see below). However, specialist research by companies involved in battery, lubricant, and materials research has climatic cells with controlled environments in the range between +40 and −40 °C. For relatively simple cold-start work, a common technique is to use a modified refrigerated ISO container (see containerized cells in Chapter 4). An alternative method is to circulate cold fluids around the engine while it is encased in a demountable, insulated "tent" fed by a supply of refrigerated air. This allows the engine to be "cold-soaked" down to the temperature required for legislative Cold Start emission tests. For general requirements for special materials and design features required in test equipment operating in low-temperature environments, see the relevant section in Chapter 17 on chassis dynamometers.

Climatic Test Cells for Vehicles

There is a requirement for this kind of facility for development work associated with various aspects of vehicle and powertrain performance under extreme climatic conditions. Subjects include drivability, cold starting, fuel waxing and solar loading, vapor lock, air-conditioning, and vehicle climatic control.

The design of air-conditioning plants for combustion air has been discussed earlier in this chapter, but the design of an air-conditioned chamber for the testing of complete vehicles is a very much larger and more specialized problem. Figure 6.14 is a schematic drawing of such a chamber, built in the form of a recirculating wind tunnel so that the vehicle on test, mounted on a chassis dynamometer, can be subjected to oncoming air flows at realistic velocities, and in this particular case over a temperature range from +40 to −30 °C.

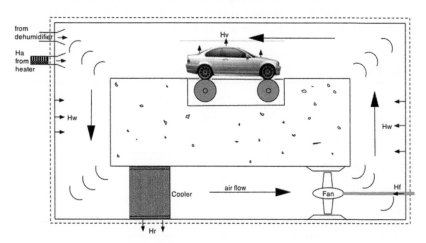

FIGURE 6.14   Schematic of climatic wind tunnel facility fitted with a chassis dynamometer.

The outstanding feature of such an installation is its very large thermal mass and, since it is usually necessary, from the nature of the tests to be performed, to vary the temperature of the vehicle and the air circulating in the tunnel over a wide range during a prolonged test period, this thermal mass is of prime significance in sizing the associated heating and cooling systems.

The thermodynamic system of the chamber is indicated in Table 6.8 and, using actual site data, the energy inflows and outflows during a particular cold test, in accordance with the methods described in Chapter 3, were as follows:

C_a Air content of chamber, return duct, etc. volume approximately 550 m³, approximate density 1.2 kg/m³, $C_p = 1.01$ kJ/kg·K, thermal mass $= 550 \times 1.2 \times 1.01 = 670$ kJ/°C.

C_v Vehicle, assume a commercial vehicle, weight 3 tonnes, specific heat of steel approximately 0.45 kJ/kg·K thermal mass, say $3000 \times 0.45 = 1350$ kJ/°C.

C_e Fan, cooler matrix, internal framing, etc., estimated at 2480 kJ/°C.

C_s Structure of chamber. This was determined by running a test with no vehicle in the chamber and a low fan speed. The rate of heat extraction by the refrigerant circuit was measured and a cooling curve plotted. Concrete is a poor conductor of heat and the coefficient of heat transfer from surfaces to air is low. Hence, during the test a temperature difference between surfaces and wall built up but eventually stabilized; at this point the rate of cooling gave a true indication of effective wall heat capacity. The test showed that the equivalent thermal mass of the chamber was about 26,000 kJ/°C, much larger than the other elements. Concrete has a specific heat of about 8000 kJ/m³ °C, indicating that the "effective" volume involved was about 26,000/8000 = 3.2 m³.

The total surface area of the chamber is about 300 m², suggesting that a surface layer of concrete about 1 cm thick effectively followed the air temperature. The thermal capacity (thermal mass) of the various elements of the system is defined in terms of the energy input required to raise the temperature of the element by 1 °C.

Adding these elements together, we arrive at a total thermal mass of 30,500 kJ/°C and since 1 kJ = 1 kW s, we can derive the rate at which the temperature of the air in the chamber may be expected to fall for the present case.

This shows that the "surplus" cooling capacity available for cooling the chamber and its contents, H_w, amounts to 221 kW. This could be expected to

TABLE 6.8 Thermodynamic System of the Chamber

In	Energy	Out	Energy
H_a make-up air	20 kW	H_r refrigerator circuit	495 kW
H_v vehicle convection and radiation	100 kW		
H_r tunnel fan power	154 kW		
H_w walls and chamber	221 kW		
Totals	495 kW		495 kW

achieve a rate of cooling of $221/30,500 = 0.0072$ °C/s, or 1 °C in 2.3 minutes. However, this is the final rate when the temperature difference between walls and air had built up to the steady-state value of about 20 °C. The observed initial cooling rate is much faster, about 1.5 °C/min. Heating presents less of a problem, since the heat released by the test vehicle and the fan assist the process rather than opposing it.

A further effect is associated with the moisture content of the tunnel air. On a warm summer's day this could amount to about 10 kg of water and during the cooling process this moisture would be deposited mostly on the cooler fins, where it would eventually freeze, blocking the passages and reducing heat transfer. To deal with this problem it is necessary to include a dehumidifier to supply dried air to the tunnel circuit.

Note that massive concrete constructions take several months to "dry out" under average temperate conditions; therefore, during the early commissioning of installations like the one described above, increased humidity and freezing of the cooler matrix can be a significant problem.

Wind Tunnels

At the extreme end of any discussion concerning test facility ventilation, the design and operation of automotive wind tunnels deserves mention. This is a specialist field and students needing to study the subject in detail are recommended to read *Aerodynamics of Road Vehicles* detailed at the end of the chapter.

Aerodynamic testing is receiving increasing investment, not only because of the continued need to use less fuel through reducing a vehicle's drag coefficient (C_d), but because of the more complex cooling flow requirements of the new hybrid vehicles, all of which tend to increase drag.

The energy required to move air in a laminar flow through an enclosed space in which the test object is placed rises almost exponentially with the speed of flow and the volume moved. However, much of the work is scalable; therefore, some of the tunnels needing to work at the highest speeds are, for example, one-third or one-half scale. In spite of this, full-size tunnels requiring 3.5 MW fans and operating at over 250 kph have been built. Designs are both closed designs such as that of climatic versions (Figure 6.13) and semi-open designs where the air is normally sucked through the test space rather than blown. Full-sized, light-vehicle tunnels have test area spaces that vary considerably in size, but typically would need to be around 16 m long by 8 m wide and 5 m high.

A particular design feature of automotive wind tunnels, not shared with aeronautical facilities, is that the test vehicle has to be in contact with a flat road surface with which it has an ever-changing force relationship, much of which depends on the aerodynamic forces working upon the vehicle. These forces, all transmitted through the road/tire contact points, have to be measured by a "wind tunnel balance" on which the vehicle sits. This balance system is made up of strain gauge flexures that are designed to respond to load. They measure multiple loads and moments by the use of multiple individual flexures, each one measuring load on just one axis. The balance is frequently mounted on a turntable so that phenomena such as side-wind stability can be measured.

One problem in real-life simulation within an automotive wind tunnel is that the "real", and aerodynamically important, situation of the air being stationary relative to the road while moving relative to the vehicle is expensive to simulate; this requires a "moving-ground" installation that a roller-based chassis dynamometer does not provide. However, moving-ground or "flat-track" chassis dynamometers have been made and are briefly described in Chapter 17 of this book.

While the forces resulting from air flow can be measured, the means by which they are created is difficult to visualize since air is invisible. Various means of air-flow visualization are used, from the simple "smoke wand" that vaporizes oil or propylene glycol to produce a dense white vapor stream, to jets of helium-filled microbubbles, all of which need to be supported by motion-capture cinema-photography units.

SUMMARY

Design of the ventilation system for a test cell is a major undertaking, and a careful and thorough analysis of expected heat loads from engine, exhaust system, dynamometer, instrumentation, cooling fans, and lights is essential if subsequent difficulties are to be avoided. Once the maximum heat load has been determined it is recommended that a diversity factor is applied so as to arrive at a realistic rating for the cell services.

Any cell using gasoline or other volatile fuels should include within the ventilation system a purge system to remove potentially explosive vapor.

The choice between full-flow ventilation and air-conditioning has to be made after a detailed review of site conditions, particularly building space, and cost.

If a separate air supply for the engine is to be used, careful consideration has to be made concerning its type and rating in order to keep costs and energy requirements in line with actual test requirements.

The ventilation of the control room should ensure that conditions there meet normal office standards.

Specialist facilities for climatic and aerodynamic work require designers and contractors that have the relevant experience in the automotive sector.

NOTATION

Atmospheric pressure	p_a (bar)
Atmospheric temperature	t_a (°C)
Density of air	ρ (m^3)
Gas constant for air	$R = 287$ J/kg·K

Specific heat of air at constant pressure	C_p (J/kg·K)
Mass rate of flow of air	m (kg/s)
Rate of heat loss, vertical surface	Q_v (W^2)
Rate of heat loss, horizontal surface	Q_h (W^2)
Temperature of surface	t_s (°C)
Rate of heat loss by radiation	Q_r (W/m^2)
Ventilation air temperature rise	ΔT (°C)
Total heat load	H_L (kW)
Ventilation air flow rate	Q_A (m^3/s)
Fan air flow rate	Q_F (m^3/s)
Velocity of air	V (m/s)
Velocity pressure	p_v (Pa)
Static pressure	p_s (Pa)
Total pressure	p_t (Pa)
Pressure loss	Δp_t (Pa)
Static air power	P_{sF} (kW)
Total air power	P_{tF} (kW)
Shaft power	P_A (kW)
Moisture content	ω (kg/kg)
Relative humidity	φ
Specific enthalpy of moist air	h (kJ/kg)
Cooling load	L_C (kW)
Heating load	L_H (kW)
Emissivity	ε
Pressure loss coefficient	K_e

REFERENCES

[1] TM 8 Design Notes for Ductwork, Chartered Institution of Building Services, London.
[2] R.J. Heinsohn, Industrial Ventilation Engineering Principles, Wiley, Chichester, 1991. ISBN-13: 978-0471637035.
[3] British Standard 848-7. Fans for General Purposes (2003).
[4] BS 7120. Specification for Rating and Performance of Air to Liquid and Liquid to Liquid Chilling Packages.

FURTHER READING

BS 599. Methods of Testing Pumps.
BS 2540. Specification for Granular Desiccant Silica Gel.
BS 5720. Code of Practice for Mechanical Ventilation and Air Conditioning in Buildings.
BS 6339. Specification for Dimensions of Circular Flanges for General Purpose Industrial Fans.
CIBS Psychrometric Chart. Chartered Institution of Building Services, London.
H.G. Freeston, Test bed installations and engine test equipment, Proc. I. Mech. E 172 (7) (1958).
J.P. Holman, Heat Transfer, McGraw-Hill Education, 2009. ISBN-13: 978-0071267694.
W.-H. Hucho (Ed.), Aerodynamics of Road Vehicles, fourth ed., SAE, 1998. ISBN-13: 978-0768000290.
E.G. Pita, Air Conditioning Principles and Systems: An Energy Approach, Wiley, Chichester, 1981. ISBN-13: 978-0135053065.

Test Cell Cooling Water and Exhaust Gas Systems

Chapter Outline

Engine Testing. DOI: 10.1016/B978-0-08-096949-7.00007-8
151

PART 1. COOLING WATER SUPPLY SYSTEMS

The cooling water system for any heat engine test facility has to provide water of suitable quality, temperature, and pressure to allow sufficient volume to pass through the equipment in order for it to have adequate cooling capacity. The water pressure and flow have to be sufficiently constant to enable the devices supplied to maintain control of temperature of engine and transmission fluids. It is essential for purchasers of water-cooled plant to carefully check the inlet water temperature specified for the required performance, since the higher the facility cooling water inlet temperature supplied by the factory, the less work the device will be capable of performing before the maximum allowable exit temperature is reached. In this book the water supplied by the test facility is considered as the *primary* circuit, while the fluid flowing through the unit under test (oil, engine coolant, etc.) are the *secondary* circuits. There are several different terms used in the English language for the same primary circuit fluid, including mains water, plant water, factory water, and raw water. Most, but not all, automatically controlled heat exchangers in the powertrain test industry have valves controlling the flow of the secondary fluid through a heat exchanger that is fed by a constant flow from the primary circuit, as is shown later in this chapter.

Water

Water is an ideal solvent that is only met in its chemically pure form in laboratories. Unless stated otherwise, this chapter refers to water as the treated, potable liquid generally available in city supplies within a developed industrial country. Water is the ideal liquid cooling medium. Its specific heat is higher than that of any other liquid, roughly twice that of liquid hydrocarbons. It is of low viscosity, widely available and, provided it does not contain sufficient quantities of dissolved salts that make it aggressive to some metals, relatively noncorrosive. The specific heat of water is usually taken as $C = 4.1868$ kJ/kg·K. (**Note**: This is the value of the "international steam table calorie" and corresponds to the specific heat at 14 °C. The specific heat of water is very slightly higher at each end of the liquid phase range: 4.21 kJ/kg·K at 0 °C and at 95 °C, but these variations may be neglected in general industrial use.)

The use of antifreeze (ethylene glycol, $H_2H_6O_2$) as an additive to water permits its operation as a coolant over a wider range of temperatures. A 50% by volume solution of ethylene glycol in water permits operation down to −33 °C. Ethylene glycol also raises the boiling point of the coolant such that a 50% solution will operate at a temperature of 135 °C with pressurization of 1.5 bar. The specific heat of ethylene glycol is about 2.28 kJ/kg·K and, since its density is 1.128 kg/l, the specific heat of a 50% by volume solution is:

$$(0.5 \times 4.1868) + (0.5 \times 2.28 \times 1.128) = 3.38 \text{ kJ/kg·K}$$

or 80% of that of water alone. Thus, the circulation rate must be increased by 25% for the same heat transfer rate and temperature rise. The relation between flow rate, q_w (liters per hour), temperature rise, ΔT, and heat transferred to the water is:

$$4.1868 q_w \Delta T = 3600H$$
$$q_w \Delta T = 860H$$

where $H =$ heat transfer rate in kW (to absorb 1 kW with a temperature rise of 10 °C, the required flow rate is thus 86 l/h).

Required Flow Rates

In the absence of a specific requirement it is good practice, for design purposes, to limit the temperature rise of the cooling medium through the engine water jacket to about 10 °C. In the case of the dynamometer the flow rate is determined by the maximum permissible cooling water outlet temperature, since it is important to avoid the deposition of scale (temporary hardness) on the internal surfaces of the machine or localized boiling.

Eddy-current dynamometers, in which the heat to be removed is transferred through the loss plates, are more sensitive in this respect than hydraulic machines, in which heat is generated directly within the cooling water. Maximum discharge (leaving) temperatures are:

Eddy current machines 60 °C
Hydraulic dynamometers 65–70 °C,[1]

provided carbonate hardness of water does not exceed 50 mg CaO/liter. For greater hardness values, limit temperatures to 50 °C.

Approximate cooling loads per kilowatt of engine power output are shown in Table 7.1. Corresponding flow rates and temperature rises are as shown in Table 7.2.

Water Quality

At an early stage in planning a new test facility, it is essential to ensure that a sufficient supply of water of appropriate quality can be made available. Control of water quality, which includes the suppression of bacteria, algae, and slime, is a complex matter and it is advisable to consult a water treatment expert who is aware of local conditions. If the available water is not of suitable quality then the project must include the provision of a water treatment plant. Most dynamometer manufacturers publish tables, prepared by a water

---

1. Approaching this absolute maximum outlet temperature range, some machines can experience flash boiling, which can lead to a degrading of control and be heard as a distinctive "crackling" noise. Running at these extreme temperature conditions will cause cavitation damage to the working chamber of a dynamometer.

TABLE 7.1 Estimated Cooling Loads

Heat Source	Output (kW/kW)
Automotive gasoline engine, water jacket	0.9
Automotive diesel engine, water jacket	0.7
Medium-speed marine diesel engine	0.4
Automotive engine oil cooler	0.1
Hydraulic or eddy-current dynamometer	0.95

chemist, which specify the water quality required for their machines. The following paragraphs are intended for the guidance of nonspecialists in the subject.

Solids in Water

Circulating water should be as free as possible from solid impurities. If water is to be taken from a river or other natural source it should be strained and filtered before entering the system. Raw surface water usually has significant turbidity caused by minute clay or silt particles that are ionized and may only be removed by specialized treatments (coagulation and flocculation). Other

TABLE 7.2 Estimated Cooling Water Flow Rates

Heat Source	In (°C)	Out (°C)	l (kWh)
Automotive gasoline engine	70	80	75
Automotive diesel engine	70	80	60
Medium-speed marine diesel engine	70	80	35
Automotive oil cooler	70	80	5
Hydraulic dynamometer	20	68	20
Eddy-current dynamometer	20	60	20

sources of impurities include drainage of dirty surface water into the sump, windblown sand entering cooling towers, and casting sand from engine water jackets. Hydraulic dynamometers are sensitive to abrasive particles and accepted figures for the permissible level of suspended solids are in the range 2–5 mg/liter. Seawater or estuarine water is used for testing large marine prime-movers with dynamometers and heat exchangers fitted with internals made from special stainless steels and marine bronzes, but it is not to be recommended for standard automotive equipment.

Water Hardness

The hardness of water is a complex property. There is a general subjective understanding of the term related to the ease of which soap lather can be created in a water sample, but the quality is not easy to measure objectively. Hard water, if its temperature exceeds about 70 °C, may deposit calcium carbide "scale", which can be very destructive to all types of dynamometer and heat exchanger. A scale deposit greatly interferes with heat transfer and commonly breaks off into the water flow, when it can jam control valves and block passages. Soft water may have characteristics that cause corrosion, so very soft water is not ideal either. Essentially, hardness is due to the presence of divalent cations, usually calcium or magnesium, in the water. When a sample of water contains more than 120 mg of these ions per liter, expressed in terms of calcium carbonate, $CaCO_3$, it is generally classified as a hard water. There are several national scales for expressing hardness, but at present no internationally agreed scale:

American and British: $1° US = 1° UK = 1$ mg $CaCO_3$ per kg water $= 1$ ppm $CaCO_3$
French: $1° F = 10$ mg $CaCO_3$ per liter water
German: $1° G = 10$ mg $CaCO_3$ per liter water
$1° dH = 10$ mg CaO per liter water $= 1.25°$ English hardness
(the old British system, 1 Clarke degree $= 1$ grain per Imperial gallon $= 14.25$ ppm $CaCO_3$).

Requirements for dynamometers are usually specified as within the range 2–5 Clarke degrees (30–70 ppm $CaCO_3$).

Water may be either acid or alkaline/basic. Water molecules, HOH or H_2O, have the ability to dissociate, or ionize, very slightly. In a perfectly neutral water equal concentrations of H^+ and OH^- are present. The pH value is a measure of the hydrogen ion concentration: its value is important in almost all phases of water treatment, including biological treatments. Acid water has a pH value of less than 7.07 and most dynamometer manufacturers call for a pH value in the range 7–9; the ideal is within the range 8–8.4.

The preparation of a full specification of the chemical and biological properties of a given water supply is a complex matter. Many compounds—phosphates, sulfates, sodium chloride, and carbonic anhydride—all contribute to the nature of the water, the anhydrides in particular being a source of dissolved oxygen that may make it aggressively corrosive. This can lead to such problems as the severe roughening of the loss plate passages in eddy-current dynamometers, which can cause failures due to local water starvation leading to plate distortion. Note that the narrow passages in eddy-current dynamometer loss plates are particularly liable to blockage arising from the use of inappropriate chemicals used in some water treatment regimes. Water treatment specifications should include the fact that, if used with water brakes, the treated water will be subjected to highly centrifugal regimes and local heating that may cause some degrading of the solution. Control of water quality also includes the suppression of bacterial infections, algae, and slime. British Standard 4959 [1] describes the additives used to prevent corrosion and scale formation, with chemical tests for the control of their concentration, and gives guidance on the maintenance and cleaning of cooling water systems. A recirculating system should include a small bleed-off to drain, to prevent deterioration of the water by concentration of undesirable compounds. A bleed rate of about 1% of system capacity per day should be adequate. If no bleed-off is included the entire system should be periodically drained, cleaned out, and refilled with fresh water. Finally, consideration should be given at the design stage of a cooling water system to the consequences of a power failure. Consider, for example, the possible effects of a sudden failure in the water supply to a hydraulic dynamometer absorbing 10 MW from a marine propulsion diesel engine operating at full speed. Even when the shutdown system operates immediately the fault is detected the engine system will take some time to come to rest, during which the brake will be operating on a mixture of air and water vapor, with the strong possibility of serious overheating. Therefore, in the case of any large engine test facility, some provision for a gravity feed of water in the event of a sudden power failure is advisable. The supply pressure to hydraulic dynamometers should be stable or the control of the machine will be affected. This implies that the supply pump must be of adequate capacity, and having as flat as possible pressure–volume characteristic in the normal operating range.

Types of Test Cell Cooling Water Circuits

Test cell cooling water circuits may be classified as follows, with increasing levels of complexity:

1. Direct mains water supplied systems containing a portable dynamometer and cooling column that allow heated water to run to waste.
2. Sump or tank-stored water systems that are "open", meaning at some point in the circuit water runs back, under gravity and at atmospheric pressure, into the sump by way of an open pipe. These systems normally incorporate self-regulating water/fluid cooling modules for closed engine

cooling systems filled with special coolant/water mix and, if required, for oil temperature control. They commonly have secondary pumps to circulate water from the sump through evaporative cooling towers when required.

3. Closed pumped circuits with an expansion, pressurization, and make-up units in the circuit. Such systems have become the most common as most modern temperature control devices and eddy-current or electrical dynamometers, unlike water brakes, do not require gravitational discharge. Closed water cooling systems are less prone to environmental problems such as the risk of Legionnaires' disease.

4. Chilled water systems (those supplying water below ambient) are almost always closed, although some may contain an unpressurized, cold, buffer tank.

Direct Mains Water to Waste Cooling

These are systems where the plant water is directly involved in the heat extraction from the process rather than from an intermediate heat exchanger. In most cases the water runs to waste on exiting the process and is most commonly used in occasional engine testing benches when portable dynamometers and cooling columns (see below) are used. While must of these systems run the water to waste it is possible to circulate the water through a gravity feed sump. One drawback of these systems is that the coolant is mains supply water and therefore cannot be dosed with any additives.

"Open" Plant Water-Cooling Circuits

The essential features of these systems are that they store water in a sump lying below floor level from which it is pumped through the various heat exchangers and a cooling tower circulation system. The sump is normally divided into hot and cold areas by a partition weir wall (see Figure 7.1).

Water is circulated from the cool side and drains back into the hot side. When the system temperature reaches the control maximum, it is pumped from the hot sump and through the cooling tower before draining back into the cool side. A rough rule for deciding sump capacity is that the water should not be turned over more than once per minute; within the restraints of cost the biggest available volume gives the best results.

Sufficient excess sump capacity, above the normal working level, should be provided to accommodate drain-back from pipework, engines, and dynamometers upon system shutdown. There is a continuous loss of water due to evaporation plus the small drainage to waste, mentioned above; therefore, make-up from the mains water supply needs to be controlled by a float valve. It is important to minimize air entrainment in the pump suction; therefore, the minimum level of the sump when the pressure and return lines are full should be sufficient to discourage the formation of an air-entraining vortex. The return flow should be by way of a submerged pipe fitted with an air vent. The

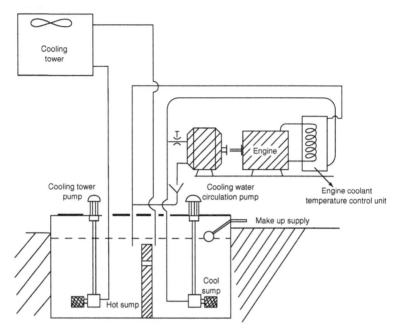

FIGURE 7.1 Simple open cooling water system incorporating a partitioned sump.

arrangement shown diagrammatically in Figure 7.1 is a classic arrangement of which thousands of similar systems are installed worldwide, but care has to be taken to keep debris such as leaves or flood water "wash-off" from entering the system. All ground-level sumps should have the top surface raised at least 100 mm above ground level to provide a lip that prevents flooding from groundwater.

A sensible design feature at sites where freezing conditions are experienced is to use pumps submerged in the sump so it can be ensured that, when not being used, the majority of pipework will be empty.

Closed Plant Water-Cooling Circuits

The design and installation of a closed water supply for a large test installation is a specialist task not to be underestimated. It may require the inclusion of a large number of test and flow balancing valves, together with air bleed points, stand-by pumps, and filters with changeover arrangements. By definition these closed systems have no sump or gravity draining from any module within the circuit. Such a system does not suffer from the evaporative losses of an open system and is less prone to contamination. Typically it uses one or more pumps to force water through the circuit, where it picks up heat that is then dispersed via closed-circuit cooling towers, then the water is returned directly to the pump inlet.

It is vital that air is taken out and kept out of the system. The whole pipe system must be provided with the means of bleeding air out at high points or any trap points in the circuit. To achieve proper circulation and cope with thermally induced changes of system volume, also to make up for any leakage, a closed system has to be fitted with an expansion tank plus some means of "make-up" and pressurization. These requirements can be met by using a form of compressed air/water accumulator connected to a pressurized make-up supply of treated water. "Balancing" of water systems is the procedure by which the required flow, through discrete parts of the circuit, is fixed by use of pressure-independent "flow-setter" valves having test points fitted for commissioning purposes. Valves are required for each subcircuit because they have their own particular thermal load and resistance, and therefore require a specific primary system flow rate. The balancing of closed cooling systems can be problematic, particularly if a facility is being brought into commission in several phases, meaning that the complete system will have to be reba- lanced at each significant system addition. None of the devices fitted within a closed plant water system should have "economizer" valves that themselves regulate the flow of the primary (plant) water, since that variation may continually unbalance the primary system. To avoid such unbalancing, devices in the circuit, the temperature control valves, should work by regulation of secondary fluid and have constant primary (plant water) flow. *Freezing protection* must be considered by the facility designer, even if the whole system is within a building. Closed, pressurized water systems can be filled with an ethylene glycol and water mix to prevent freezing but, as mentioned at the beginning of this chapter, the cooling efficiency of the mixture is inversely proportional to the concentration of glycol. It should also be remembered that some materials in seals and pipes, such as natural rubber derivatives, may deteriorate and fail if exposed for long periods to glycols at elevated temperatures. The pipework of open water systems can be trace heated. Such systems consist of special heating tape being wound around the pipes in a long pitched spiral under insulating material. The control is usually fully automated such that the heating current is regulated according to ambient temperature.

Engine Coolant Temperature Control: Cooling Columns

If special engine coolants are not required, a cooling column is a simple and economical solution commonly used in the USA (see Figure 7.2). It can be portable and located close to the engine under test. The column allows the engine outlet temperature to run up to its designed level; at this operating point a thermostatic valve opens, allowing cold water to enter the bottom of the column and hot water to run to waste or the sump from the top. The top of the column is fitted with a standard automotive radiator cap for correct engine pressurization and use when filling the engine circuit.

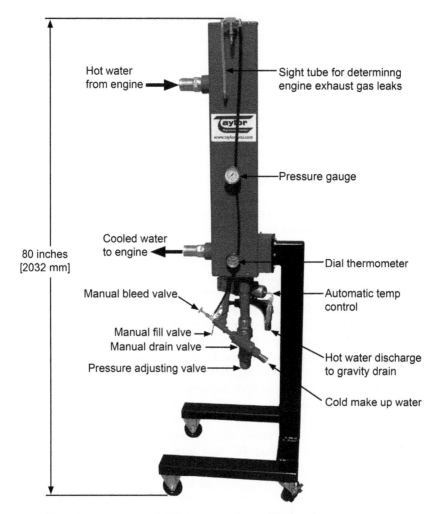

FIGURE 7.2 Engine cooling column used widely in the USA, often in conjuction with direct engine-mounted dynamometers (see Chapter 10). *(Reproduced with the permission of Taylor Dynamometer, Milwaukee.)*

Closed Engine Coolant and Oil Temperature Control Modules

The inclusion in the test engine of its own coolant thermostat will depend on the design of experiment (DoE) and the type of engine system. Where the thermostat body has to be present due to the physical layout of the pipework, or because an active (pumped and heated) coolant control unit is being used, it can be rendered ineffective by drilling holes in the diaphragm. Whether or not the engine under test is fitted with its own thermostat, precise control of coolant

TABLE 7.3 Typical Control Ranges and Limits for Modern Coolant Control Units

Device	Control Range and Tolerance
Pressurized engine coolant, passive cooling only. Steady state at engine inlet	$70-125 \pm 1$ °C
Engine oil, passive, cooling only. Steady state at engine outlet	$70-145 \pm 2$ °C
Pumped, heated, and cooled with engine bypass flow. Dynamic performance is superior to other types	$70-145 \pm 0.75$ °C

temperature is not easily achieved with a standard "passive"[2] service module unless it is closely designed to match the thermal characteristics of the engine with which it is associated. Coolant temperature control units can be installed to control either the temperature of the coolant entering or leaving the engine; it is the position of the PRT giving feedback to the controller that determines what is controlled. Control of the "coolant in" temperature is always superior to that of "coolant out"; typical control ranges and limits figures for modern plant are listed in Table 7.3.

Even here it may be difficult to achieve stable temperatures at light load. The instability of temperature control is increased if the engine is much smaller than that for which the cooling circuit is designed. The capacity of the heat exchanger is the governing factor and it may be advisable, when a wide range of engine powers is to be accommodated, to provide several coolers with a range of capacities.

There are many passive, closed system engine coolant temperature control units on the market, most working on the principle of a closed-loop, control valve controlling flow of coolant through a heat exchanger, and they can be broken down into the following types:

- Type 1: Mobile pedestal type
- Type 2: Special engine pallet-mounted systems
- Type 3: User-specific, wall-mounted systems
- Type 4: Complex, heated and pumped, fixed pedestal type with conditioned fluid bypass close to the engine.

---

2. Passive in this case means that the coolant service module is not fitted with either a coolant circulating pump or an auxiliary heater circuit.

Figure 7.3 is an illustration of a typical service module incorporating heat exchangers for jacket coolant and lubricating oil, while Figure. 7.4 shows a simplified schematic of the circuit. The combined header tank and heat exchanger is a particularly useful feature. This has a filler cap and relief valve and acts in every way as the equivalent of a conventional engine radiator and ensures that the correct pressure is maintained.

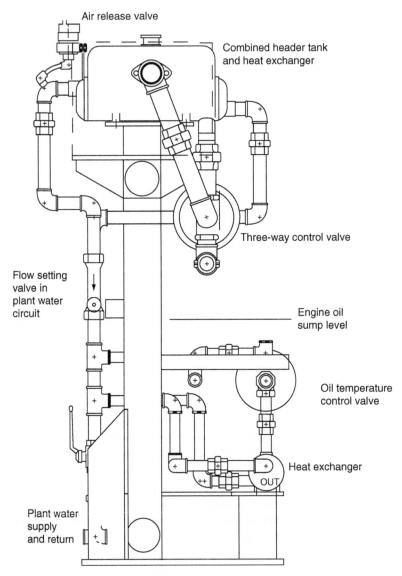

FIGURE 7.3 Diagram of movable coolant conditioning unit with combined heat exchanger/header tank.

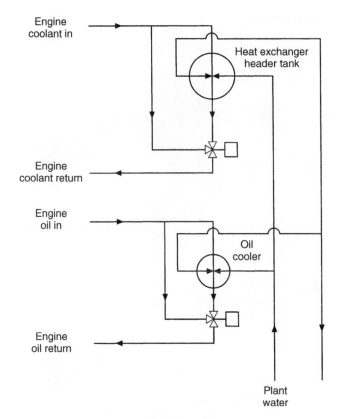

FIGURE 7.4a Circuit diagram of unit in Figure 7.3.

For ease of maintenance, it should be possible to withdraw exchanger tube stacks without major dismantling of the system, and simple means for draining both oil and coolant circuits should be provided. The most usual arrangement is to control the temperature by means of a three-way thermostatically controlled valve in the engine fluid system. As discussed earlier, the alternative, where temperature is controlled by regulating the primary cooling water flow, will work, but gives an inherently lower rate of response to load changes. Types 2 and 3 listed above are often designed and built by the user, particularly the pallet-mounted systems, which may use specific ex-vehicle parts for such items as the header tank and expansion vessel.

Type 4 is the most complex and incorporates a coolant circulation pump, heaters, fluid-to-fluid cooler, and complex control strategies to deal with low engine loads and transient testing. Such devices, while considerably more expensive than passive types, give superior control response to rapid changes of heat load and allow a wider range of engines to be tested with the same

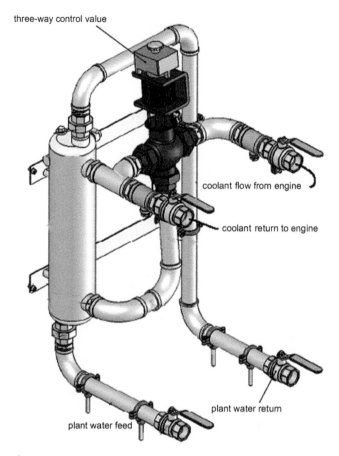

three-way control value

coolant flow from engine

coolant return to engine

plant water return

plant water feed

FIGURE 7.4b   Example of the same circuit shown at the top of Figure 7.4a used for a simple wall-mounted engine coolant control module using a self-contained electrical controller operating a three-way valve.

device; also, much faster warm-up times and pre-heating of the engine are possible.

If engines are to be tested without their own coolant pumps, the module must be fitted with a circulating pump, commonly of the type used in central heating systems; this is ideal and easy to incorporate in the circuit (Figure 7.5).

Temperature Control and the Effects of System Thermal Inertia

None of the heat exchanger units will operate satisfactorily if not integrated well with the engine and cell pipework.

A control time-lag is a common fault of many poorly installed coolant control systems, particularly the passive types, and has several causes. The sum of these phenomena is often referred to as the "thermal inertia" of the

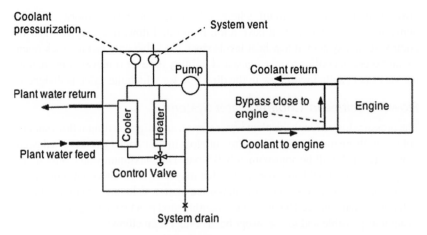

FIGURE 7.5 Diagram of an engine coolant control system fitted with full temperature control, its own circulation pump, and a fluid bypass close to the engine.

cooling system. To reduce thermal inertia there are three recommended strategies:

1. Reduce the distance and fluid friction-head between cooler, the control valves, and engine.
2. Circulate the conditioned coolant between engine and cooler/heater in a pumped circuit with an engine bypass as in type 4 units described above and shown in Figure 7.5.
3. Interconnecting pipes should be insulated against heat loss/gain.

Strategy 1 is best served by arranging a pallet-mounted cooling module close to the engine. In cases such as anechoic cells, where the heat exchanger is inevitably remote from the engine, strategy 2 is required to speed up the rate of circulation by an auxiliary pump mounted outside the cell to reduce lag and noise.

Three types of device are commonly adopted for the control of the secondary fluids through heat exchangers:

- Independent packaged controllers using electrically operated valves such as those produced by Honeywell
- Indirect control by instruments having internal control "intelligence" and given set points from a central (engine test) control system
- Direct control from the engine test controller via auxiliary PID software routines and pneumatically or electrically operated valves.

A proportional and integral (P and I) controller will give satisfactory results in most cases but it is important that variable-flow valves should be correctly sized, since they will not function satisfactorily at flow rates much lower than the optimum for which they are designed. In design terms the valves have to have "authority" over the range of flows that the coolers require. A common error, which creates a circuit where valves do not have authority at low demands, is to fit valves that have a too high rated flow. It is very difficult to control such a system at low heat levels since the valve only has to crack open for an excess of cooling to take place and stability is never achieved. Ensure that the valve is installed with flow in the direction for which the value is disigned!

Flow Velocities in Cooling Water Systems

The pressure loss through the plant water circuit, at design maximum flow, should not normally be more than 1.5 bar. Pressure loss in the coolant circuits at a design flow of 12 m^3/h will be approximately 0.5 bar. The maximum water velocity in a supply system should not exceed 3 m/s but may need to reach a minimum of 1.5 m/s in some systems to sweep away deposited matter. Velocities in gravity drains will not normally exceed 0.6 m/s. It is essential good practice to keep pipe runs as straight as possible and to use swept bends rather than elbows.

Design of Water-to-Water Heat Exchangers

Manufacturers of heat exchangers and devices using cooling water invariably provide simple design procedures for establishing flow rates and pressure drop for a given heat exchanger performance. If it should be required to design a heat exchanger from first principles, Ref. [2] gives a detailed design procedure.

Chilled Water Circuits

In powertrain test facilities the term "chilled water" normally covers water stored and circulated between 4 and 8 °C. Chilled water, supplementing plant water, may be required for the following processes:

- Control room air-conditioning
- Fuel temperature control
- Combustion air temperature control
- Thermal shock coolant conditioning module.

These systems have markedly different thermal loads and operational regimes. A common fault of badly designed chilled water systems is "crosstalk"; this is when good control of one process is lost because of a sudden change of supply pressure or temperature caused by events occurring in a separate process within a shared supply. If a common chilled water system supplies several processes having differing thermal loads, each subsystem should have its own control and an adequately sized buffer tank should be incorporated in the system. In a closed system a thermally stratified buffer tank can be used where the system's return water enters the top of the tank and from there is drawn into the chiller, which returns the treated flow to the bottom of the tank from whence it is drawn and circulated.

Control of liquid fuel temperature is critical to repeatability of results in engine testing; the control has a thermal rating much lower than other typical cell requirements. These small (1–2 kW) circuits are best served by having a dedicated chilled water supply; suitable units that are designed to provide chilling for beer supplies in hotels and restaurants are commercially available.

Engine Thermal Shock Testing

To accelerate durability testing of engines and engine components, including cylinder-head gaskets, many manufacturers carry out thermal shock tests that may take several forms, but commonly require the sudden exchange of hot coolant within a running engine for cold fluid. The term "deep thermal shock" is usually reserved for tests having a ΔT of around 100 °C; such tests are also called, descriptively, "head cracking" tests. All such sudden changes in fluid temperatures cause significant differential movement between seal faces. Thermal shock tests are normally carried out in normal (not climatic) test cells and are achieved by having a source of cold fluids that can be switched directly into the engine cooling circuit via three-way valves inserted into the coolant system inlet and outlet pipes. Due to the speed required to chill down the engine and the cyclic nature of the tests, it is not usual to use secondary heat exchangers between the chilled fluid and the engine coolant. The energy requirements for such tests clearly depend on the size of engine and the cycle time between hot stabilized running and the chilled engine condition, but they can be very considerable and often require cells with specially adapted fluid services that include a large buffer tank of chilled coolant. Mobile thermal shock systems, complete with chiller and buffer tank, are used to allow any cell fitted with suitable actuating valves in the engine coolant circuit to be used for thermal shock testing; however, their capacity is less than that of a purposely adapted cell. Note that self-sealing couplings commonly used to connect mobile fluid treatment modules may create a high flow resistance, which can considerably add to the thermal inertia of the circuit. Globe valves and cam-locked couplings are a cheaper and a less flow-restrictive substitute.

The control for thermal shock and thermal cycling tests requires special test sequence subroutines and rapidly functioning valve actuators. There will be a need to operate the fluid control valves in the primary circuit in strict sequence in order to direct engine coolant flows to or from chilled water buffer stores. The stage "end condition" is usually defined by the attainment of a given (low) temperature in the engine cylinder head. Since this type of testing may be expected to induce a failure in some part of the engine system, the failure monitoring system must be designed accordingly.

Commissioning of Cooling Water Circuits

Before dynamometers and any other test cell instrumentation are connected to the building's cooling water system, it must be cleaned and the water treated appropriately. Good practice dictates that during water system commissioning the system is first fully flushed. Devices that have to be supplied with water should be temporarily bypassed by pipes that allow water to be circulated rather

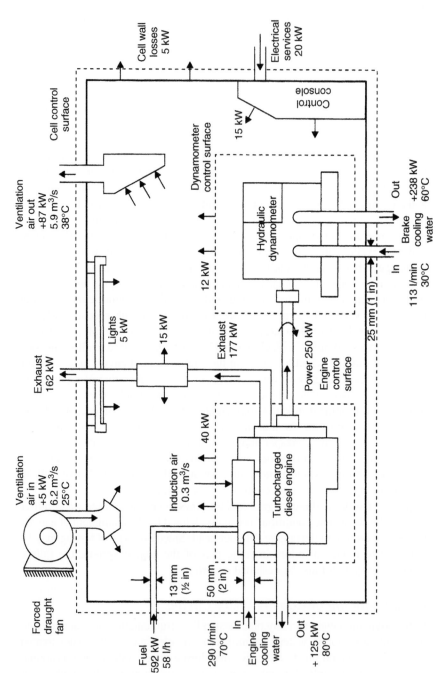

FIGURE 7.6 Energy balance and energy flow diagram for example 250 kW cell.

than to contaminate the instrument circuit. If this is not possible or if large heat exchangers have to be flushed, then temporary strainers should be put in the circuit. Only when temporary strainers and filters are not picking up debris should they be removed and the instruments connected into the circuit. In locations where water is freely available it is recommended that the first system fill of untreated mains supplied water used for flushing is dumped and the settling tanks cleaned if necessary. Debris will include both magnetic and nonmagnetic material such as jointing compound, PTFE tape, welding slag, and building dust, all injurious to test plant.

It is cost- and time-effective to take trouble in getting the coolant up to specification at this stage rather than dealing with the consequences of valve malfunction or medium-term corrosion/erosion problems later.

After flushing the system should then be filled with clean water that has been treated to balance the hardness, pH level, and biocide level to the required specification.

Drawing up the Energy Balance and Sizing the Water System

In Chapter 3 it is recommended that at an early stage in the design of a new test cell a diagram similar to Figure 7.6 should be drawn up to show all the flows: air, fuel, water, exhaust gas, electricity, and heat into and out of the cell.

The process is best illustrated by an example, and the full-load regime for a 250 kW turbocharged diesel engine has been chosen driving a hydraulic dynamometer (rather than an air-cooled AC machine). The design of the ventilation system for the cell is dealt with in Chapter 6. The cooling water flow for the hydraulic dynamometer may be calculated on the basis of the following assumptions:

- 95% of power absorbed appears in the cooling water
- Cooling water inlet temperature 30 °C
- Outlet temperature not to exceed 70 °C.

Then flow to brake:

$$\frac{250 \times 0.95 \times 60}{4.186 \times 40} = 851 \text{ l/min}$$

Calculated pipe sizes may be as follows:

Fuel, velocity 0.2 m/s, 10.8 mm diameter, say 0.5 in
Engine cooling water, 3 m/s, 45 mm diameter, say 2 in
Brake cooling water, 3 m/s, 28 mm diameter, say 1.5 in.

PART 2. EXHAUST GAS SYSTEMS

The layout and special design features of exhaust systems forming part of full gaseous emission analysis systems is covered in Chapter 16.

It is possible to run into operational safety problems with test cell exhaust systems; there have been accidents because of these systems, some of them fatal. Due to the toxicity of exhaust gases any pipe or duct carrying them should exit enclosed building spaces as soon as practically possible. Particular care should be taken in the detailed design and construction of ducting containing diluted exhaust gases under positive pressure from a fan. Where possible the ducting within a building space of such a system should be on the suction side of the fan. Fan-assisted exhaust systems with the discharge duct within a building space may be acceptable in large, open factory spaces where the fan is on the cell roof and the duct is taken directly, preferably vertically, out of the shell building with as few, well-sealed, joints as possible. Within the test cell the most significant risk to operation staff is contact with hot metal exhaust piping. Fixed cell sections can be permanently lagged while the sections after the engine, particularly those running at low level in the cell, should be guarded or wrapped with thermal bandage to avoid burns to staff and vulnerable transducer cables. Many modern automotive engines are fitted with emission control systems that require that the actual vehicle exhaust system should be employed during engine tests. This can impose problems of layout even in cells of large floor area. If vehicle systems are to be modified it should be remembered that any change in the length of the primary pipe is particularly undesirable, since this can lead to changes in the pattern of exhaust pulses in the system. This can affect the volumetric efficiency and power output of the engine and, in the case of two-stroke engines, it may prove impossible to run the engine at all with a wrongly proportioned exhaust system. An increase in the length of pipe beyond the after-teatment devices is less critical, but care should be taken to limit the back-pressure imposed upon the engine to the design figure. Engine test cell exhaust systems should always be designed to give a much lower back-pressure to the maximum gas flow than that required or specified for any test engine; control of back-pressure can then be adjusted by a permanently installed control valve. Such valves, usually of butterfly type, fitted into the oversized pipework of the cell exhaust system, need to be fitted with stainless steel internals and spindle seals that work at elevated temperatures. Automatic control and remote actuation of these valves is only fitted in cases of critically tuned systems or where engines of widely different sizes are run, otherwise they are manually adjusted and locked in position. Turbocharged engines in particular can have complex exhaust systems and run at such high temperatures that large areas of manifold and exhaust pipe can appear incandescent and this can represent a large heat load on the ventilating system. They may also give false triggering of a fire system if inappropriate sensors have been fitted. Unless the system has been specifically designed to run the exhaust pipe in a floor duct and housekeeping is of a high order, it is not generally a good idea because of the fire risk caused by fallen debris and unseen corrosion.

The practice sometimes adopted, of discharging the engine exhaust into the main ventilation extraction duct, has several disadvantages:

- To avoid rapid acid attack and corrosion, the duct has to be made of a suitable stainless steel. Galvanized sheet ducting and particularly spiral-wound galvanized tubing will have a very limited life when used in a diluted exhaust system and is *not* recommended.
- The air flow must be increased to a level greater than that necessary for basic ventilation, to maintain an acceptable duct temperature. On a cold day this can lead to chilling in the cell.
- Soot deposits and corrosive, staining condensate in the fan and ducting are unsightly and make maintenance difficult.
- Other difficulties can arise, such as noise, variability in exhaust back-pressure, etc.

If "tail-pipe" testers of the type used in vehicle servicing and inspection are to be employed then easy and safe access to a suitable extraction tapping is necessary.

Test cell exhaust layouts may be classified as follows:

- Individual cell, close coupled
- Individual cell, vehicle exhaust system, scavenged duct
- Multiple cells with common scavenged duct
- Specially designed emission cells (see Chapter 16).

Individual Cell, Close Coupled Exhaust

Such an arrangement is shown schematically in Figure 7.7a. It may be regarded as the "standard" arrangement for a general-purpose test bed. The exhaust manifold is coupled to a flexible stainless steel pipe, of fairly large diameter, to

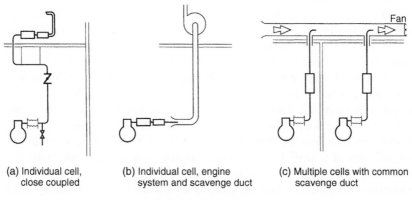

(a) Individual cell, close coupled

(b) Individual cell, engine system and scavenge duct

(c) Multiple cells with common scavenge duct

FIGURE 7.7 Exhaust systems.

minimize pressure waves, and led by way of a back-pressure regulating valve to a pipe system suspended from the cell roof. Condensate, which is highly corrosive, tends to collect in these pipes, which should be laid to a fall with suitable drainage arrangements. In cells designed to occasionally use smoke or particulate analysis it is usually a requirement that the sample is taken from the exhaust pipe at particular points; some devices have quite specific requirements regarding size, position and angle of probe insertion, and it is desirable, in order to ensure a representative sample, that there should be six diameters length of straight pipe both upstream and downstream of the probe. It is therefore good practice, when designing the exhaust system, to arrange a straight horizontal run of exhaust pipe; this pipe is easily replaced by a pipe with specific probe tappings.

Individual Cell, Scavenged Duct

When it is considered necessary to use the vehicle exhaust system, two options are available: one is to take the pipe outside the building through a panel in the cell wall, the other is to put the tail pipe into a scavenge air system, as shown schematically in Figure 7.7b. In this case the tailpipe is simply inserted into a bell mouth through which cell air is drawn; the flow rate should be at least twice the maximum exhaust flow, preferably more. This outflow should be included in the calculations of cell ventilation air flow. The scavenger flow is induced by a fan, preferably external to the building, usually centrifugal, which must be capable of handling the combined air and exhaust flow at temperatures that, if close to the cell, may reach 150 °C.

Multiple Cells, Common Scavenged Duct

This arrangement (Figure 7.7c) is found in many large installations. Note that in these layouts silencers have to be fitted near the engine before the tail pipe runs into the dilution duct.

We can take as an example an installation of three test cells, each like that schematically illustrated in Figure 7.6, running a 250 kW turbocharged diesel engine for which maximum exhaust flow rate is 1365 kg/hour. To ensure adequate dilution and a sufficiently low temperature in all circumstances, we need to cater for the possibility of all three engines running at full power simultaneously and a scavenging air flow rate of about 10,000 kg/h, say 2.3 m^3/s would be appropriate. Table 6.4 (Chapter 6) indicates a flow velocity in the range of 15–20 m/s and hence a duct size in the region of 400 mm × 300 mm, or 400 mm diameter. As before, the scavenging fan must be suitable for temperatures of at least 150 °C. This arrangement is recommended only for diesel engines. In the case of spark-ignition engines there is always the possibility that unburned fuel, say from an engine that is being motored, could accumulate in the ducting and then be ignited by the exhaust from another engine. This possibility may seem remote, but accidents of this kind are by no means unknown. Note that, in Figure 7.7b and c, the fan controls must be interlocked with the cell control systems so that engines can only be run when the duct is being evacuated.

Where the engine's exhaust system is not used, the section of exhaust tubing adjacent to the engine must be flexible enough to allow the engine to move on its mountings and a stainless steel bellows section is to be recommended. Exhaust tubing used in this area should be regarded as expendable and the workshop should be equipped to make up replacements. As a final point, carbon steel silencers that are much oversized for the capacity of the engine will never get really hot and can be rapidly destroyed by corrosion accelerated by condensate. The dual use of an exhaust gas extraction duct to act as a cell purge system is not now recommended because fans that are compliant with European ATEX rules are more expensive than noncompliant units and they have to be integrated with the hydrocarbon level alarm system; both are good reasons for a "stand-alone" purge system.

Cooling of Exhaust Gases

There may be an operational requirement to reduce the temperature of exhaust gas exiting the cell. This is usually achieved by using a stainless steel, water-jacketed cooler having several gas tubes that give a low resistance to gas flow. These occasionally used devices are often made specifically for the project in which they are used and there are three important design considerations to keep in mind:

- If water flow is cut off to the exhaust gas cooler, a steam explosion can be caused; therefore, primary and secondary safety devices should be fitted.
- High internal stresses can be generated by the differential heating of the cooler elements; therefore, some expansion of the outer casing should be built into the design.
- Since the water jacket pressure will be higher than exhaust gas pressure, tube leakage could fill the engine cylinders unless the system is regularly checked.

Direct water spray has been used in vertical sections of exhaust systems in order to wash and cool exhaust flow into a common evacuated exhaust main. The resulting waste water can be highly corrosive and staining; therefore, adequate drainage and proper disposal should be provided.

Estimation of Exhaust Gas Flows

The volume of raw and diluted exhaust gas flow into and out of the test cell's fixed exhaust system is highly variable so, for system design purposes, the maximum flow needs to be based on the fuel consumption of the largest scheduled engine at full rated power; an example of the data required and the calculation for a naturally aspirated gasoline engine of 130 kW output is shown in Table 7.4.

TABLE 7.4 Example Estimation of Exhaust Gas Flow for Test Cell Running a 130 kW Gasoline Engine, With and Without Dilution Air Entrained From Test Cell

	Value	Units
Fuel		
Mass rate	31.5	kg/h
Volume rate	42.0	liter/h
Specific fuel consumption	0.242	kg/kWh
Combustion Air		
Mass rate	463	kg/h
Intake density at 1 bar	1.22	kg/m^3
Volume rate	378	m^3/h
Exhaust Gas		
Mass rate (air + fuel)	495	kg/h
Density, after manifold 1.5 bar	1.22	kg/m^3
Volume rate	779	m^3/h
Density, out of cell 1.3 bar	0.586	kg/m^3
Volume rate of cell	844	m^3/h
With Exhaust Dilution in Cell		
Dilution ratio	1.5	
Mass flow at intake	742	kg/h
Density at intake (cell at −50 Pa)	1.16	kg/m^3
Volume rate at intake	637	m^3/h
Total mass flow (exh. + dil. air)	1236	kg/h
Mixture temperature after intake	218	°C
Mixture temperature out of cell	60	°C
Mixture density out of cell	1.06	kg/m^3
Mixture volume flow rate out of cell	1168	m^3/h
	0.32	m^3/s

Exhaust Noise

The noise from test cell exhaust systems can travel considerable distances and be the subject of complaints from neighboring premises, particularly if running takes place at night or during weekends. The design of test cell exhaust systems is largely dictated by the requirement that the performance of engines under test should not be adversely affected and by restraints on noise emission, both internally and external to the building.

Essentially there are two types of device for reducing the noise level in ducts: resonators and absorption mufflers. A resonator, sometimes known as a reactive muffler, is shown in Figure 7.8. It consists of a cylindrical vessel divided by partitions into two or more compartments. The exhaust gas travels through the resonator by way of perforated pipes, which themselves help to dissipate noise. The device is designed to give a degree of attenuation, which may reach 50 dB, over a range of frequencies (see Figure 7.9 for a typical example).

Absorption mufflers, which are the type most commonly used in test facilities, consist essentially of a chamber lined with sound-absorbent material through which the exhaust gases are passed in a perforated pipe. Absorption mufflers give broadband damping but are less effective than resonators in the low-frequency range. However, they offer less resistance to flow.

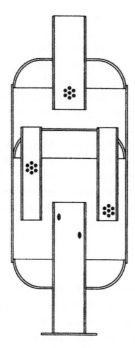

FIGURE 7.8 Exhaust resonator or reactive muffler.

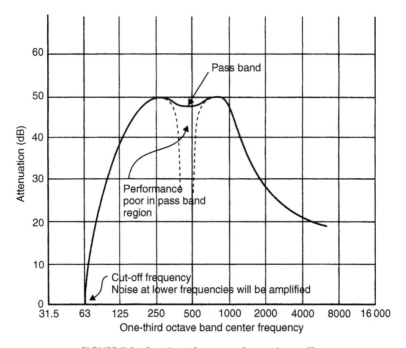

FIGURE 7.9 Sample performance of a reactive muffler.

Selection of the most suitable designs for a given situation is a matter for the specialist. Both types of silencer are subject to corrosion if not run at a temperature above the dew point of the exhaust gas, and condensation in an absorption muffler is particularly to be avoided.

As with a number of engine-attached devices, the exhaust silencer may be considered as part of the engine rigging when it is engine model specific, or it may be considered as part of the cell, or both may be used. Modern practice tends to use the vehicle exhaust system complete with silencers within the cell then extract the gas, mixed with a proportion of cell air, into a duct fitted, before exiting the building, via absorption attenuators. When fans used in the extraction are rated to work up to gas/air temperatures of 200 °C the effect of temperature on the density of the mixture and therefore the fan power must be remembered in the design process.

Tail Pipes

The noise from the final pipe section is directional and therefore often points skywards, although the ideal would be to terminate in a wide radius 90° bend away from the prevailing wind. The pipe end can, if incorrectly positioned and with a plain 90° end, suffer from wind-induced pressure effects; this has led to difficulties in getting correlation of results between cells. Remember that the condensate of exhaust gases is very corrosive, and that rain, snow, etc. should not be allowed to run into undrainable catch points or the engine.

Exhaust Cowls on Buildings

When deciding on the position of the exhaust termination outside a building, it is important to consider the possibility of recirculation of exhaust fumes into ventilation inlets and also to avoid the imposition of back-pressure through wind flow. Prevention of recirculation requires the careful relative positioning of exhausts and inlet ducts in relation to each other and to the prevailing local (building affected) wind direction. The facility owner and architect may have strong opinions about the number and design of exhaust cowls. In the case of multi-cell facilities the individual outlets are often consolidated into one or more chimneys, forming an architectural feature. However, Figure 7.10a shows a simple termination tube suitable for use on

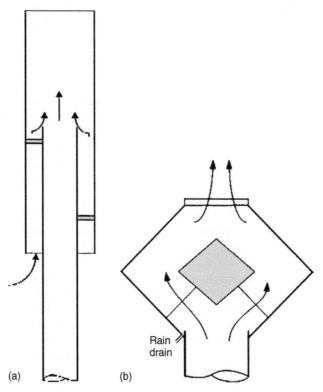

(a)   (b)

FIGURE 7.10 (a) A simple rain-excluding dilution tube fitted to a single-cell raw exhaust outlet. (b) One design of a building cowl for the termination of a multi-cell diluted exhaust system.

a single cell. This is a raw exhaust outlet that incorporates a shroud tube that acts as a plume dilution device and rain excluder. Figure 7.10b shows a type of rain-excluding cowl commonly used on multiple-cell diluted exhaust outlets. In both cases the exit tubes terminate vertically to minimize horizontal noise spread.

PART 3. TURBOCHARGER TESTING

The turbocharger is an important component of a modern diesel engine and increasingly so of gasoline engines. For many years turbochargers have been of a fixed internal geometry installed within a system wherein overpressurization of the engine has been prevented by the use of an internal, or external, waste gate (valve). This device acts as an air pressure relief valve by diverting a variable amount of exhaust gas flow from the turbine and sending it to waste, thus reducing the compressor speed and lowering boost air pressure. Recent advances in materials and engine design mean that the modern turbocharger is now often fitted with a variable-geometry turbine (VGT) that is electronically controlled through the ECU so that its performance is optimized to deal with the wide range of operating conditions and to support the correct function of exhaust gas recirculation (EGR) and other emission control strategies. Variable-geometry turbine systems may be servo-electric or servo-hydraulic and do not require waste gates, but in most applications still require antisurge valves. Antisurge valves operate as a pressure relief valve in the boost air circuit and are designed to deal with the sudden rise in air pressure that arises between the compressor and the engine when the engine throttle is suddenly closed from near, or at, wide-open throttle (WOT) operation. A complication of the operation of antisurge valves is that the air relieved has to be returned to the compressor inlet and after the mass-airflow sensor, which has already registered its passing, to prevent incorrect fueling. The moving blades and parts of the actuation mechanism of VGT units have to work in the highest temperature zone of the turbine, because of the demands made upon the materials used; until recently, VGT designs have been confined to diesel engines, which have lower exhaust gas temperatures than gasoline engines. A problem for vehicle designers is the cost of VGT designs, which is often well over double that of fixed-geometry designs; therefore, in truck installations, where space may not be as constrained as in light vehicles, the use of two-stage or twin-turbo designs may be more cost-effective. A two-stage boost system can be based on two separate turbochargers, one HP and one LP, permanently connected in series, or on two units fitted in parallel with some modulation of the gas flow to one or both units, to give optimized boost and engine power. Turbocharger testing takes two forms:

- Testing the turbocharger and its associated actuators and sensors, in a customer built test stand, as a separate module without any other part of an engine. In this case a combustion-gas generator is required to drive the turbine.

- Testing the turbocharger mounted on an engine as an integrated system. In this case the engine provides the gas generation to drive the turbine.

Turbocharger Test Stands

A range of test stands that are designed for testing the automotive range of turbochargers are commercially available from companies such as Kratzer Automation [3]; these units typically use natural gas in hot-gas generators whose energy output is variable over the full operating cycle of the intended engine installation. Although gas burners provide a virtually particulate-free gas flow, it is important for some tests to simulate operation in poor gas conditions; therefore, such rigs have to be capable of simulating less than optimal gas conditions.

Non-engine-based turbocharger test rigs are cost-effectively used by manufacturers for the full range of development, quality assurance, and endurance testing, including:

- Testing over the extreme thermal range of the turbine unit and thermal shock testing.
- Testing over the full range of the pressure–volume envelope to confirm compressor surge and choke lines and the unit's efficiency "onion rings" or "efficiency islands".
- Checking for turbine blade resonance, fatigue testing.
- Seal and bearing module performance testing and blow-by tests (oil leakage into either induction or turbine side has serious consequences).
- Mapping of both turbine and compressor speed–pressure–temperature performance.
- Lubrication system and oil suitability testing, including resistance to "oil-coking" following shutdown while at high turbine temperature.
- Running at over-speed and to destruction in order to check failure modes and debris retention.

Testing of Turbochargers Using an Engine

This must be considered as distinct from testing a turbocharged engine and an example of an instrumented test unit is shown in Figure 7.11. In the latter case the operation of the unit is one more variable to be mapped during engine calibration and, where appropriate, will be the subject of its own "mini-map" covering the operation of the servo-controller of the VGT vanes within the engine fueling map. For example, in engine calibration work the interaction of EGR and turbocharger geometry has to be optimized, but when testing the

FIGURE 7.11 A variable-geometry turbocharger being tested while fitted to an automotive truck engine. For test, the turbine housing has been fitted with three pressure probes (wrapped transducers), three displacement transducers (two thin-threaded transducers visible, the third hidden), and thermocouples. The VG actuator is electromechanical and fitted with water cooling. *(Photo courtesy of Cummins Turbo Technologies Ltd.)*

turbine using an engine, for speed–pressure sweeps, etc., the EGR will interfere with system control and may be switched out of circuit. Similarly, it will be found that running the engine and dynamometer in manual control mode (position/position) will give the most repeatable results in much turbocharger testing because it eliminates the majority of engine and cell control system influences. Gas flow, from the exhaust of individual engine cylinders into the turbine, is complex and subject to pressure pulsing; therefore, the pressure regime of the whole exhaust manifold and turbine has to be understood in order to optimize its design. Such test work, which is practically impossible to accurately simulate on a test bench, has to be done in an engine test cell. Such testing has led to the development of twin-scroll turbine designs, which divide the gas flows into the turbine housing from pairs of cylinders in order to reduce pressure-pulse crosstalk evident only on a fully instrumented engine.

The matching of turbocharger with a particular design of engine can only truly be checked in the engine test cell, and testing includes:

- Integrated turbine and compressor performance during engine power curve sequences.
- Turbocharger "lag" during step changes in power demand, over the full operating range.
- Testing of two-stage or twin-turbo operation.
- Intercooler operation: matching performance and control.
- Engine-based lubrication, cooling and seal operation, including testing on inclined engine test stands.

- Effects of various exhaust pipe configurations on back-pressure, turbine, and engine performance.
- Thermal soak effects such as following sudden shutdown after WOT running, etc.

Turbine Speed and Blade Vibration Sensing

While many of the larger automotive turbochargers are fitted with a "once per revolution" speed sensing facility, during development testing it is usually desirable to measure blade passing frequency. This allows the tester to compare the actual blade arrival time at the sensor with that predicted by the rotational speed and the number of turbine blades; any difference will be caused by blade deflection and vibration (resonance). Optical laser, capacitive, and eddy current probes can be used for the blade arrival time measurements [4].

Special Applications

Large *two-stroke marine* diesel engines have to use multiple turbochargers in order to keep the devices to manageable size; single units have now reached the point where 27,300 kW of engine power can be supported per turbocharger.[3] Motor sport applications are usually not as restricted in their use of turbocharging by the same emission legislation as road-legal vehicles, but they do have to comply with the FIA regulations relevant to their particular specialization. In order to cap maximum power, World Rally Car (WRC) rules currently (2010, due to change in 2011–12) dictate that the two-liter turbocharged engines are fitted with a 34-mm-diameter air restrictor before the turbocharger inlet; this limits the air flow to about 10 m^3/min, which limits maximum rpm. Typical boost pressure in WRC engines is 4–5 bar, compared with a maximum of up to 1 bar for a road car; this produces a high torque output having a rather flat curve from 3000 rpm to the 7500 rpm maximum. In WRC work, it is vital for the turbine to be kept running at optimum speed whatever the driver is doing—for example, when the car is "flying" with no power transmitted through the wheels; this creates some unusual test sequences, ignition timing, and engine fueling strategies, all part of an integrated Anti Lag System (ALS). Where emission controls are not required, throttled turbocharged engines can be controlled using air pressure rather than mass air flow (MAF), thus avoiding the ECU being confused when the antisurge valve opens, as mentioned above. In the case of WRC engine testing, a "fresh air valve" is mapped to bleed charge air into the turbine inlet below a set throttle position as part of the ALS and to keep the turbine running at its target speed.

---

3. MAN TCA88-25 turbochargers on the MAN-B&W 8K98MC two-stroke diesel engine.

Health and Safety Implications of Turbocharger Testing

Because of the high rotational speeds of the turbine-compressor shaft within a turbocharger and because they can be seen to glow bright red-hot in the test cell, there is often a perception that they represent an unusually high risk to cell operators and the cell fabric. In the rotating parts, the turbine blades in particular are on low mass and operate within a housing designed for their containment in the case of failure. Once again the perception of danger is higher than the reality and in-service failures of turbochargers are, almost always, due to poor lubrication and engine system maintenance. However, there are hazards associated with turbocharger testing that have to be taken into consideration, including but not exclusively:

- There is a worldwide market in after-market, counterfeit, turbochargers and spare parts that are inherently dangerous to use; therefore, all engine testers should ensure the units being run are genuine OEM units.
- Bearing failures have serious implications and will be caused by impaired oil supply in the case of conventional bearings or air supply in the case of "foil" bearings.
- Seal failures can lead to contamination of engine systems and even cause a diesel engine to "run away" (see "Emergency stop" section in Chapter 12).
- Turbines and near exhaust pipes may be incandescent while the engine is running at full power; it is vital that any nearby cables and pipes are well secured and suitably shielded. Take note of the warnings concerning the use of spot-cooling fans contained in Chapter 6 and also note that false fire alarms may be trigged by flame detectors fitted as part of an automatic fire suppression system.
- A cool-down period with the engine running at idle for 5 minutes is always recommended in any test sequence involving turbocharged engines. No special guarding against debris from a catastrophic turbine failure is recommended unless such an event is planned or anticipated during an over-speed test; the comments about the choice of cell windows given in Chapter 4 are relevant.

REFERENCES

[1] BS 4959. Recommendations for Corrosion and Scale Prevention in Engine Cooling Systems.
[2] W.H. McAdams, Heat Transmission, McGraw-Hill Maidenhead, 1973. ISBN-13: 978-0070447998.
[3] http://www.kratzer-automation.com
[4] http://www.agilis.com/
[5] P.A. Vesilind et al., Introduction to Environmental Engineering, third ed., CL-Engineering 2010. ISBN-13: 978-01495295853.

Fuel and Oil Storage, Supply and Treatment

INTRODUCTION

The storage and transport of volatile liquid and gaseous fuels worldwide is now subject to extensive and ever-developing regulation. The legislation surrounding the bulk storage of fuels is made more complicated because the safety-related risks of fire and explosion tend to be covered by one group of rules, while the risks of environmental pollution from leakage tend to be covered by another group of rules. Each set of rules may be administered by their own officials, who occasionally have policies concerning any one site that are in conflict with each other.

Engine Testing. DOI: 10.1016/B978-0-08-096949-7.00008-X

When planning the construction or modification of any bulk fuel storage facility it is absolutely vital to contact the responsible local official(s) early in the process, so that the initial design meets with concept approval.

While much of the legislation quoted in this chapter is British or European, the practices they require are valid and are recommended as good practice in most countries in the world. In the UK the Health and Safety Executive (HSE) is responsible for the regulation of risks to health and safety arising from work activity, while the 433 local authorities (LA) in the UK have responsibility for the rules governing storage of petroleum and volatile fuels; each local authority will have a petroleum officer.

Many of the rules and the licensing practices imposed on bulk fuel storage are designed to cover large farm (agricultural) or transport company diesel fuel systems and retail filling stations, where fuel is moved (dispensed) from the store tank into vehicles. Test facilities that dispense fuel into vehicles will fall under such rules; those that only handle fuel within a closed (reticulation) pipe system may not.

The most important reference legislations in the UK are:

- The Petroleum (Consolidation) Act 1928
- The Health and Safety at Work etc. Act 1974
- Dangerous Substances and Explosive Atmospheres Regulations in 2002
- ATEX Directive 94/9/EC
- Control of Pollution (Oil Storage) (England) Regulations 2001.

Copies of all this legislation and the codes of practice supporting them can be found on the internet.

Individual countries and administrative areas within the EEC have a level of local regulation, as is the case in different administrative areas in the USA.

In the UK, if fuel including petrol, diesel, vegetable, synthetic or mineral oil is stored in a container with a storage capacity of over 200 liters (44 gallons), then the owner may need to comply with the Control of Pollution (Oil Storage) (England) Regulations 2001.

The Explosive Atmospheres Regulations 2002 SI 2002/2776 issued to all local authorities in the UK having responsibility for fuel storage sites are worth study by any responsible individual wherever based (http://www.hse.gov.uk/lau/lacs/65-49.htm).

It should be clear to any senior test facility planner or manager that unless their site complies with the laws and practices that are in force in their region concerning fuel storage, their site insurers will not cover the, potentially ruinous, costs of a fuel fire or a major leakage causing environmental damage.

BULK FUEL STORAGE AND SUPPLY SYSTEMS

There was a time when it was standard practice for diesel fuel to be stored in a bunded tank above ground and for gasoline (petroleum) to be stored in a single skinned tank buried below ground. It still common practice in retail filling stations for all tanks to be buried. In the case of test facilities, if built within a secure perimeter, there is an increasing preference for fuel tanks and fuel piping to be installed above ground, where any leakage can be quickly detected. Local geology and proximity of groundwater sources or streams will play a part in determining the suitability of a fuel storage site.

The height below the datum ground level and any movement of the local water table[1] is of critical importance. In areas where the water table is high and variable, it is not unknown for buried tanks to "float" to the surface, with consequential fracturing of connections and fuel leakage.

The Approved Codes of Practice in the UK and in the USA require any underground fuel storage tanks (USTs) to be provided either with secondary containment or a leak detection system covering both tanks and associated pipework. Steel tanks also have to be fitted with a cathodic corrosion protection system.

The European regulations such as the ATEX Directive 94/9/EC and the EPA regulations in the USA, which in the 1980s required a major UST replacement program, encouraged many owners to convert to above-ground storage. However, in many sites burial was required by local officials when the installation was considered unsightly, exposed to high solar heat load, vulnerable to vehicular collision, vulnerable to vandalization or terrorist attack. These criteria are important, which is why a secure perimeter and careful layout is required when planning this part of a test facility.

If a full fuel storage license is required in the UK, the following information will usually be required (in the rest of the world attention to the same detail is recommended):

1. Site location map to a scale of 1:1250 or 1:2500 indicating all site boundaries. Two copies of site layout to scale 1:100 clearly indicating the intended layout of fuel storage and distribution system.
2. The layout plan must show the following:
   a. Location of storage tanks and tank capacities.
   b. Route taken by road delivery vehicles (this must be a "drive-through" not "cul-de-sac").
   c. Position of tanker fill points and their identification.
   d. Location of pipework including all vent pipes, etc.
   e. Location and type description of metering pumps, dispensers, etc.
   f. All site drainage and its discharge location.
   g. Petrol interceptor location and drainage discharge point.

---

1. The water table may be visualized as the surface of the subsurface materials that are saturated with groundwater in a given locality; it is not simply the depth to which groundwater settles in a hole. Note that some saturated ground materials are dangerously unstable. Local topography and geology will determine the details, which sometimes can vary over surprisingly small areas of ground.

h. All other buildings within the site and their use.
i. All neighboring buildings within 6 m from the boundary.
j. Position of LPG storage, if applicable.
k. Location of car wash and drainage if applicable.
l. Main electrical intake point and distribution board.
m. Position of all fire-fighting appliances.

Very similar restrictions and requirements to those described above for the UK exist in most countries in the world, and any engineers taking responsibility for anything beyond the smallest and simplest test cell should make themselves familiar with all relevant regulations of this kind, both national and local.

Any contractor used to install or modify fuel systems should be selected on the basis of proven competence and, where applicable, relevant licensing to carry out such work and the required post-installation testing.

Both the legislative requirements and the way in which they are interpreted by local officials can vary widely, even in a single country. Where engine test installations are well established there should be a good understanding of the requirements, whereas in localities where such systems are novel, the fire and planning officers may have no experience of the industry and can react with concern; in these cases they may require tactful guidance.

Figure 8.1 shows a typical arrangement for a fuel oil or gasoline bunded storage tank according to British Standard 799.

The risk of oil being lost during tank filling or from ancillary pipework is higher than tank rupture; the UK Control of Pollution (Oil Storage) Regulations 2001 recognize this fact and require that tanks have a list of ancillary equipment such as sight tubes, taps, and valves retained within a secondary containment system. The use of double-skinned tanks is strongly recommended and

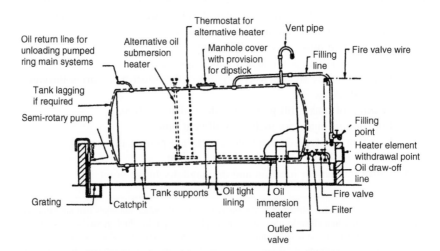

FIGURE 8.1 Typical above-ground bunded fuel tank with fittings.

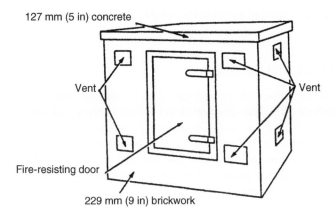

FIGURE 8.2 Design of fuel drum store, minus statutory warning signs.

increasingly required by regulation when used in underground installations [1]. Double-skinned tanks and fuel pipes should also be fitted with an interstitial leak monitoring device.

For the storage of fuel in drums or other containers, a secure petroleum store such as that shown in Figure 8.2 is to be recommended. This design typically meets the requirements of the local UK authorities, but its location and the volume of fuel allowed to be stored may be subject to site-specific regulation. The lower part forms a bunt, not more than 0.6 m deep, capable of containing the total volume of fuel authorized to be kept in the store. Ventilators at high and low level are to be covered by fine wire gauze mesh and protective grilles. Proprietary designs should allow access to a wheeled drum lifter/trolley. Packaged units built from 20-foot ISO shipping containers, designed for use at large road building sites, are probably a cost-effective solution for some facilities.

Finally, when planning bulk fuel storage, precautions should be taken against the theft of fuel and malicious damage (vandalism); where the risk is considered to be high the increased cost of burial is clearly justified.

DECOMMISSIONING OF UNDERGROUND FUEL TANKS

Sites with disused underground fuel tanks have a double problem: the land above is virtually unusable and the risk of environmental pollution increases with time. One policy has been to pump the tank full with a cement-based grout, but that makes the tank too heavy to remove at some future date and it is extremely difficult to fill the void entirely. A better method, after as much of the fuel as possible has been pumped out, is to have the tank filled with chemical-absorbing foam created from an amino-plastic resin, which completely fills the space, soaks up hydrocarbon residues, and doesn't significantly increase the mass of the tank. Specialist contractors need to carry out such decommissioning and certify the site's safety.

AUDITING OF FUEL USE

There have been cases where long-term fuel leakage into the environment or loss through theft has been discovered, not by management monitoring systems or facility staff, but by accountants trying to balance quantities of fuel purchased with that costed out to test contracts. Simple flow meters measuring the accumulated flow into each cell, that are read at the start and end of each project, are all that is required and are a recommended part of the quality plan for any engine test organization.

FUEL PIPES

The expensive consequences of fuel leaks means that modern best practice is for fuel lines to be run high above ground, where they can be seen but not accidentally damaged by normal site activities.

The use of standard (nongalvanized) drawn steel tubing for fuel lines is entirely satisfactory providing that they remain full of liquid fuel, but if they are likely to spend appreciable periods partially drained then the use of stainless steel is recommended. The use of threaded fittings introduces some risk of leakage, although in drawn steel tube and provided there is no significant, thermally induced, pipe movement they can be satisfactory when used with modern sealants. The use of any kind of fibrous "pipe jointing" should be absolutely forbidden as fiber contamination is difficult to clear.

Threading of stainless steel tube with pipe treading machines using normal dies often gives poor results and is not recommended. Preferably, all stainless steel fuel lines, and certainly all underground lines, should be constructed with orbital welded joints, or with the use of compression fittings approved for the fuels concerned (some fuels such as "winter diesel" appear, from experience, to be particularly penetrating). External fuel oil lines should be lagged and must be trace heated if temperatures are likely to fall to a level at which fuel "waxing" may take place.

Within the test cell flexible lines may be required in short sections to allow for engine rigging or connection to fixed instrumentation. Such tubing and fittings must be specifically made and certified to handle the range of fuels being used; the generally recommended specification is for metal braided, electrically conductive, Teflon hoses. All conductive fuel lines should be linked and form part of the building's equipotential earthing scheme, as described in Chapter 5.

If the site has a fuel bowser for filling vehicles then, as stated elsewhere, the whole site has to comply with the rules covering retail fuel outlets, including the type of bowser, its interconnections, warning signs, and electrical supply system.

Underground Fuel Pipes

Whatever their type of construction it is vital that any underground pipe is buried sufficiently deep and is sufficiently protected from earth movement and compaction from vehicles running over them. When running under any type of road or walkway, the route should be marked by permanent signage so that protective measures can be taken in the case of movement of construction plant, etc.

Increasingly, in Europe and the USA, buried fuel lines are required to be of a double-walled design with the facility to check leakage from the primary tube into interstitial space (interstitial monitoring). Double-walled pipe systems are now usually made of extruded high-density plastics with the outer sleeve composition chosen for high abrasion resistance and the inner sleeve composition for very low permeability to fuels. Such pipes and fittings have to be electrofusion welded, have a maximum pressure rating for the inner containment sleeve of 10 bar, and a 30-year minimum design life.

Where single-wall steel fuel lines are used subfloor, particularly where they run under and through concrete floor slabs, they should be wrapped with water-repellent bandage (in the UK, sold as "Denzo®️ tape") and laid on well-compacted fine gravel within a trench. The wrapping prevents the concrete from trapping the pipes and allows them to move without local stressing as they expand and contract. It is good practice to run underground fuel lines in a sealed trench of cast concrete sections with a load-bearing lid; the route of the trench has to be clearly marked and vehicles kept clear. Such ducts should be fitted with hydrocarbon "sniffers" connected to an alarm system to detect fuel leakage before ground contamination takes place.

The "fuel farm" for a modern R&D facility may be provided with a number of storage tanks feeding several ring mains; these must be clearly and permanently labeled with their contents at all filling and draw-off points.

Air-locking can be a problem in the long, quite small bore piping used in test facility fuel systems; it can be particularly difficult to cure, so careful planning and installation is needed to avoid creation of vapor or air traps.

Pipes containing gravity-fed gasoline (from a header tank) that are run for any distance, external to the building and exposed to high solar load, are vulnerable to vapor locking and may require tall vent pipes[2] to atmosphere to be fitted at the point at which the downward legs to cells are fitted.

STORAGE AND TREATMENT OF RESIDUAL FUELS

Many large stationary diesel engines and the majority of slow-speed marine engines operate on heavy residual fuels that require special treatment before

---

2. The termination height of any vent needs to be some meters higher than the pressure head of fuel in the system under any circumstances.

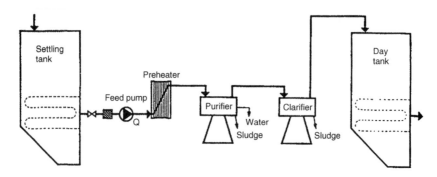

FIGURE 8.3   Fuel supply and treatment system for residual fuel oil.

use and delivery to the fuel injection system. In this fuel classification there are two commonly used grades of fuel:

- Number 5 fuel oil, sometimes known as Bunker B.
- Number 6 fuel oil, which is a high-viscosity residual oil sometimes known as residual fuel oil (RFO) or Bunker C; it may contain various impurities plus 2% water and 0.5% mineral particles (soil).

Any test system running engines with these fuels needs to incorporate the special features of the fuel supply and treatment systems typically installed in ships propelled by slow-speed diesel engines. Bunker C oil needs to be preheated to between 110 and 125 °C; therefore, the bulk storage tanks will require some form of heating coils within them to enable the heavy oil to be pumped into a heated settling tank. It is always necessary with fuels of this type to remove sludge and water before use in the engine. The problem here is that the density of residual fuels can approach that of water, making separation very difficult. The accepted procedure is to raise the temperature of the oil, thus reducing its density, and then to feed it through a purifier and clarifier in series. These are centrifugal devices, the first of which removes most of the water while the second completes the cleaning process [2]. Figure 8.3 shows a schematic arrangement.

It is also necessary to provide changeover arrangements in the test cell so that the engine can be started and shut down on a light fuel oil. An emergency stop resulting in a prolonged shutdown when there is heavy residual oil in the engine fuel rails creates a very unpleasant problem.

STORAGE OF BIOFUELS

Test facilities involved in the preparation and mixing of biofuels may have special storage problems to consider. Most biodiesel fuels are a mixture of mineral- and vegetable-derived liquids; they are usually designated by the preface B, followed by the percentage of the biomass-derived content; thus,

B20 contains 20% of nonpetroleum fuel. Biofuels tend to be hydrophilic; therefore, storage tanks and plain steel pipelines can be subject to corrosion from condensation.

Laboratories working with the constituents of biodiesel mixtures will require storage for animal- or vegetable-derived oils in bulk or drums; these may require low-grade heating and stirring in some climatic conditions. However, any water in the fuel will tend to encourage the growth of microbe colonies in heated fuel tanks, which can form soft masses that plug filters.

Ethanol/gasoline mixtures are designated with the preface E, followed by the percentage of ethanol. E5 and E10 (commonly called "gasohol") are commercially available in various parts of Europe. E20 and E25 are standard fuel mixes in Brazil, from where no particular technical problems of storage have been reported.

The storage and handling of 100% ethanol and methanol raises problems of security; the former is highly intoxicating and the latter highly toxic, and anyone likely to drink the intoxicant probably lacks the chemical knowledge to distinguish between the two.

REFERENCE FUEL DRUMS

Where frequent use is made of special or reference fuels, supplied by the drum, special provision must be made for their transportation from the secure fuel store, such as that shown in Figure 8.2, and protection while in use. Special drum containers made from high-density plastic, designed to be transported by fork-lift, are available and these may be parked immediately against an outside wall of the test cell requiring the supply. Connection and supply into the cell system can be achieved by using an automotive 12 V pump system designed to screw into the drum and deliver fuel through a connection point outside the cell fitted with identical fire isolation interlocks as used in the permanently plumbed supplies.

NATURAL GAS (NG), LIQUEFIED NATURAL GAS (LNG), AND COMPRESSED NATURAL GAS (CNG)

Natural gas consists of about 90% methane, which has a boiling point at atmospheric pressure of $-163\ °C$. Engine test installations requiring natural gas for dual fuel engines usually draw this from a mains supply at just above atmospheric pressure, so high-pressure storage arrangements are not necessary. In the UK the distribution of NG, to individual commercial users, is by way of a national grid and generally covered by the Gas (Third Party Access) Regulations 2004. Fire hazards are moderate when compared with LPG installations.

Engine test cells using LNG and NG have to comply with gas industry regulations, which may include explosion relief panels in the cell and that part of the engine exhaust system within the building. Early contact and negotiation with the local authority and supply company is vital.

LIQUEFIED PETROLEUM GAS (LPG)

LPG is often referred to as autogas or auto propane. In engine test facilities it will be stored under approximately 7 bar (100 psi) pressure as a liquid in an external bulk storage tank. LPG is known as one of the most "searching" of gases—that is, it can escape through gaps that would hold water and other gases. It has a density of 1.5 that of air and concentrations of 2–10% LPG in air make a flammable mix. Test cells using LPG must be fitted with suitable hydrocarbon detectors at the lowest cell level and the ATEX precautions for the avoidance of explosive atmospheres, discussed elsewhere in this book, have to be taken. It should be clear that any LPG installation has to employ specialist contractors during the planning, installation, and maintenance stages of its life.

Useful guidance concerning the safe storage and handling of LPG is to be found in the "UK LP Gas Association Code of Practice 1: Bulk LPG Storage at Fixed Installations, Part 1: Design, Installation and Operation of Vessels Located Above Ground" (http://www.uklpg.org/shop/codes-of-practice/).

FUEL SUPPLY TO THE TEST CELL

Fuel supplies to a cell may be provided either under the static pressure head from a day tank, a pumped supply from a reference fuel drum, or by a pressurized reticulation system fed from the central fuel farm.

Day tanks should be fuel specific to prevent cross-contamination, so there may be a requirement for several. Modern safety practice usually, but not exclusively, dictates that gasoline day tanks should be kept on the outside wall rather than inside the test facility. Diesel day tanks are most frequently inside the shell building and thus protected, where necessary, from low (waxing) temperatures.

In all cases day tanks must be fitted with a dump valve for operation in case of fire; this allows fuel contained in the vulnerable above-ground tanks and pipes to be returned to the fuel farm, or a specific underground dump tank. Besides having to be vented to atmosphere, day tanks also have to be fitted with a monitored overflow and spill return system in the case of a malfunction of any part of the supply system. Figure 8.4 shows a recommended day-tank system diagrammatically.

The static head is commonly at 4.5 m or above, but may need to be calculated specifically to achieve the 0.5–0.8 bar inlet pressure required by some industrial standard fuel consumption and treatment instruments.

When the tanks are exposed to ambient weather conditions they should be shielded from direct sunlight and, in the case of diesel fuels, lagged or trace-heated.

If a pumped system is used it must be remembered that, for much of the operating life, the fuel demand from the cells will be below the full rated flow of

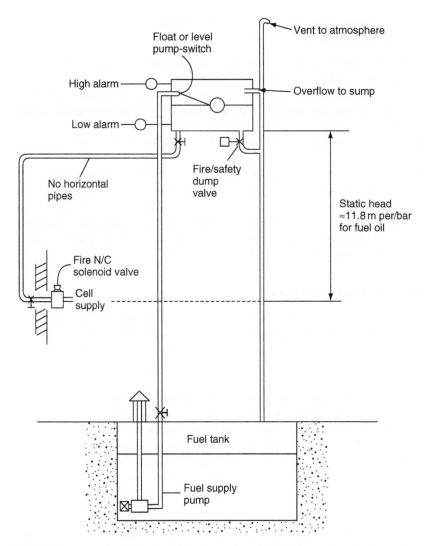

FIGURE 8.4 A schematic showing the elements of a typical fuel day-tank system, in this case combined with a subterranean bulk tank.

the pump; therefore, the system must be able to operate under bypass, or stall, conditions without cavitation or undue heating.

The use of positive displacement, pneumatically operated pumps, incorporating a rubberized air/fluid diaphragm, has a number of practical advantages: they do not require an electrical supply and they are designed to be able to stall at full (regulated air) pressure, thus maintaining a constant fuel pressure supply to the test cells.

In designing systems to meet this wide range of requirements, some general principles should be remembered:

- Pumps handling gasoline must have a positive static suction head to prevent cavitation problems on the suction side.
- Fuel pipes should have low flow resistance, where smooth bends, rather than sharp elbows, are used and no sudden changes in internal cross-section exist.
- Fuel lines should be pressure tested by competent staff before filling with fuel.
- Each fuel line penetrating the cell wall should be provided with a normally closed solenoid-operated valve interlocked with both the cell control system and the fire protection circuits. In some parts of the USA, local regulations require that the solenoid valves are supplemented by mechanical, normally spring-closed, valves that are held open by a soft-solder plug; in the case of a fire the plug melts and the supply is shut off.

IN-CELL FUEL SYSTEMS

The fuel system in the test cell will vary widely in complexity. In some cases the system may be limited to a single fuel line connected to the engine's fuel pump, but a special-purpose test cell may call for the supply of many different fuels, all passing through fuel temperature control and fuel consumption measurement devices.

For the occasional test, using a portable engine dynamometer system, a simple fuel tank and tubing of the type used for outboard marine engines, with capacity not more than 10 liters, may be all that is necessary. These marine devices are certified as safe to use in their designated roles of containing and supplying fuel to an engine; the use of other containers is not safe and would endanger the insurance of the premises in which they are used.

The following points should be considered in the planning of in-cell systems:

- It may be necessary to provide several separate fuel supplies to a cell. A typical provision would be three lines for diesel fuel, "standard", and "super" unleaded gasoline respectively. Problems can arise from carry-over, with consequent danger of "poisoning" exhaust catalyzers. The capacity of the fuel held in the cell system, including such items as filters, needs to be kept to a minimum. To minimize cross-contamination it is desirable, although often difficult, to locate the common connection as close to the engine as possible. A common layout provides for an inlet manifold of fuels to be fitted below the in-cell fuel conditioning and measuring systems. Selection of fuel can be achieved by way of a flexible line fitted with a self-sealing connector from the common system to the desired manifold mounted connector.

- It is good practice to have a cumulative fuel meter in each line for general audit and for contract charging.
- Air entrainment and vapor locking can be a problem in test cell fuel systems so every care should be taken in avoiding the creation of air traps. An air-eliminating valve should be fitted at the highest point in the system, with an unrestricted vent to atmosphere external to the cell and at a height and in a position that prevents fuel escape.
- Disconnection of, or breaking into fuel lines, with consequent spillage or ingress of air, should, as far as possible, be avoided. It is sensible to mount all control components on a permanent wall-mounted panel or within a special casing, with switching by way of interlocked and *clearly marked valves*. The run of the final flexible fuel lines to the engine should not interfere with operator access. A common arrangement is to run the lines through an overhead boom. Self-sealing couplings should be used for engine and other frequently used connections.
- It is essential to fit oil traps to cell drain connections to avoid the possibility of discharging oil or fuel into the foul water drains.
- Aim for a flow rate during normal operation to be between 0.2 and a maximum recommended fuel line velocity: 1.0 m/s.
- Where self-sealing, quick connectors are used to switch fuels from supply lines to cell circuits, it is recommended, where possible, to use "fail-safe" connectors that physically prevent the accidental poisoning of, for example, a gasoline circuit with diesel oil.

ENGINE FUEL PRESSURE CONTROL

There are three fuel pressure control problems that may need to be solved when designing or operating an engine test cell:

1. Pressure of fuel supply to the fuel conditioning system or consumption measurement instrument. Devices such as the AVL fuel balance require a maximum pressure of 0.8 bar at the instrument inlet; in the case of pumped systems this may require a pressure-reducing regulator to be fitted before the instrument.
2. Pressure of fuel supplied to the engine. Typically systems fitted in normal automotive cells are adjustable to give pressures at the engine's own high-pressure pump system inlet of between 0.05 and 4 bar.
3. Pressure of the fuel being returned from the engine. When connected to a fuel conditioning and consumption system the fuel return line in the test cell may create a greater back-pressure than required. In the vehicle the return line pressure may be between zero and 0.5 bar, whereas cell systems may require over 1.5 bar to force fuel through conditioning equipment and to overcome static head differences between the engine and instruments; therefore, a pressure-reducing circuit including a pump may be

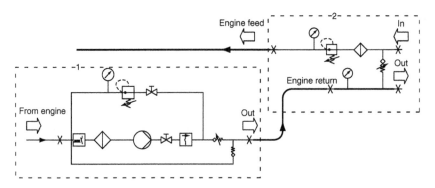

FIGURE 8.5 Regulation of engine supply and return pressures. Within dotted lines, (1) is the fuel return pressure control and (2) the fuel feed to the engine from the fuel consumption measurement system. *(AVL List.)*

required. Figure 8.5 shows a circuit of a system that allows for the independent regulation of engine supply and return pressures. Care must be exercised in the design of such circuits to avoid the fuel pressure being taken below the point at which vapor bubbles form.

ENGINE FUEL TEMPERATURE CONTROL

All the materials used in the control of fuel temperature, such as heat exchangers, piping, and sealing materials, must be checked with the manufacturer as to their suitability for use with all specified fuels.

If fuel supplied to the engine fuel rail inlet has to be maintained at one standard temperature, as is usually the case in quality audit cells, then a relatively simple control system may use hot water circulated at controlled temperature through the water-to-fuel heat exchanger. Such an arrangement can give good temperature control for quite short and undemanding test sequences despite wide variations in fuel flow rate.

The control of the fluid fuel temperature within the engine circuit is complicated by the fuel rail and spill-back strategy adopted by the engine designer; because of this and variations in the engine rigging pipe work, no one circuit design can be recommended. Commercially available fuel conditioning devices can only specify the temperature control at the unit discharge: the system integrator has to ensure the heat gains and losses within the connecting pipework do not invalidate experiments.

It is particularly important to minimize the distance between the temperature-controlling element and the engine so that, if running is interrupted, the engine receives fuel at the control temperature with the minimum of delay.

Commercially available temperature control units typically have the following basic specifications:

- Setting range of fuel temperature: 10–45 °C.
- Deviation between set value and actual value at instrument output: <1 °C.

- Power input: 0.4 kW or 2 kW (with heating).
- Cooling load: typically 1.5 kW.

It should be noted that the stability of temperature control below ambient will be dependent on the stability of the chilled water supply, which should have a control loop independent of other, larger, cooling loads in any shared system.

A constant fuel temperature over the duration of an experiment is a prerequisite for accurate consumption measurement (see Chapter 14).

As with all systems covered in this book, the complexity and cost of fuel temperature control will depend on the operation range and accuracy specified.

To confirm that temperature change across the fuel measurement circuit has a significant influence on fuel consumption and fuel consumption measurement, especially during low engine power periods, Figure 8.6 should be considered. The graph lines of Figure 8.6 show that making a cyclic step change in the set point of a fuel temperature controller of only ± 0.2 °C within the measurement circuit produces an oscillation in fuel temperature at the engine inlet of ± 0.1 °C, and a variation in the measured fuel consumption value of ± 0.05 kg/h (top line) is created. This equals a variation of $\pm 2\%$ of the actual fuel consumption value and shows that instrument manufacturers' claims of fuel consumption accuracy are only valid under conditions of absolutely stable fuel temperatures.

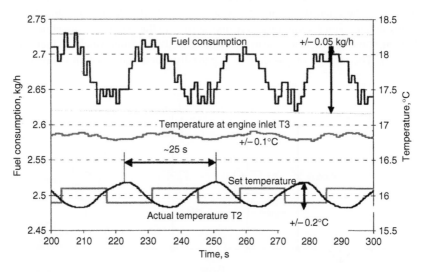

FIGURE 8.6 Effects of fuel temperature changes on the fuel consumption measurement values. T2 is the controlled outlet of the fuel conditioning device superimposed on the step demand changes. T3 is the temperature at the engine inlet. The difference between the two is due to system damping. *(AVL List.)*

ENGINE OIL COOLING SYSTEMS

Certain important layout requirements apply to oil cooling units. Unless the cell is operating a "dry-sump" lubrication system on the engine, the entire lubrication oil temperature control circuit must lie below engine sump level so that there is no risk of flooding the sump. It may be necessary to provide heaters in the circuit for rapid warm-up of the engine; in this case the device must be designed or chosen to ensure that the skin temperature of the heating element cannot reach temperatures at which oil "cracking" can occur.

Figure 8.7 shows schematically a separate lubricating oil cooling and conditioning unit.

Where very accurate transient temperature control is necessary, the use of separate pallet-mounted cooling modules located close to the engine may be required; otherwise permanently located oil and cooling water modules offer the best solution for most engine testing. A once common position for such modules was behind the dynamometer, where both units were fed from the external cooling water system and the engine connection hoses were run under the dynamometer to a connection point near the shaft end. The use of vehicle exhaust systems has made this layout less possible and transferred the modules in modern cells to one of the side walls of the engine pallet.

PROPERTIES OF GASOLINE AND "SHELF LIFE"

The calorific value of automotive gasoline can be taken, for energy flow calculation in the test cell and *not* in critical test calculations, as 43.7 MJ/kg, but as sold at the retail pump in any country of the world gasoline is variable in many details. Some variabilities are appropriate to the geographic location and season, some of which are due to the original feedstock from which the fuel was refined. The storage of gasoline in a partially filled container, in conditions

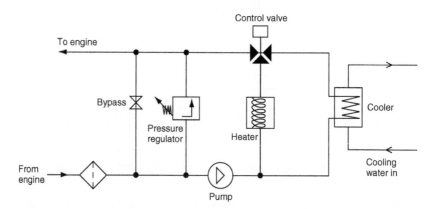

FIGURE 8.7 Schematic of oil temperature control unit, sensing points and control connections omitted.

where it is exposed to direct sunlight, will cause it to degrade rapidly. Reference fuels of all types should be stored in dark, cool conditions and in containers that are 90–95% full; their shelf life should not be assumed to be more than a few months.

Engine development work is often concerned with "chasing" very small improvements in fuel consumption; these differences can easily be swamped by variations in the calorific value of the fuels. Similarly, comparisons of identical engines manufactured at different sites will be invalid if they are tested on different batches of the "same" retail fuels or fuels that have been allowed to deteriorate at differential rates.

For example, the lower calorific values (LCVs) of hexane and benzene, typical constituents of gasoline, are respectively 44.8 and 40.2 MJ/kg. A typical value for a "branded" retail gasoline would be 43.7 MJ/kg, but the hexane/benzene content could easily vary by ±2% in different parts of the world, while the presence of alcohol as a constituent can depress the calorific value substantially.

Octane number is the single most important gasoline specification, since it governs the onset of detonation or "knock" in the engine. This fault condition, if allowed to continue, will rapidly destroy an engine. Knock limits the power, compression ratio, and hence the fuel economy of an engine. Too low an octane number also causes run-on when the engine is switched off.

Three versions of the octane number are used: research octane number (RON), motor octane number (MON), and front-end octane number (R100). RON is determined in a specially designed (American) research engine: the Cooperative Fuel Research (CFR) engine. In this engine the knock susceptibility of the fuel is graded by matching the performance with a mixture of reference fuels, iso-octane with an RON = 100 and n-heptane with an RON = 0.

The RON test conditions are now rather mild where modern engines are concerned and the MON test imposes more severe conditions but also uses the CFR engine. The difference between the RON and the MON for a given fuel is a measure of its "sensitivity". This can range from about 2 to 12, depending on the nature of the crude and the distillation process. In the UK it is usual to specify RON, while in the US the average of the RON and the (lower) MON is preferred. R100 is the RON determined for the fuel fraction boiling at below 100 °C.

Volatility is the next most important property and is a compromise. Low volatility leads to low evaporative losses, better hot start, less vapor lock, and in older engine designs, less carburettor icing. High volatility leads to better cold starting, faster warm-up, and hence better short-trip economy, also to smoother acceleration.

PROPERTIES OF DIESEL FUELS

The calorific value of automotive diesel is generally taken, for energy flow calculation in the test cell, as 43,000 kJ/kg, but just as the range of size of the diesel engine is much greater than that of the spark-ignition engine, from 1–2 to 50,000 kW, the range of fuel quality is correspondingly great.

Cetane number is the most important diesel fuel specification. It is an indication of the extent of ignition delay: the higher the cetane number, the shorter the ignition delay, the smoother the combustion, and the cleaner the exhaust. Cold starting is also easier the higher the cetane number.

Viscosity covers an extremely wide range. BS2869 specifies two grades of vehicle engine fuels, Class Al and Class A2, having viscosities in the range 1.5–5.0 and 1.5–5.5 cSt respectively at 40 °C. BS MA 100 deals with fuels for marine engines. It specifies nine grades, Class M1, equivalent to Class A1, with increasing viscosities up to Class M9, which has a viscosity of 130 cSt maximum at 80 °C.

FUEL DYES AND COLORING

Fuel oil and gasoline are often colored by the addition of dyes at the blending stage of production. Fuels are artificially colored to enable recognition by the taxation authorities, or by the user for safety reasons to minimize the risk of using the wrong fuel type.

Fuel "laundering" whereby such fuels as "red diesel", originally bought at a low tax rate, is de-colored and resold is highly illegal; for this reason alone test facilities should only buy bulk fuel from fully certified suppliers.

Reference fuels of the types used in critical engine testing are not dyed. Table 8.1 shows examples of some national and international standards of fuel dyeing.

TABLE 8.1 Some Examples of the Use of Fuel Dyes in Common Use for Purposes of Safety and Prevention of Tax Avoidance

Country or Group	Fuel	Dye Color
European Union	All automotive diesel	Solvent, Yellow 124
UK	Agricultural gas oil, low tax	Quinizarin, Red
USA	High sulfur, low tax diesel	Solvent, Red 164
France	Marine diesel, low tax	Solvent, Blue 35
Australia	Regular unleaded gasoline Premium unleaded gasoline	Purple Yellow
Worldwide	Aviation gasoline 100LL	Blue
Worldwide	Aviation gasoline 100/130	Green
Worldwide	Aviation gasoline 80/87	Red

SAFETY PRINCIPLES IN HANDLING AND STORING VOLATILE FUELS

The mnemonic "VICES" is used by the HSE in the UK and is worth remembering:

- V = ventilation. Adequate ventilation rapidly disperses flammable vapors.
- I = ignition. All ignition sources should be removed from fuel handling areas.
- C = containment. The fuels must be held in containers suitable for their containment with secondary devices such as trays to catch spillage and absorbent materials to hold and clean up any leakage.
- E = eliminate. Is it possible to eliminate or reduce some of the fuel containment?
- S = separation. Fuels should be stored in areas well separated from other storage or work areas or areas where they are exposed to accidental damage (delivery trucks, etc.).

REFERENCES

[1] J.R. Hughes, S. Swindells, Storage and Handling of Petroleum Liquids, Griffin, London, 1987. ISBN-13: 978-0471629665.
[2] Recommendations for Cleaning and Pretreatment of Heavy Fuel Oil. Alfa Laval, London.

FURTHER READING

BS 2869 Part 2. Specification for Fuel Oil for Agricultural and Industrial Engines and Burners.
BS 3016. Specification for Pressure Regulators and Automatic Changeover Devices for Liquefied Petroleum Gases.
BS 4040. Specification for Leaded Petrol (Gasoline) for Motor Vehicles.
BS 4250 Part 1. Specification for Commercial Butane and Propane.
BS 5355. Specification for Filling Ratios and Developed Pressures for Liquefiable and Permanent Gases.
BS 6843 Parts 0 to 3. Classification of Petroleum Fuels.
BS 7405. Guide to Selection and Application of Flowmeters for the Measurement of Fluid Flow in Closed Conduits.
BS EN 228. Specification of Unleaded Petrol (Gasolene) for Motor Vehicles.
BS EN 589. Specification for Automotive LPG.
Dangerous Substances and Explosive Atmospheres Regulations (DSEAR) 2002.

Vibration and Noise

PART 1: VIBRATION AND ITS ISOLATION

Note: Vibration is considered in this chapter with particular reference to the design and operation of engine test facilities, engine mountings, and the isolation of engine-induced disturbances. Torsional vibration within shaft systems is covered as a separate subject in Chapter 11.

Vibration

Usually the engine itself is, or should be, the only significant source of vibration and noise in the engine and powertrain test cell. Secondary sources of noise and vibration within the test cell such as the ventilation system, pumps, and fluid circulation systems or the dynamometer are usually swamped by the effects of the engine.

Engine Testing. DOI: 10.1016/B978-0-08-096949-7.00009-1

There are several aspects to the problem of engine vibration and noise:

- The engine must be mounted in such a way that neither it, nor connections to it, can be damaged by excessive movement or excessive constraint.
- Transmission of engine-induced vibration to and via services connected to the cell structure or to adjoining buildings must be minimized.
- Noise levels in the cell should be contained as far as possible and the design of alarm systems should take into account ambient noise levels in all related work areas.

Fundamentals: Sources of Vibration

Since the vast majority of engines likely to be met with are single or multicylinder in-line vertical engines, we shall being by concentrating on this configuration.

Any unit under test (UUT) may be regarded as having six degrees of freedom of vibration about orthogonal axes through its center of gravity: linear vibrations along each axis and rotations about each axis (see Figure 9.1). In practice only three of these modes are usually of importance in engine testing:

- Vertical oscillations on the X axis due to unbalanced vertical forces
- Rotation about the Y axis due to cyclic variations in torque
- Rotation about the Z axis due to unbalanced vertical forces in different transverse planes.

Torque variations will be considered later. In general the rotating masses are carefully balanced but periodic forces due to the reciprocating masses cannot be avoided. The crank, connecting rod, and piston assembly shown in Figure 9.2 is subject to a periodic force in the line of action of the piston, given approximately by:

$$f = m_{\mathrm{p}}\omega_{\mathrm{c}}^2 r \cos \theta + \frac{m_{\mathrm{p}}\omega_{\mathrm{c}}^2 r \cos 2\theta}{n} \tag{9.1}$$

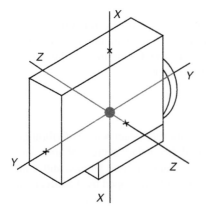

FIGURE 9.1 Internal combustion engine: principal axes and degrees of freedom.

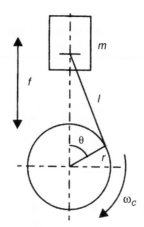

FIGURE 9.2 Connecting rod crank mechanism: unbalanced forces.

where $n = l/r$ in Figure 9.2. Here m_p represents the sum of the mass of the piston plus, by convention, one-third of the mass of the connecting rod (the remaining two-thirds is usually regarded as being concentrated at the crankpin center).

The first term of equation (9.1) represents the first-order inertia force. It is equivalent to the component of centrifugal force on the line of action generated by a mass m_p concentrated at the crankpin and rotating at engine speed. The second term arises from the obliquity of the connecting rod and is equivalent to the component of force in the line of action generated by a mass $m/4n$ at the crankpin radius but rotating at twice engine speed.

Inertia forces of higher order ($3\times$, $4\times$, etc., crankshaft speed) are also generated but may usually be ignored.

It is possible to balance any desired proportion of the first-order inertia force by balance weights on the crankshaft, but these then lead to an equivalent reciprocating force on the Z axis, which may be even more objectionable.

Inertia forces may be represented by vectors rotating at crankshaft speed and twice crankshaft speed. Table 9.1 shows the first- and second-order vectors for engines having from one to six cylinders. Note the following features:

- In a single-cylinder engine, both first- and second-order forces are unbalanced.
- For larger numbers of cylinders first-order forces are balanced.
- For two- and four-cylinder engines, the second-order forces are unbalanced and additive.

This last feature is an undesirable characteristic of a four-cylinder engine and in some cases has been eliminated by counter-rotating weights driven at twice crankshaft speed.

TABLE 9.1 First- and Second-Order Forces in Engines of One to Six Cylinders

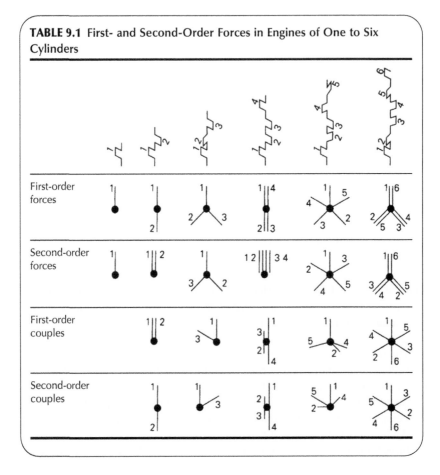

The other consequence of reciprocating unbalance is the generation of rocking couples about the transverse or Z axis and these are also shown in Figure 9.1:

- There are no couples in a single-cylinder engine.
- In a two-cylinder engine there is a first-order couple.
- In a three-cylinder engine there are first- and second-order couples.
- Four- and six-cylinder engines are fully balanced.
- In a five-cylinder engine there is a small first-order and a larger second-order couple.

Six-cylinder engines, which are well known for smooth running, are balanced in all modes.

Variations in engine turning moment are discussed in Chapter 11. These variations lead to equal and opposite reactions on the engine, which tend to cause rotation of the whole engine about the crankshaft axis. The order of these disturbances, i.e. the ratio of the frequency of the disturbance to the engine speed, is a function of the engine cycle and the number of cylinders. For a four-stroke engine the lowest order is equal to half the number of cylinders: in a single cylinder there is a disturbing couple at half engine speed while in a six-cylinder engine the lowest disturbing frequency is at three times engine speed. In a two-stroke engine the lowest order is equal to the number of cylinders.

Design of Engine Mountings and Test-Bed Foundations

The main problem in engine mounting design is that of ensuring that the motions of the engine and the forces transmitted to the surroundings, as a result of the unavoidable forces and couples briefly described above, are kept to manageable levels. In the case of vehicle engines it is sometimes the practice to make use of the same flexible mounts and the same location points as in the vehicle; *this does not, however, guarantee a satisfactory solution.* In the vehicle the mountings are carried on a comparatively light structure, while in the test cell they may be attached to a comparatively massive pallet or even to a seismic block. Also, in the test cell the engine may be fitted with additional equipment and various service connections. All of these factors alter the dynamics of the system when compared with the situation of the engine in service and can cause fatigue failures of both the engine support brackets and those of auxiliary devices such as the alternator.

Truck diesel engines usually present less of a problem than small automotive engines, as they generally have fairly massive and well-spaced supports at the flywheel end. Stationary engines will in most cases be carried on four or more flexible mountings in a single plane at or near the horizontal centerline of the crank shaft, as in Figure 9.3.

We shall consider the simplest case, an engine of mass m kg carried on undamped mountings of combined stiffness k N/m (Figure 9.3). The differential

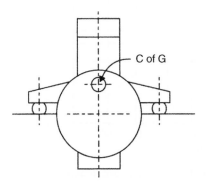

FIGURE 9.3 Engine carried on four flexible mountings.

C of G

equation defining the motion of the mass equates the force exerted by the mounting springs with the acceleration of the mass:

$$\frac{m d^2 x}{dt^2} + kx = 0 \tag{9.2}$$

A solution is:

$$x = \text{constant} \times \sin\sqrt{\frac{k}{m}} \times t \tag{9.3}$$

$$\frac{k}{m} = \omega_0^2 \qquad \text{natural frequency} = \eta_0 = \frac{\omega_0}{2\pi} = \frac{1}{2\pi}\sqrt{\frac{k}{m}}$$

The static deflection under the force of gravity $= mg/k$, which leads to a very convenient expression for the natural frequency of vibration:

$$\eta_0 = \frac{1}{2\pi}\sqrt{\frac{g}{\text{static deflection}}} \tag{9.4a}$$

or, if static deflection is in millimeters:

$$\eta_0 = \frac{15.76}{\sqrt{\text{static deflection}}} \tag{9.4b}$$

This relationship is plotted in Figure 9.4.

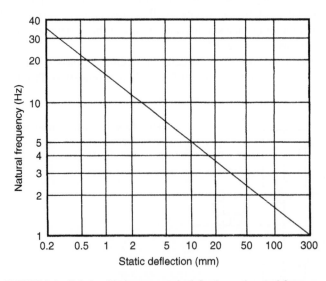

FIGURE 9.4 Relationship between static deflection and natural frequency.

Next consider the case where the mass m is subjected to an exciting force of amplitude f and frequency $\omega/2\pi$. The equation of motion now reads:

$$m\frac{\mathrm{d}^2x}{\mathrm{d}t^2} + kx = f\sin\omega t$$

The solution includes a transient element. For the steady-state condition, amplitude of oscillation is given by:

$$x = \frac{f/k}{(1 - \omega^2/\omega_0^2)} \tag{9.5}$$

Here f/k is the static deflection of the mountings under an applied load f. This expression is plotted in Figure 9.5 in terms of the amplitude ratio, x divided by static deflection. It has the well-known feature that the amplitude becomes theoretically infinite at resonance, $\omega = \omega_0$.

If the mountings combine springs with an element of viscous damping, the equation of motion becomes:

$$m\frac{\mathrm{d}^2x}{\mathrm{d}t^2} + c\frac{\mathrm{d}x}{\mathrm{d}t} + kx = f\sin\omega t$$

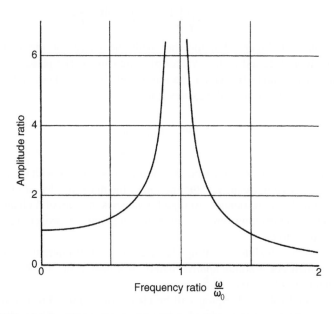

FIGURE 9.5 Relationship between frequency and amplitude ratio (transmissibility), undamped vibration.

where c is a damping coefficient. The steady-state solution is:

$$x = \frac{f/k}{\sqrt{\left(1 - \frac{\omega^2}{\omega_0^2}\right)^2 + \frac{\omega^2 c^2}{mk\omega_0^2}}} \sin(\omega t - A) \qquad (9.6a)$$

If we define a dimensionless damping ratio:

$$C^2 = \frac{c^2}{4mk}$$

the above equation may be written as:

$$x = \frac{f/k}{\sqrt{\left(1 - \frac{\omega^2}{\omega_0^2}\right)^2 + 4C^2 \frac{\omega^2}{\omega_0^2}}} \sin(\omega t - A) \qquad (9.6b)$$

(if $C = 1$ we have the condition of critical damping when, if the mass is displaced and released, it will return eventually to its original position without overshoot).

The amplitude of the oscillation is given by the first part of this expression:

$$\text{Amplitude} = \frac{f/k}{\sqrt{\left(1 - \frac{\omega^2}{\omega_0^2}\right)^2 + 4C^2 \frac{\omega^2}{\omega_0^2}}}$$

This is plotted in Figure 9.6, together with the curve for the undamped condition, Figure 9.5, and various values of C are shown. The phase angle A is a measure of the angle by which the motion of the mass lags or leads the exciting force. It is given by the expression:

$$A = \tan^{-1} \frac{2C}{\frac{\omega_0^\circ}{\omega} - \frac{\omega}{\omega_0}} \qquad (9.7)$$

At very low frequencies A is zero and the mass moves in phase with the exciting force. With increasing frequency, the motion of the mass lags by an increasing angle, reaching 90° at resonance. At high frequencies the sign of A changes and the mass leads the exciting force by an increasing angle, approaching 180° at high ratios of ω to ω_0.

Natural rubber flexible mountings have an element of internal (hysteresis) damping, which corresponds approximately to a degree of viscous damping corresponding to $C = 0.05$.

The essential role of damping will be clear from Figure 9.6: it limits the potentially damaging amplitude of vibration at resonance. The ordinate in Figure 9.6 is often described as the transmissibility of the mounting system: it is

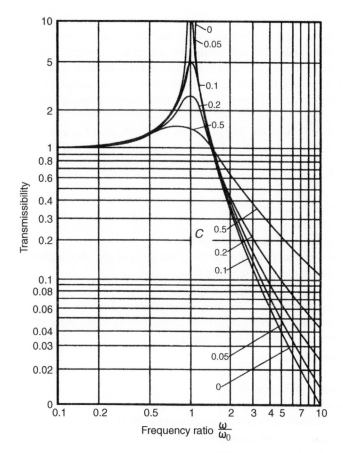

FIGURE 9.6 Relationship between transmissibility (amplitude ratio) and frequency, damped oscillations for different values of damping ratio C (logarithmic plot).

a measure of the extent to which the disturbing force f is reduced by the action of the flexible mounts. It is considered good practice to design the system so that the minimum speed at which the machine is to run is not less than three times the natural frequency, corresponding to a transmissibility of about 0.15. It should be noticed that once the frequency ratio exceeds about 2, the presence of damping actually has an adverse effect on the isolation of disturbing forces.

Practical Considerations in the Design of Engine and Test-Bed Mountings

In the above simple treatment we have only considered oscillations in the vertical direction. In practice, as has already been pointed out, an engine carried on flexible mountings has six degrees of freedom (Figure 9.1). While in many cases a simple analysis of vibrations in the vertical direction will give a satisfactory result, under test cell conditions a complete computer analysis of the various modes of vibration and the coupling between them may be advisable. This is particularly the case with tall engines with mounting points at a low level, when cyclic variations in torque may induce transverse rolling of the engine.

Reference [1] lists the design factors to be considered in planning a system for the isolation and control of vibration and transmitted noise:

- Specification of force isolation:
  - as attenuation (dB)
  - as transmissibility
  - as efficiency
  - as noise level in adjacent rooms.
- Natural frequency range to achieve the level of isolation required.
- Load distribution of the machine:
  - is it equal on each mounting?
  - is the center of gravity low enough for stability?
  - exposure to forces arising from connecting services, exhaust system, etc.
- Vibration amplitudes—low frequency:
  - normal operation
  - fault conditions
  - starting and stopping
  - is a seismic block or sub-base needed?
- Higher-frequency structure-borne noise (100 Hz+):
  - is there a specification?
  - details of building structure
  - sufficient data on engine and associated plant.
- Transient forces:
  - shocks, earthquakes, machine failures.
- Environment:
  - temperature
  - humidity
  - fuel and oil spills.

Detailed design of engine mountings for test-bed installation is a highly specialized matter. In general the aim is to avoid "coupled" vibrations, e.g. the generation of pitching forces due to unbalanced forces in the vertical direction, or the generation of rolling moments due to the torque reaction forces exerted by the engine. These can create resonances at much higher frequencies than the simple frequency of vertical oscillation calculated in the following section and lead to consequent trouble, particularly with the engine-to-dynamometer connecting shaft.

Massive Foundations and Air-Sprung Bedplates

The analysis and prediction of the extent of transmitted vibration to the surroundings is a highly specialized field. The starting point is the observation that a heavy block embedded in the earth has a natural frequency of vibration that generally lies within the range 1000–2000 c.p.m. There is thus a possibility of vibration being transmitted to the surroundings if exciting forces, generally associated with the reciprocating masses in the engine, lie within this frequency range. An example would be a four-cylinder four-stroke engine running at 750 rev/min: we see from Table 9.1 that such an engine generates substantial second-order forces at twice engine speed, or 1500 c.p.m. Figure 9.7 gives an indication of acceptable levels of transmitted vibration from the point of view of physical comfort.

Figure 9.8 is a sketch of a typical small seismic block. Reinforced concrete weighs roughly 2500 kg/m^3 and this block would weigh about 4500 kg. Note that the surrounding tread plates must be isolated from the block, also that it is essential to electrically earth (ground) the mounting rails. The block is shown carried on four combined steel spring and rubber isolators, each having a stiffness of 100 kg/mm (Figure 9.9). From equation (9.4a) the natural frequency of vertical oscillation of the bare block would be 4.70 Hz, or

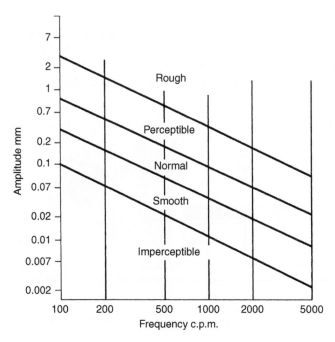

FIGURE 9.7 Perception of vibration. *(Redrawn from Ref. [2].)*

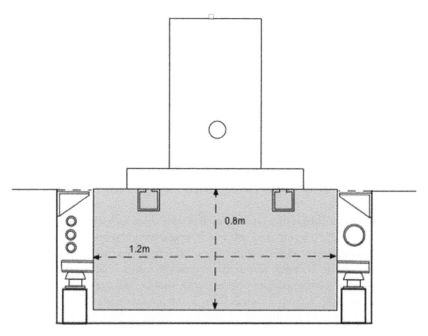

FIGURE 9.8 Spring-mounted seismic block with embedded box-section rails in top face and service pipes running down each side of the pit. Pit must have a drainage point or pump sump.

282 c.p.m., so the block would be a suitable base for an engine running at about 900 rev/min or faster. If the engine weight were, say, 500 kg the natural frequency of block + engine would be reduced to 4.46 Hz, a negligible change. An ideal design target for the natural frequency is considered to be 3 Hz.

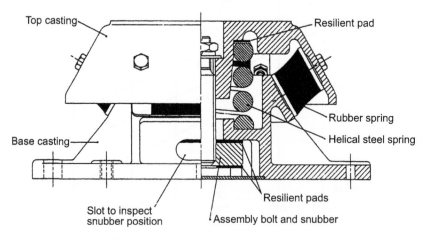

FIGURE 9.9 Combined spring and rubber flexible mount.

Heavy concrete foundations (seismic blocks) carried on a flexible membrane are expensive to construct, calling for deep excavations, complex shuttering, and elaborate arrangements, such as tee-slotted bases, for bolting down the engines. With the wide range of different types of flexible mounting now available, it is questionable whether, except in special circumstances, such as a requirement to install test facilities in proximity to offices, their use is economically justified. The trough surrounding the block may be of incidental use for installing services; if the gap is small, then there should be means of draining out contaminated fluid spills.

It is now common practice for automotive engines to be rigged on vehicle-type engine mounts, then on trolley systems, which are themselves mounted on isolation feet; therefore, less engine vibration is transmitted to the building floor. In these cases a more modern alternative to the deep seismic block is shown in Figure 9.10a and is sometimes used where the soil conditions are suitable. Here the test bed sits on a thickened and isolated section of the floor cast in situ on the compacted native ground. The gap between the floor and block is almost filled with expanded polystyrene boards and sealed at floor level with a flexible, fuel-resistant, sealant; a damp-proof membrane should be inserted under both floor and pit or block.

Where the subsoil is not suitable for the arrangement shown in Figure 9.10a, then a pit is required, cast to support a concrete block that sits on a mat or pads

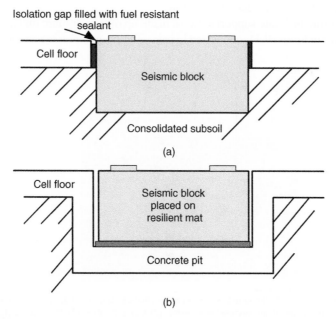

FIGURE 9.10 (a) Isolated foundation block for test stand set onto firm compacted subsoil. (b) Seismic block onto resilient matting in a shallow concrete pit.

of a material such as cork/nitrile rubber composite, which is resistant to fluid contamination (Figure 9.10b). The latter design, while still used in some industries, has the problem of fluids and foreign objects dropping into the narrow annular space between block and pit unless provision is made to avoid it.

Air-Sprung Bedplates

The modern replacement for the cast concrete seismic block is a large cast-iron bedplate supported by air springs, as shown in Figure 9.11. By using air springs with automatic level control a constant, load-independent horizontal position of the isolated base plate can be maintained; control accuracy of between ±0.1 and ±0.01 mm can be achieved depending on the model of device used.

Such precision of surface-level maintenance is completely unnecessary for most automotive engine test cells, but the self-leveling action is valuable when UUTs of different weights are changed on the bedplate. These systems can be tuned from "soft" where the single bedplate may be rocked by operators moving on it, to almost rigid.

The first set-up of a multisection bed can take some time, particularly if the quality of the base floor is of low-quality finish and uneven.

Air-sprung bedplate installations have natural frequencies of around 3 Hz, providing good decoupling between the test bed and the building at low frequencies but at considerable expense.

Air-spring bedplate support systems made up of four or more units, such as that shown in Figure 9.11, connected in pairs within a compressed air circuit are

FIGURE 9.11 A BILZ® air spring, which is part of a self-leveling bedplate support and damping system. (1) Top plate of air suspension element. (2) Control valve. (3) Pressure safety valve. (4) Level control. (5) Air hose. The units are usually connected in pairs and supplied with filtered air. (6) Bar supply pressure. Bedplate control accuracies of down to ±0.01 mm are claimed.

now used in large powertrain cells. When the air supply is switched off the block or plate will settle down and rest on packers; this allows removal for maintenance (item 5 in Figure 9.12). The air springs require a reliable, low-flow, condensate-free air supply that is not always easy to provide to the lowest point in the cell system. Sufficient room must be left around the bedplate to allow maintenance access to the air-spring units and access to hatch plates for any centrally located.

In cases where there is no advantage in having the bedplate face at ground level (as is required for wheeled trolley systems), the bedplate can be mounted on the flat cell floor. It is possible that plant and engines mounted on rubber viscous mounts or air-spring systems could, unintentionally, become electrically isolated from the remainder of the facility because of the rubber elements; it is vital that a common grounding scheme is included in such facility designs (see Chapter 5).

A special application concerns the use of seismic blocks for supporting engines in anechoic cells. It is, in principle, good practice to mount engines undergoing noise testing as rigidly as possible, since this reduces noise radiated as the result of movement of the whole engine on its mountings. Lowering of the center of gravity is similarly helpful, since the engines have to be mounted with a shaft center height of at least 1 meter to allow for microphone placement.

Except in anechoic cells, it is common practice to mount both engine and dynamometer on a common block; if they are separated the relative movement

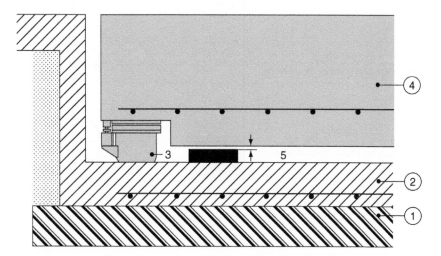

FIGURE 9.12 Seismic block (or cast-iron bedplate not shown) (4) mounted on air springs (3) within a shallow concrete pit (2) on consolidated subsoil (1). When the air supply is off, the self-leveling air springs allow the block to settle on support blocks (5), the rise and fall typically being 4–6 mm. Maintenance access to the springs is not shown.

Vibrations of a pendulum or spring and mass

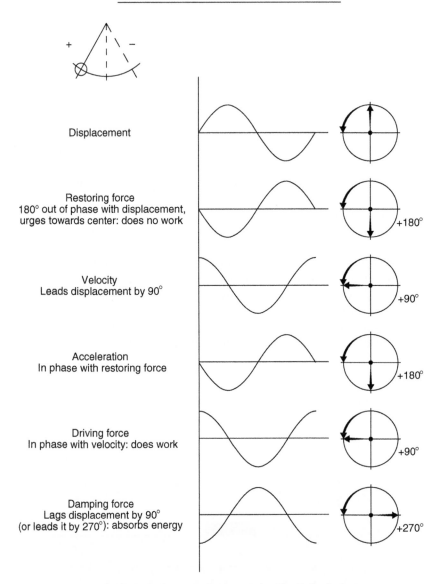

Displacement

Restoring force
180° out of phase with displacement,
urges towards center: does no work
+180°

Velocity
Leads displacement by 90°
+90°

Acceleration
In phase with restoring force
+180°

Driving force
In phase with velocity: does work
+90°

Damping force
Lags displacement by 90°
(or leads it by 270°): absorbs energy
+270°

FIGURE 9.13   Diagram of phase relationship of vibrations.

between the two must be within the capacity of the connecting shaft and its guard.

Finally, it should be remarked that there is available on the market a bewildering array of different designs of vibration isolator or flexible mounting, based on steel springs, air springs, natural or synthetic rubber of widely differing properties used in compression or shear, and combinations of these materials. For the nonspecialist the manufacturer's advice should be sought, with the specific test facility requirement clearly specified.

Summary of Vibration Section

This section should be read in conjunction with Chapter 10, which deals with the associated problem of torsional vibrations of engine and dynamometer. The two aspects—torsional vibration and other vibrations of the engine on its mountings—cannot be considered in complete isolation. A diagrammatic summary of the vibrations of a sprung mass that may be found useful is given in Figure 9.13.

Vibration Notation

Mass of piston + one-third connecting rod	m_p (kg)
Angular velocity of crankshaft	ω_c (s^{-1})
Crank radius	r (m)
Crank angle from TDC	θ
Connecting rod length/r	n
Unbalanced exciting force	f (N)
Mass of engine	m (kg)
Combined stiffness of mountings	k (N/m)
Amplitude of vibration	x (m)
Angular velocity of vibration	ω (rad/s)
Angular velocity at resonance	ω_0 (rad/s)
Natural frequency	n_0 (Hz)
Phase angle	A (rad)
Damping coefficient	c (N s/m)
Damping ratio	C
Acceleration due to gravity	g (m/s^2)

PART 2: NOISE AND ITS ATTENUATION IN TEST FACILITIES

Note: This section deals with some of the fundamentals of noise, its generation and attenuation in powertrain test facilities.

The testing of automotive modules for noise, vibration, and harshness (NVH) and the design of anechoic cells for this and EMC test facilities are dealt with in more detail in Chapter 18. Exhaust silencers are covered in Chapter 7. Ventilation silencers for noise "breakout" and fan noise are covered in Chapter 6.

Sound Intensity

The starting point in the definition of the various quantitative aspects of noise measurement is the concept of sound intensity, defined as:

$$I = \frac{p^2}{\rho c} \text{ W/m}^2$$

where p^2 is the mean square value of the acoustic pressure, i.e. the pressure variation due to the sound wave, ρ the density of air, and c the velocity of sound in air.

Intensity is measured in a scale of decibels (dB):

$$\text{dB} = 10 \log_{10}\left(\frac{I}{I_0}\right) = 20 \log_{10}\left(\frac{\bar{p}}{p_0}\right)$$

where I_0 corresponds to the average lower threshold of audibility, taken by convention as $I_0 = 10^{-12}$ W/m^2, an extremely low rate of energy propagation.

From these definitions it is easily shown that a doubling of the sound intensity corresponds to an increase of about 3 dB ($\log_{10} 2 = 0.301$). A tenfold increase gives an increase of 10 dB, while an increase of 30 dB corresponds to a factor of 1000 in sound intensity. It will be apparent to the reader that intensity varies through an enormous range. The value on the decibel scale is often referred to as the sound pressure level (SPL).

In general, sound is propagated spherically from its source and the inverse square law applies. Doubling the distance results in a reduction in SPL of about 6 dB ($\log_{10} 4 = 0.602$).

The human ear is sensitive to frequencies in the range from roughly 16 Hz to 20 kHz, but the perceived level of a sound depends heavily on its frequency structure. The well-known Fletcher–Munson curves (Figure 9.14) were obtained by averaging the performance of a large number of subjects who were asked to decide when the apparent loudness of a pure tone was the same as that of a reference tone of frequency 1 kHz.

Loudness is measured in a scale of *phons*, which is only identical with the decibel scale at the reference frequency. The decline in the sensitivity of the ear is greatest at low frequencies. Thus, at 50 Hz an SPL of nearly 60 dB is needed to create a sensation of loudness of 30 phons.

Acoustic data are usually specified in frequency bands one octave wide. The standard mid-band frequencies are:

31.5   62.5   125   250   500   1000   2000   4000   8000   16,000 Hz

e.g. the second octave spans 44–88 Hz. The two outer octaves are rarely used in noise analysis.

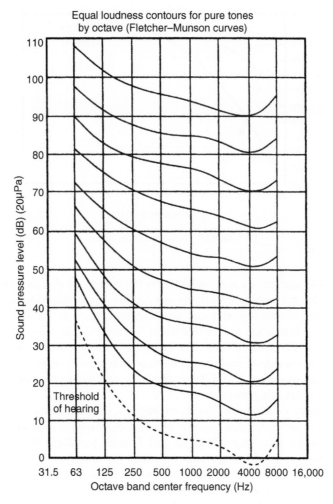

Equal loudness contours for pure tones
by octave (Fletcher–Munson curves)

FIGURE 9.14 Fletcher–Munson curves of equal loudness.

Noise Measurements

Most instruments for measuring sound contain weighted networks that give a response to frequency which approximates to the Fletcher–Munson curves. In other words, their response to frequency is a reciprocal of the Fletcher–Munson relationship (Figure 9.14). For most applications the A-weighting curve (Figure 9.15) gives satisfactory results and the corresponding SPL readings are given in dBA. B- and C-weightings are sometimes used for high sound levels, while a special D-weighting is used primarily for aircraft noise measurements.

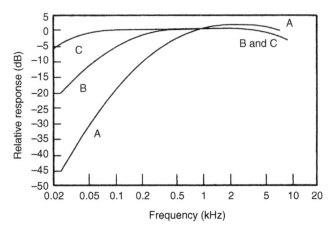

FIGURE 9.15 Noise weighting curves.

The dBA value gives a general "feel" for the intensity and discomfort level of a noise, but for analytical work the unweighted results should be used. The simplest type of sound-level meter for diagnostic work is the octave band analyzer. This instrument can provide flat or A-weighted indications of SPL for each octave in the standard range. For more detailed study of noise emissions, an instrument capable of analysis in one-third octave bands is more effective. With such an instrument it may, for example, be possible to pinpoint a particular pair of gears as a noise source.

For serious development work on engines, transmissions, or vehicle bodies, much more detailed analysis of noise emissions is provided by the discrete Fourier transform (DFT) or fast Fourier transform (FFT) digital spectrum analyzer. The mathematics on which the operation of these instruments is based is somewhat complex, but fortunately they may be used effectively without a detailed understanding of the theory involved. It is well known that any periodic function, such as the cyclic variation of torque in an internal combustion engine, may be resolved into a fundamental frequency and a series of harmonics. General noise from an engine or transmission does not repeat in this way and it is accordingly necessary to record a sample of the noise over a finite interval of time and to process this data to give a spectrum of SPL against frequency. The Fourier transform algorithm allows this to be done.

Permitted Levels of Noise in the Workplace

Regulations concerning the maximum and average levels of noise to which workers are exposed exist in most countries of the world. In the UK (the Control of Noise at Work Regulations 2005) and Europe the levels of exposure to noise (peak sound pressure) of workers, averaged over a working day or week, at which an employer has to take remedial action are:

- Lower exposure action values:
  - daily or weekly exposure of 80 dB
  - peak sound pressure of 135 dB.
- Upper exposure action values:
  - daily or weekly exposure of 85 dB
  - peak sound pressure of 137 dB.
- There are also absolute levels of noise exposure that must not be exceeded without the supply and use of personal protection equipment:
  - daily or weekly exposure of 87 dB
  - peak sound pressure of 140 dB.

Noise levels actually within an engine test cell nearly always exceed the levels permitted by statutes, while the control room noise level must be kept under observation and appropriate measures taken. The employer's obligations under prevailing regulations include the undertaking of an assessment when any employee is likely to be exposed to the first action level. This means that hearing protection equipment must be available to all staff having to enter a running cell, where sound levels will often exceed 85 dBA, and that control room noise levels should not exceed 80 dBA.

Noise-canceling earphones, in place of the noise-attenuating type, should be considered for use in high noise areas where vocal communication between staff is required.

Noise External to Test Facility and Planning Regulations

Almost all planning authorities will impose some form of restriction on the building of a new engine test laboratory that will seek to restrict the noise pollution caused. In the case of an existing industrial site, these restrictions will often take the form of banning any increase in sound levels at the nominated boundaries of the site. It is vital, therefore, that in the early planning and design stages a set of boundary noise readings are taken by both the system integrator and the user. The survey should be based on readings taken at important points, such as the boundary position nearest to residential buildings, then marked on a site map with GPS readings. The readings should be taken in as normal a working situation as possible, without exceptional occurrences, and at several times throughout the period the facility is to be used. Such a datum set of readings should be taken before work starts and agreed as relevant with any interested party; they will then provide a key reference for any dispute post-commissioning.

Noise Reverberation in the Test Cell Environment

The measured value of SPL in an environment such as an engine test cell gives no information as to the power of the source: a noisy machine in a cell having

TABLE 9.2 A Sample of Absorption Coefficients for Materials Found in Test Cell Construction, at Three Frequencies

Wall Material	125 Hz	500 Hz	4 kHz
Poured concrete, unpainted	0.01	0.04	0.1
Brick, painted	0.01	0.02	0.03
Glass (large cell window)	0.18	0.04	0.02
Perforated metal (13% open) over 50 mm fiberglass	0.25	0.99	0.92

good sound-absorbent surfaces may generate the same SPL as a much quieter machine surrounded by sound-reflective walls. The absorption coefficient is a measure of the sound power absorbed when a sound impinges once upon a surface. It is quite strongly dependent on frequency and tends to fall as frequency falls below about 500 Hz.

Information on absorption coefficients for a wide range of structural materials and sound insulators is given in IHVE Guide B 12. A few approximate values are given in Table 9.2 and indicate the highly reverberatory properties of untreated brick and concrete.

It should be remembered that the degree to which sound is absorbed by its surroundings makes no difference to the intensity of the sound received directly from the engine.

"Crosstalk" between test cell and control room and other adjacent rooms can occur through any openings in the partition walls and through air-conditioning ducts and other service pipes, when there is a common system.

Noise Notation

R.m.s. value of acoustic pressure	$\bar{p}$ (N/m^2)
Density of air	ρ (kg/m^3)
Velocity of sound in air	c (m/s)
Sound intensity	I (W/m^2)
Threshold sound intensity	I_0 (W/m^2)

REFERENCES

[1] A.N. Maw, The design of resilient mounting systems to control machinery noise in buildings, Plastics, Rubber and Composites Processing and Applications 18 (1992) 9–16. ISSN: 0959-8111.

[2] W. Ker-Wilson, Vibration Engineering, Griffin, London, 1959. ISBN-10: 0852640234 (out of print in 2011).

USEFUL TEXTS

[1] BS 3425. Method for the Measurement of Noise Emitted by Motor Vehicles.

[2] BS 3539. Specification for Sound Level Meters for the Measurement of Noise Emitted by Motor Vehicles.

[3] BS 4196 Parts 0 to 8. Sound Power Levels of Noise Sources.

[4] Control of Noise at Work Regulations, Health and Safety Executive, London, 2005.

[5] B. Fader, Industrial Noise Control: Fundamentals and Applications, Wiley, Chichester, 1981. ISBN-13: 978–0824790288.

[6] I.H.V.E. Guide B12: Sound Control. Chartered Institution of Building Services, London.

[7] W.T. Thompson, Theory of Vibration, Prentice-Hall, London, 1988. ISBN-13: 978-0412783906.

[8] J.D. Turner, A.J. Pretlove, Acoustics for Engineers, Macmillan Education, London, 1991. ISBN-13: 978-0333521434.

[9] R.H. Warring, Handbook of Noise and Vibration Control, fifth ed., Trade and Technical Press, Morden, UK, 1983. ISBN-13: 978-0471395997.

FURTHER READING

BS 2475. Specification for Octave and One-third Octave Band-Pass Filters.

BS 3045. Method of Expression of Physical and Subjective Magnitudes of Sound or Noise in Air.

BS 4198. Method for Calculating Loudness.

BS 4675 Parts 1 and 2. Mechanical Vibration in Rotating Machinery.

BS 5330. Method of Testing for Estimating the Risk of Hearing Handicap due to Noise Exposure.

W.T. Thomson, Theory of Vibration with Applications, fourth ed., Taylor & Francis, 1996. ISBN-13: 978-0412546204.

Dynamometers: The Measurement of Torque, Speed, and Power

Engine Testing. DOI: 10.1016/B978-0-08-096949-7.00010-8

INTRODUCTION

While the majority of this chapter relates to the measurement of torque and the calculation of power through the use of a dynamometer coupled to the prime mover being tested it is, of course, possible to measure the torque being developed within the powertrain when the prime mover is operating within its intended vehicle. The chassis dynamometer, covered in Chapter 17, measures the torque developed at the tire/roll interface rather than the engine flywheel. But recent developments in torque transducers mean that they can be installed within the powertrain system, often transmitting data wirelessly, and these widen the range of data collection available to the test engineer. However, under real-life operating conditions, precise control of speed and torque, even on a chassis dynamometer, is not realistically possible to the degree required for most engine development work and such testing is usually reserved to record "track data" for later study or replay under more controlled conditions.

The dynamometers used in prime-mover testing, of which there are several different types, all resist, and thus measure, the torque produced by the unit under test. The accuracy with which a dynamometer absorbs and measures torque and controls rotational speed is fundamental to the power measurements and all other derived performance figures made in the test cell. Perhaps the most difficult question facing the engineer setting up a test facility is the choice of the most suitable type and size of dynamometer. The matching of the power and torque-producing characteristics of the engine, motor, or turbine with the power and torque absorption characteristics of the dynamometer is vital for accuracy of data and control: the problem is discussed later in this chapter.

In transmission module testing the unit under test is not the torque-producing prime mover; that will be an electric motor that forms part of the dynamometry system and that, as in the case of "cold engine" testing, may itself be a dynamometer.

All modern commercially available dynamometers measure speed, if not as a displayed data channel, then as part of their control system.

The SI unit of torque is the newton meter, sometimes hyphenated as newton-meter and written symbolically as Nm or N·m. One newton meter is equal to the torque resulting from a force of one newton applied perpendicularly to a moment arm that is one meter long:

$$T = r \times F$$

In SI units the relationship between torque and power is:

$$P_{kW} = \frac{\tau_{N \cdot m} \times \omega_{rpm}}{9549}$$

where P_{kW} is the power in kilowatts, $\tau_{N \cdot m}$ is the torque in newton meters, and ω_{rpm} is the rotational velocity in revolutions per minute.

In Imperial units, still used in some parts of the English-speaking world, torque is measured in foot-pounds, which is a force of 1 pound force[1] (lbf) through a displacement (from the pivot) of 1 foot. In Imperial units the relationship between torque and power is:

$$P_{hp} = \frac{\tau_{ft \cdot lbf} \times \omega_{rpm}}{5252}$$

The geometry of a trunnion-mounted dynamometer and the calibration devices of all dynamometers embodies the torque formula. The value r is the effective length of the arm from which the force F is measured. To meet the requirement of the force acting perpendicular to that arm, its centers of connection must be horizontal, as shown in Figure 10.1.

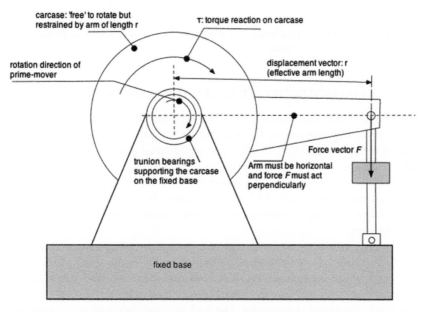

FIGURE 10.1 Diagram of a typical modern trunnion-mounted eddy-current dynamometer fitted with a load cell. Torque can be calibrated and measured in either rotational direction with the load cell in either compression or tension.

---

1. Confusion can arise, when using calibration "weights", between the unit of mass named a "pound" and the unit of force that is a pound-foot. The implications of this are discussed in the section dealing with torque calibration.

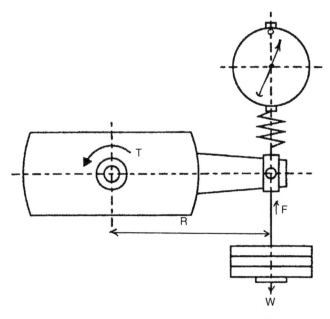

FIGURE 10.2 Diagram of Froude-type, trunnion-mounted, sluice-gate dynamometer measuring torque with dead weights and spring balance.

TRUNNION-MOUNTED (CRADLE[2]) DYNAMOMETERS

The essential feature of trunnion-mounted or cradled dynamometers is that the power-absorbing element of the machine is mounted on trunnion bearings coaxial with the machine shaft. The torque generated has the tendency to turn the carcase in the trunnions, but it is resisted and thus measured by some kind of transducer acting tangentially at a known radius from the machine axis (shown diagrammatically in Figures 10.1 and 10.2). Until the beginning of the twenty-first century the great majority of dynamometers used this method of torque measurement.

Modern trunnion-mounted machines, shown diagrammatically in Figure 10.1, use a force transducer, almost invariably of the strain gauge type, together with an appropriate bridge circuit and amplifier. The strain gauge transducer or "load cell" has the advantage of being extremely stiff, so that no positional adjustment is necessary. The backlash and "stiction" free mounting of the transducer between carcase and base is absolutely critical.

In machine designs predating the mid 1970s, and still in use, such as the Froude sluice-gate machine shown diagrammatically in Figure 10.2, the torque

---

2. These are different names for the same physical system based on measuring the reaction, against the ground, of a dynamometer's torque-developing carcase.

measurement was achieved by physically balancing a combination of dead weights and a spring balance against the torque absorbed. As the stiffness of the balance was low, it was necessary to adjust its position depending on the torque, to ensure that the force measured was accurately tangential.

The trunnion bearings are either a combination of a ball-bearing (for axial location) and a roller-bearing or hydrostatic type. Hydrostatic designs, requiring a pressurized oil feed, are more complex and expensive than designs using rolling-element bearings, but higher calibration accuracies are usually claimed.

Trunnion bearings operate under unfavorable conditions, with no perceptible angular movement, and the rolling element type are consequently prone to brinelling, or local indentation of the races, and to fretting. This is aggravated by vibration that may be transmitted from the engine, and periodical inspection and turning of the outer bearing race is recommended in order to avoid poor calibration.

A Schenck dynamometer design replaces the trunnion bearings by two radial flexures, thus eliminating possible friction and wear, but at the expense of the introduction of torsional stiffness, of reduced capacity to withstand axial loads and of possible ambiguity regarding the true center of rotation, particularly under side-loading.

Many modern AC and most older DC electric motor-based dynamometers are also built as trunnion-mounted machines fitted with load cells, as shown in Figure 10.3.

MEASUREMENT OF TORQUE USING IN-LINE SHAFTS OR TORQUE FLANGES

The alternative method of measuring the torque produced in the prime-mover/dynamometer system is by using an in-line *torque shaft* (Figure 10.4) or *torque flange* (Figure 10.5). A torque shaft is mounted in the drive shaft between engine and brake device. It consists essentially of a double-flanged, pedestal bearing assembly containing torque shaft fitted with strain gauges. Designs are available both with slip rings and with RF signal transmission. Figure 10.4 is a brushless torque shaft unit intended for rigid mounting.

More common in automotive testing is the "disk" type torque transducer, commonly known as a *torque flange*, which is a device that is bolted directly to the input flange of the brake and transmits data to a static antenna encircling it.

A perceived advantage of the in-line torque measurement arrangement is that it avoids the necessity, discussed below, of applying torque corrections under transient conditions of torque measurement. However, not only are such corrections, using known constants, trivial with modern computer control systems, but there are important problems that may reduce the inherent accuracy of this arrangement.

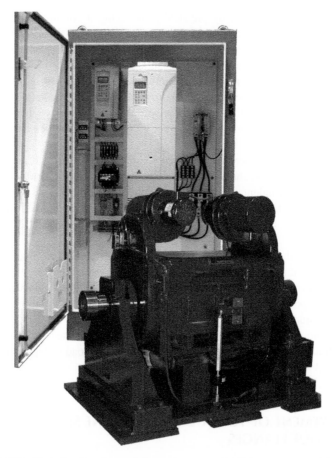

FIGURE 10.3   Typical modern fan-cooled, trunnion-mounted, AC dynamometer and the cabinet housing its power electrics.

For steady-state testing, a well designed and maintained trunnion machine will give more consistently verifiable and accurate torque measurements than the in-line systems; the justifications for this bold statement can be listed as:

- The in-line torque sensor has to be oversized for the rating of its dynamometer and being oversized the resolution of the signal is lower. The transducer has to be overrated because it has to be capable of dealing with the instantaneous torque peaks of the engine, which are not experienced by the load cell of a trunnion bearing machine.
- The in-line transducer forms part of the drive shaft system and requires very careful installation to avoid the imposition of bending or axial stresses on the torsion-sensing element from other components or its own clamping device.
- The in-line device is difficult to protect from temperature fluctuations within and around the driveline.

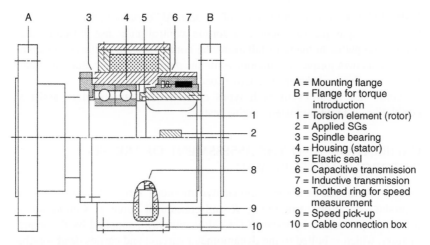

A = Mounting flange
B = Flange for torque
    introduction
1 = Torsion element (rotor)
2 = Applied SGs
3 = Spindle bearing
4 = Housing (stator)
5 = Elastic seal
6 = Capacitive transmission
7 = Inductive transmission
8 = Toothed ring for speed
    measurement
9 = Speed pick-up
10 = Cable connection box

FIGURE 10.4 Brushless torque shaft for mounting in shaft line between engine and "brake".

- Calibration checking of these devices is not as easy as for a trunnion-mounted machine; it requires a means of locking the dynamometer shaft in addition to the fixing of a calibration arm in a horizontal position without imposing bending stresses.
- Unlike the cradled machine and load cell, it is not possible to verify the measured torque of an in-line device during operation.

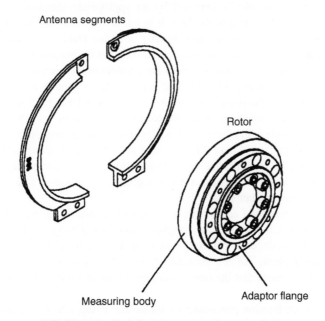

FIGURE 10.5 Shaft-line components of a torque flange.

It should be noted that, in the case of modern AC dynamometer systems, the tasks of torque measurement and torque control may use different data acquisition paths. In many installations the control system of the AC machine may use its own torque calculation, while the data acquisition values are taken from an in-line transducer or reactive load cell.

Some test facilities fit both types of torque measurement to assist in correlation of highly transient engine tests.

CALIBRATION AND THE ASSESSMENT OF ERRORS IN TORQUE MEASUREMENT

We have seen that in a conventional dynamometer torque T is measured as a product of torque arm radius R and transducer force F. Calibration is invariably performed by means of a *calibration arm*, supplied by the manufacturer, which is bolted to the dynamometer carcase and carries dead weights that apply a load at a certified radius. The manufacturer certifies the distance between the axis of the weight hanger bearing and the axis of the dynamometer. There is no way, apart from building an elaborate fixture, in which the dynamometer user can check the accuracy of this dimension; therefore, the arm should have its own calibration certificate and be stamped with its effective length. For R&D machines of high accuracy the arm should be stamped and matched with the specific machine.

The "dead weights" should in fact be more correctly termed "standard masses" and they should be certified by an appropriate standards authority. Whatever the certified weight marked, it is not always remembered that their weight does not necessarily equate directly with the force they exert. While mass, measured in kilograms or pounds, is a constant throughout the world, the force they exert on the calibration arm is the product of their mass and the local value of "g", which is not constant. The value of "g" is usually assumed to be 9.81 m/s^2, but this value is only correct at sea level and a latitude of about 47°N. It increases towards the poles and falls towards the equator, with local variations. As an example, a machine calibrated with its matched weights in London, where $g = 9.81$ m/s^2, will read 0.13% high if recalibrated in Sydney, Australia, and 0.09% low if recalibrated in St Petersburg without correcting for the different local values of g. These are not negligible variations if one is hoping for accuracies better than 1%.

The actual process of calibrating a dynamometer with dead weights, when treated rigorously, is not entirely straightforward. We are confronted with the facts that no transducer is perfectly linear in its response, and no linkage is perfectly frictionless. We are then faced with the problem of adjusting the system so as to ensure that the (inevitable) errors are at a minimum throughout the range.

A suitable calibration procedure for a machine using a typical strain-gauge load cell for torque measurement is as follows.

The dynamometer should not be coupled to the engine. After the system has been energized long enough to warm up, the load cell output is zeroed with the machine in its normal no-load running condition (cooling water on, etc.) and the calibration arm weight balanced by equal and opposite force. Dead weights are then added to produce approximately the rated maximum torque of the machine. This torque is calculated and the digital indicator set to this value.

The weights are removed, the zero reading noted, and weights are added again, but this time preferably in 10 equal increments, the cell readings being noted. The weights are removed in reverse order and the incremental readings again noted.

The procedure described above means that the load cell indicator was set to read zero before any load was applied (it did not necessarily return to zero after the weights had been added and removed), while it was adjusted to read the correct maximum torque when the appropriate weights had been added.

Let us assume we apply this procedure to a machine having a nominal rating of 600 Nm torque, and that we have six equal weights, each calculated to impose a torque of 100 Nm on the calibration arm. Table 10.1 shows the indicated torque readings for both increasing and decreasing loads, together with the calculated torques applied by the weights. The corresponding errors, or the differences between torque applied by the calibration weights and the indicated torque readings, are plotted in Figures 10.6 and 10.7.

The machine is claimed to be accurate to within ±0.25% of nominal rating and in this example the machine meets the claimed limits of accuracy and may be regarded as satisfactorily calibrated.

TABLE 10.1 Dynamometer Calibration (Example Taken From Actual Machine)

Mass (kg)	Applied Torque (Nm)	Reading (Nm)	Error (Nm)	Error (% reading)	Error (% full scale)
0	0	0.0	0.0	0.0	0.0
10	100	99.5	−0.5	−0.5	−0.083
30	300	299.0	−1.0	−0.33	−0.167
50	500	500.0	0.0	0.0	0.0
60	600	600.0	0.0	0.0	0.0
40	400	400.5	+0.5	+0.125	+0.083
20	200	200.0	0.0	0.0	0.0
0	0	0.0	0.0	0.0	0.0

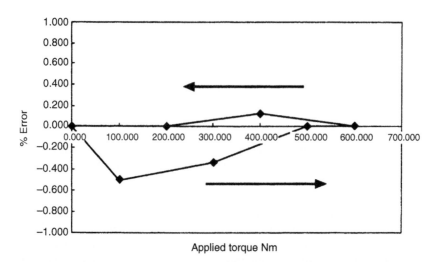

FIGURE 10.6 Dynamometer calibration error as percentage of applied torque plotted for both incremental addition and subtraction of weights on the calibration arm.

It is usually assumed, though it is not necessarily the case, that hysteresis effects, manifested as differences between observed torque with rising load and with falling load, are eliminated when the machine is running, due to vibration, and it is a commonly accepted practice when calibrating to knock the machine carcase lightly with a soft mallet after each load change to achieve the same result.

It is certainly not wise to assume that the ball joints invariably used in calibration arm and torque transducer links are frictionless. These bearings are

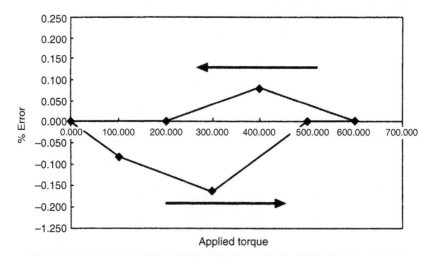

FIGURE 10.7 Dynamometer calibration error as percentage full scale (% FS) as per Figure 10.6.

designed for working pressures on the projected area of the contact in the range 15–20 MN/m^2 and a "stick slip" coefficient of friction at the ball surface of, at a minimum, 0.1 is to be expected. This clearly affects the effective arm length (in either direction) and must be relaxed by vibration.

Some large water-brake dynamometers are fitted with torque multiplication levers, reducing the size of the calibration masses. In increasingly litigious times and ever more stringent health and safety legislation, the frequent handling of multiple 20 or 25 kg weights may not be advisable. It is possible to carry out torque calibration by way of "master" load cells or proving rings.[3] These devices have to be mounted in a jig attached to the dynamometer and give a verifiable measurement of the force being applied on the target load cell by means of a hydraulic actuator. Such systems produce a more complex "audit trail" to refer the calibration back to national standards.

It is important, when calibrating an eddy-current machine, that the water pressure in the casing should be at operational level, since pressure in the transfer pipes can give rise to a parasitic torque. Similarly, any disturbance to the number and route of electrical cables to the carcase must be avoided once calibration is completed. Finally, it is possible, particularly with electrical dynamometers with forced cooling, to develop small parasitic torques due to air discharged nonradially from the casing. It is an easy matter to check this by running the machine uncoupled under its own power and noting any change in indicated torque.

Experience shows that a high-grade dynamometer such as would be used for research work, after careful calibration, may be expected to give a torque indication that does not differ from the absolute value by more than about ±0.1% of the full load torque rating of the machine.

Systematic errors such as inaccuracy of torque arm length or wrong assumptions regarding the value of "g" will certainly diminish as the torque is reduced, but other errors will be little affected. It is safer in most test facilities to assume a band of uncertainty of constant width; this implies, for example, that a machine rated at 400 Nm torque with an accuracy of ±0.25% will have an error band of ±1 N. At 10% of rated torque, this implies that the true value may lie between 39 and 41 Nm.

All load cells used by reputable dynamometer manufacturers will compensate for changes in temperature, though their rate of response to a change may vary. They will not, however, be able to compensate for internal temperature gradients induced, for example, by air blasts from ventilation fans or radiant heat from exhaust pipes.

The subject of calibration and accuracy of dynamometer torque measurement has been dealt with in some detail, but this is probably the most critical measurement that the test engineer is called upon to make, and one for which

---

3. A proving ring is a hollow steel alloy ring whose distortion under a rated and certified range of compressive loads is known and measured by means of an internal gauge.

a high standard of accuracy is expected but not easily achieved. Calibration and certification of the dynamometer and its associated system should be carried out at the very least once a year, and following any system change or major component replacement.

Finally, it should be mentioned that the best calibration procedures will not compensate for errors introduced "downstream" of the raw data flow by software or, more rarely, hardware "corrections". The author has been involved in several investigations into apparent test equipment inaccuracies only to find that unknown and undocumented power correction factors are running in the embedded software of dynamometer drive units or the test cell controllers. In one case investigated by the author, an incorrectly programmed correction factor installed in the control system of a quality audit test bed had been the source of undetected and optimistic power figures for several years.

TORQUE MEASUREMENT UNDER ACCELERATING AND DECELERATING CONDITIONS

With the increasing interest in transient testing it is essential to be aware of the effect of speed changes on the "apparent" torque measured by a trunnion-mounted machine.

The basic principle is as follows:

Inertia of dynamometer rotor	I kgm^2
Rate of increase in speed	ω rad/s^2
	N rpm/s
Input torque to dynamometer	T_1 Nm
Torque registered by dynamometer	T_2 Nm

$$T_1 - T_2 = I\omega = \frac{2\pi NI}{60} \text{ Nm}$$
$$= 0.1047NI \text{ Nm}$$

To illustrate the significance of this correction, a typical eddy-current dynamometer capable of absorbing 150 kW with a maximum torque of 500 Nm has a rotor inertia of 0.11 kgm^2. A DC regenerative machine of equivalent rating has a rotational inertia of 0.60 kgm^2.

If these machines are coupled to an engine that is accelerating at the comparatively slow rate of 100 rpm/s, the first machine will read the torque low *during the transient phase* by an amount:

$$T_1 - T_2 = 0.1047 \times 100 \times 0.11 = 1.15 \text{ Nm}$$

while the second will read low by 6.3 Nm.

If the engine is decelerating the machines will read high by the equivalent amount.

Much larger rates of speed change are demanded in some transient test sequences and this can represent a serious variation of torque indication, particularly when using high inertia dynamometers.

With modern computer processing data acquisition systems corrections for these and other transient effects can be made with software supplied by test plant manufacturers.

MEASUREMENT OF ROTATIONAL SPEED

Rotational speed of the dynamometer is measured either by a system using an optical encoder system or a toothed wheel and a pulse sensor within its associated electronics and display. While the pulse pick-up system is robust and, providing the wheel to transducer gap is correctly set and maintained, reliable, the optical encoders, which use the sensing of very fine lines etched on a small disk, need more care in mounting and operation. Since the commonly used optical encoders transmit over 1000 pulses per revolution, misalignment of its drive may show up as a sinusoidal speed change; therefore, they should be mounted as part of an accurately machined assembly, physically registered to the machine housing.

It should be remembered that all bidirectional dynamometers and modern electrical machines operate in four quadrants (Figure 10.8); it is therefore necessary to measure not only speed but also direction of rotation. Encoder systems can use separate tracks of their engraved disks to sense rotational direction. It is extremely important that the operator uses a common and clearly understood convention describing direction of rotation throughout the facility, particularly in laboratories operating reversible prime movers.

As with torque measurement, specialized instrumentation systems may use separate transducers for the measurement of engine speed and for the control of

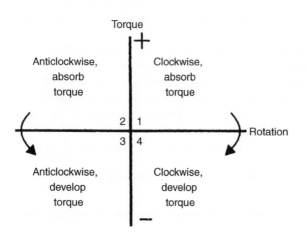

FIGURE 10.8 Dynamometer operating quadrants (observer looking at dynamometer coupling flange).

the dynamometer. In many cases engine speed is monitored separately to dynamometer speed. The control system can use these two signals, with an appropriate time-lag to deal with shaft system twist, in order to shut down automatically in the case of a shaft failure. Measurement of crank position during rotation is discussed in Chapter 15.

Measurement of power, which is the product of torque and speed, raises the important question of sampling time. Engines never run totally steadily and the torque transducer and speed signals invariably fluctuate. An instantaneous reading of speed will not be identical with a longer-term average. Choice of sampling time and of the number of samples to be averaged is a matter of experimental design and compromise.

ONE, TWO, OR FOUR QUADRANT?

Figure 10.8 illustrates diagrammatically the four "quadrants" in which a dynamometer may be required to operate. Most engines run counter-clockwise when looking on the flywheel; therefore, testing takes place in the first quadrant (looking on the dynamometer coupling). On occasions it is necessary for a test installation using a unidirectional water brake to accept engines running in either direction; one solution is to mount the dynamometer on a turntable, with couplings at both ends, allowing running in either direction depending on the dynamometer's orientation. Most very large and some "medium-speed" marine engines are directly reversible.[4]

Hydraulic dynamometers are usually designed for one direction of rotation, though they may be run in reverse at low fill state without damage. When designed specifically for bidirectional rotation they are larger than a single-direction machine of equivalent power, and torque control may not be as precise as that of the unidirectional designs. The torque measuring system must of course operate in both directions. Eddy-current machines are inherently reversible.

When it is required to operate in the third and fourth quadrants (i.e. for the dynamometer to produce power as well as to absorb it), the choice is effectively limited to DC or AC electric motor-based machines. These machines are generally reversible when powered by drives capable of import and export of electrical power and therefore operate in all four quadrants. There are (rare) dynamometer systems based on hydrostatic drives that operate in four quadrants, as do hybrid eddy-current/electric-motor units.

Table 10.2 summarizes the performance of machines in this respect.

4. Designs like the Pielstick engine achieve direct reversing by hydraulically powered, lateral movement of the camshafts.

TABLE 10.2 Operating Quadrants of Dynamometer Designs

Type of Machine	Quadrant
Hydraulic sluice plate	1 or 2
Variable-fill hydraulic	1 or 2
"Bolt-on" variable-fill hydraulic	1 or 2
Disk-type hydraulic	1 and 2
Hydrostatic	1, 2, 3, 4
DC electrical	1, 2, 3, 4
AC electrical	1, 2, 3, 4
Eddy current	1 and 2
Friction brake	1 and 2
Air brake	1 and 2
Hybrid	1, 2, 3, 4

DYNAMOMETER TORQUE/SPEED AND POWER/SPEED CHARACTERISTICS

The different types of dynamometer have significantly different torque–speed and power–speed curves, and this can affect the choice made for a given application. Due to the requirement of legislative calibration test sequences, for many automotive powertrain engineers in the twenty-first century, engaged in emission homologation or engine mapping, the only viable choice of dynamometer is a four-quadrant AC unit, matched with an IBGT-based vector drive and control electronics. Typical characteristics of the torque–speed and power–speed curves of machines of this type are shown in Figure 10.9.

For engineers involved in testing prime movers in the power ranges above 1 MW, then the water-brake dynamometer is the only viable choice.

If the reader compares the power and torque envelope shapes in Figure 10.9 with those of hydraulic water brakes shown in Figures 10.10 and 10.11, the important differences in the performance curves, within a single absorbing quadrant, between hydraulic, eddy-current, and "motoring" dynamometers will be clear. These differences become important when matching engines with dynamometers and give the four-quadrant machines two advantages:

- The AC or DC dynamometer can produce full, or near full, torque at zero speed while the hydraulic or eddy-current unit has a significant "gap" on

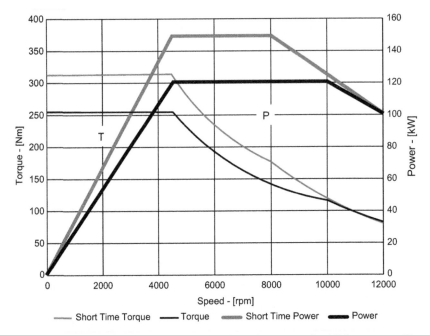

FIGURE 10.9   Typical performance curve shape of a modern, automotive AC dynamometer. The maximum torque rating is based on the safe operating temperature of the machine rather than maximum rotor stresses; thus, many machine units will have a higher, short-time, torque rating, such as +10% for 5 minutes in every hour.

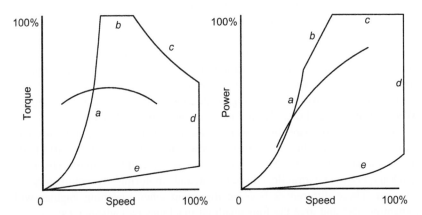

FIGURE 10.10   Typical hydraulic dynamometer torque (left) and power (right) curves. The torque and power curves of an engine producing greater low-speed torque than the dynamometer can absorb has been superimposed.

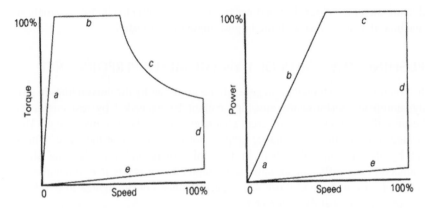

FIGURE 10.11 Performance curve shapes of an eddy-current dynamometer.

the left-hand side of its torque curve. A common problem in matching a high-torque diesel engine with some dynamometers is shown in Figure 10.11, where the engine's torque curve comes outside the dynamometer's torque envelope.

- The two-quadrant design dynamometers have a rising "minimum torque with speed" characteristic. This is due to rotor windage in the case of eddy-current machines and torque generated by the required minimum water content in the hydraulic units; this internal resistance to rotation appears as parasitic torque in the measuring system. AC and DC machines can "motor-out" such inherent internal resistance.

Figure 10.10 shows the power curve of an engine superimposed on that of the dynamometer. It can be seen that, while the maximum torque, power, and rotational speed of the engine are well within the capabilities of the dynamometer, the torque produced by the engine during the first 20% of its running range are to the left of the inherent torque curve of the dynamometer.

Figure 10.11 shows the performance curves for an eddy-current machine, which lie between those of the previous two machines. The key parts of the performance envelopes are:

a. Low-speed torque corresponding to maximum permitted coil excitation.
b. Performance limited by maximum permitted shaft torque.
c. Performance limited by maximum permitted power, which is a function of cooling water throughput and its maximum permitted temperature rise.
d. Maximum permitted speed.
e. Minimum torque corresponding to residual magnetization, windage, and friction.

In choosing a dynamometer for an engine or range of engines, it is essential to superimpose the maximum torque– and power–speed curves onto the dynamometer envelope. For best accuracy it is desirable to choose the smallest machine that will cope with the largest engine to be tested.

PUSHING THE LIMITS OF DYNAMOMETER PERFORMANCE

It should be understood that, in general, the limits set by the curved lines in any dynamometer performance envelope cannot be exceeded by test engineers because they represent the inherent characteristics of the machine design. The straight lines of the performance graphs, however, represent the "safe" limits set by the machine designer. In principle, therefore, provided the shaft connection has sufficient strength (see Chapter 11), it is possible, if not always advisable, to go beyond those (safe) performance cut-off figures to overload and overspeed most dynamometers.

Hydraulic dynamometers are generally robust machines, some, such as the now obsolete Froude G4 derivatives much loved still by some motor sport testers in the USA, legendarily so.

Most designs are well able to deal with a moderate degree of overload and some overspeed; the critical dynamometer parameter if running at overload is water exit temperature.

Froude hydraulic dynamometers were power rated for a water flow that gave a ΔT of 28 °C. The maximum safe outlet temperature[5] is 60 °C, meaning that to draw full power the inlet water must be under 33 °C and for any power overload to be sustained the inlet water temperature must be commensurately lowered in order to maintain the 60 °C outlet temperature.

Overspeeding is not so advisable, particularly when running with high exit temperatures, as it may cause cavitation damage that would quickly destroy the working chambers. This destructive phenomenon can sometimes be detected by a distinctive crackling noise coming from the carcase and is caused by localized boiling and bubble collapse.

Special high-speed "cropped rotor" designs of some hydraulic dyna-mometers are made for testing race engines and gas turbines, etc. The power curve of this type of machine is flattened (lower torque) and shifted right (higher speed), with a much larger "hole" in the curve on the left (no torque at low speed). The Froude model designation indicates the angle of rotor crop, so the G4 unit having a full rotor has a high-speed variant with a 90° cropped rotor designated as G490.

AC dynamometers are usually rated with a stated overload capacity for a short duration, e.g. 5 minutes in every hour. In high ambient temperatures this overload capacity may reduce and cause the tripping of the machine through the motor winding temperature protection circuit.

---

5. There are conditions and designs where this is increased to 70 °C for a short time, but all normal operations should be based on the safe lower limit.

Pushing DC dynamometers beyond their rated limits is not advisable as this can lead to damage to commutators.

Overloading eddy-current machines of any design is ill-advised as over-heating and consequential distortion of eddy-current loss plates or coil may be caused. Like any dynamometer that converts kinetic energy to heat and removes it by flowing water, the ΔT across the machine is critical to avoid internal boiling; therefore, if maximum power or power overload is required, ensure that plenty of water is flowing through the machine with as low an inlet temperature as possible.

WATER QUANTITY REQUIRED TO ABSORB A GIVEN POWER IN WATER BRAKES

The volume and pressure of water required by water-brake dynamometers varies according to their design; therefore, the manufacturer's published requirements should be obeyed.

There are a number of "rules of thumb" that have been used for many years, but these tend only to relate to water flow per unit time and absorbed engine power; the missing parameter in these cases is the change in temperature (delta T or ΔT) across the machine. The exit temperature of any water brake has to remain under the critical temperature at which microboiling and cavitation begins; therefore, any rule for dynamometer water flow has to state the calculated ΔT.

The best calculation known to the author was produced some years ago, so has stood the test of time, and states:

For a ΔT of 22 °C the water requirement for a variable-fill hydraulic dynamometer equals 39.2 l/kWh or about 6.4 imp. gall/bhp·h. The figure of 22 °C was chosen because it was considered to be the sensible limit of the ΔT achieved when using an air-blast radiator.

Note that this is a higher flow than the one used in the USA for portable dynamometers that would give a commensurately higher water exit tempera-ture for any given inlet.

It is important to understand that the maximum permissible outlet temperature determines the maximum permitted inlet temperature, which in turn determines the maximum power that a given machine can absorb under those site conditions.

ENGINE CRANKING AND STARTING

Starting an engine when it is connected to a dynamometer may present the cell designer and operator with a number of problems, and is a factor to be remembered when selecting the dynamometer. In the absence of an engine-mounted starter, a complete system to start and crank the engine must be available that compromises neither the torsional characteristics nor the torque measurement accuracy (see Chapter 11 for the requirements of rigging or providing starting systems).

CHOICE OF DYNAMOMETER

The first question concerning the choice of a dynamometer is: "Does it need to be a motoring (four-quadrant) machine?"

If so, then an AC motor-based system may be the best choice. For transmission testing and in any facility doing highly dynamic engine or transmission testing, the choice would be the same, although there are some specialist fields, such as EMC cells (see Chapter 18), where DC machines should be used because of the electromagnetic environment created.

So a question that may arise in the mind of the reader is: "Why consider any other type of dynamometer than a modern AC unit?"

It must be remembered that the technology involved in variable-speed, high-power, four-quadrant AC or DC dynamometers does not come without inherent problems. Exporting electrical power into a factory or national grid requires that a number of technical and legislative problems are solved (see Chapter 5), and frequently requires the purchase of new electrical transformers and special power cables, none of which are cheap.

The final choice of dynamometer for a given application may be influenced by some of the following factors (also see Table 10.3):

1. The speed of response required by the test sequences being run. The terms steady state, transient, dynamic, or high dynamic are very subjective but each industrial user will be aware of the speed of response required to match the needs of their test sequences.

2. Annual running time. From a technical viewpoint this is only a problem if the machine will spend long periods out of use; then the possibilities of corrosion must be considered, particularly in the case of hydraulic or wet gap eddy-current machines. Can the machine be drained readily? Should the use of corrosion inhibitors be considered? From a cost accounting viewpoint it makes no sense having expensive plant that is sitting idle and not earning review, which is why comparatively cheap and robust disk-type eddy-current machines are widely used in engine test facilities with intermittent work patterns.

3. Overloads. If significant transient and/or occasional overloading of the machine is possible, as in some motor sport tuning facilities, a hydraulic machine may be preferable, in view of its greater tolerance of such conditions. The torque-measuring system needs to have an adequate overload capacity.

4. Large and frequent changes in load. This can and does give rise to problems with eddy-current machines, due to rapid expansion and contraction of the loss plates causing distortion that leads to water-seal failure or closure of the rotor/stator air gap.

5. Wide range of engine sizes to be tested. An area where AC or DC machines may be preferable. When using two-quadrant machines it may be difficult to achieve good control and adequate accuracy when testing the smallest engines and the minimum dynamometer torque may also be inconveniently high. The greater the range of engine, the greater is the chance of hitting torsional vibration problems in any one dynamometer/shaft combination (see Chapter 11).

6. How are engines to be started? If a nonmotoring dynamometer is favored it may be necessary to fit a separate starter to the dynamometer shaft. This represents an additional maintenance commitment and may increase inertia.

7. Is there an adequate supply of cooling water of satisfactory quality? Hard water and some inappropriate water treatments that are susceptible to the high centrifuging action of water brakes will result in blocked cooling passages. The calcite deposits from hard water may break away and cause mechanical jamming of control valves. Plant water supplied at over 30 °C will compromise sustained full-power use in many hydraulic and some eddy-current designs.

8. Is the pressure of the water supply subject to sudden variations? Sudden pressure changes or regular pulsations will affect the stability of control of hydraulic dynamometers. Eddy-current and indirectly cooled machines are unaffected providing inlet flow does not fall below emergency "low-flow" trip levels.

9. The electrical supply system is a major consideration when using AC and DC machines, and the reader needs to read Chapter 5 in full. With the exception of air brakes and manually controlled hydraulic machines, all dynamometers are affected by electrical interference and voltage changes.

10. Is it proposed to use an automatic shaft-docking system for coupling engine and dynamometer? Are there any special features or heavy overhung or axial loads associated with the coupling system? Such features should be discussed with the dynamometer manufacturer before making a decision. Some machines, notably the Schenck flexure plate mounting system, are not suited to taking axial loads as are experienced in most shaft-docking systems.

CLASSIFICATION OF DYNAMOMETER TYPES AND THEIR WORKING PRINCIPLES

Hydrokinetic or "Hydraulic" Dynamometers (Water Brakes)

With the exception of the disk dynamometer, all hydrokinetic machines work on similar principles (see Figure 10.12). A shaft carries a cylindrical rotor that

TABLE 10.3 Dynamometers: Advantages and Disadvantages of Available Types

Dynamometer Type	Advantages	Disadvantages
Froude sluice plate Examples: Froude DPX_n and DPY_n models	Obsolete, but many cheap and reconditioned models in use worldwide. Robust	Slow response to change in load. Manual control that is not easy to automate
Variable-fill water brakes Current examples: Froude "F" types, AVL "Omega" range, Horiba DT range	Units can match the most powerful prime movers built. Capable of medium-speed load change, automated control, robust, and tolerant of overload	"Open" water system required. Can suffer from cavitation or corrosion damage
"Bolt-on" variable-fill water brakes Current examples: units made by Piper (UK), Taylor, Go-Power (USA)	Cheap and simple installation. Up to 1000 kW	Lower accuracy of measurement and control than most fixed machines
Disk-type hydraulic	Suitable for high speeds such as required in small turbine testing	Poor low-speed performance
Hydrostatic	For special applications, provides four-quadrant performance	Mechanically complex, noisy, and expensive. System contains large volumes of high-pressure oil
DC electrical motor Produced by most major test equipment manufacturers	Mature technology. Four-quadrant performance. Limited in automotive top-speed range	High inertia, commutator and brushes require maintenance, harmonic distortion of supply possible
Asychronous motor (AC) Produced by most major test equipment manufacturers	Now mature technology, lower inertia for same rating than DC. Four-quadrant performance. Higher speed range than DC	Expensive. Large drive cabinet needs suitable housing. Care must be taken in environment of the drive unit and the connection into the power system. Some RF emission
Permanent magnet motor Produced by AVL, MTS, and others	Lowest inertia, most dynamic four-quadrant performance. Small size in cell	Very expensive. Large drive cabinet needs suitable housing

TABLE 10.3 Dynamometers: Advantages and Disadvantages of Available Types—cont'd

Dynamometer Type	Advantages	Disadvantages
Eddy current, water cooled	Low inertia (disk-type air gap) well adapted to computer control. Mechanically simple	Vulnerable to poor cooling supply. Not suitable for sustained rapid changes in power (thermal cycling)
Friction brake	Special purpose applications for very high torques at low speed	Limited speed range
Air brake	Cheap. Very little support services needed	Noisy limited control accuracy
Hybrid	Possible cost advantage over sole electrical machine	Complexity of construction and control

revolves in a watertight casing. Toroidal recesses formed half in the rotor and half in the casing or stator are divided into pockets by radial vanes set at an angle to the axis of the rotor. When the rotor is driven centrifugal force sets up an intensive toroidal circulation, as indicated by the arrows in Figure 10.12a. The effect is to transfer momentum from rotor to stator and hence to develop a torque resistant to the rotation of the shaft, balanced by an equal and opposite torque reaction on the casing.

A forced vortex of toroidal form is generated as a consequence of this motion, leading to high rates of turbulent shear in the water and the dissipation of power in the form of heat to the water. The center of the vortex is vented to atmosphere by way of passages in the rotor and the virtue of the design is that power is absorbed with minimal damage to the moving surfaces, either from erosion or from the effects of cavitation.

As rotational speed rises the increase in water velocity and centrifugal stresses in the complex rotor becomes critical; therefore, for high-speed, lower peak torque designs the rotor may be cut off at half height of the pockets (along the line a–a in Figure 10.12); this is called a 90° crop. The "missing" rotor water pocket is then formed within the stator so the water's toroidal circulation is maintained, but with a much reduced rotor diameter capable of much higher rotation speeds and a significant change in the shape of the power envelope.

All hydrokinetic machines generate torque in the same manner but the means of varying the torque generation varies between the variable-fill types and the constant-fill types.

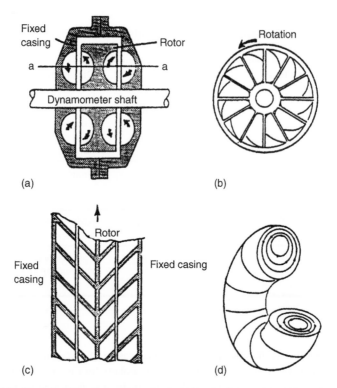

FIGURE 10.12 Principles of operation of a hydrokinetic dynamometer: (a) section through dynamometer; (b) end view of rotor; (c) development of section a–a of rotor and casing; (d) representation of toroidal vortex. A 90° cropped rotor (see text) is cut off along a–a.

Constant-Fill Machines

The classic, and now largely obsolete, sluice-plate design is shown in Figure 10.13. In this machine torque is varied by inserting or withdrawing pairs of thin sluice plates between rotor and stator, thus controlling the extent of the development of the toroidal vortices.

Variable-Fill Machines

In these machines (Figure 10.14) the torque absorbed is varied by adjusting the quantity (mass) of water in circulation within the casing. This is achieved by a valve, usually but not exclusively fitted to the carcase at the water outlet and associated with control systems of widely varying complexity. The particular advantage of the variable-fill machine is that the torque may be varied much more rapidly than is the case with sluice-plate control. Amongst this family of machines are the largest dynamometers ever made, with rotors of around 5 m diameter.

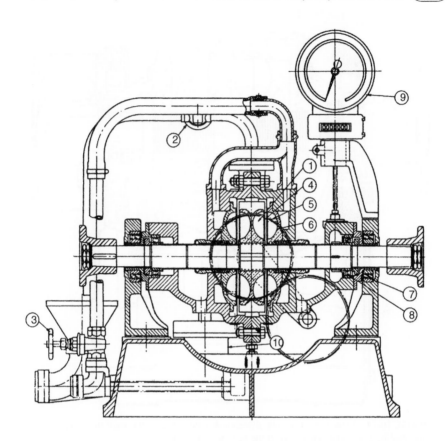

Typical cross-section through casing of Froude dynamometer, type DPX

(1) Rotor   (6) Casing liners

(2) Water outlet valve   (7) Casing trunnion bearing

(3) Water inlet valve   (8) Shaft bearing

(4) Sluice plates for load control   (9) Tachometer

(5) Water inlet holes in vanes

FIGURE 10.13 Froude sluice-plate dynamometer.

There are several designs of water control valve and valve actuating mechanisms depending on the range and magnitude of the loads absorbed and the speed of change of load required. Most machines have a balanced butterfly valve that is hydraulically actuated and requiring an electrically driven power-pack. There are some American units fitted with characterized rotary valves needing only electrical stepper motor actuation. For the fastest response it is

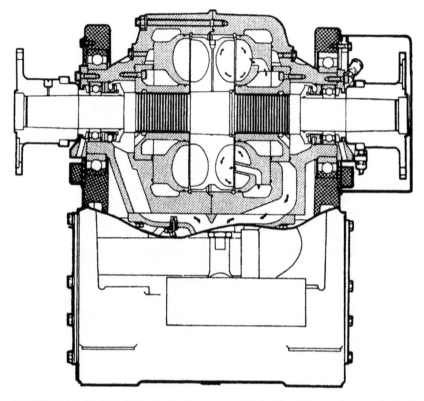

FIGURE 10.14 Variable-fill hydraulic dynamometer fitted with a full rotor (no cropping) and controlled by fast-acting outlet valve at the bottom of the stator.

necessary to have adequate water pressure available to fill the casing rapidly; to satisfy major steps in power it may be necessary to fit both inlet and outlet control valves with an integrated control system.

All these designs require a free, gravitational discharge of water so are not suitable for closed, pressurized water systems.

"Bolt-On" Variable-Fill Water Brakes

These machines, built for many years in the USA and more recently in Europe, are much favored by the truck and bus engine overhaul industry. They operate on the same principle as those described above, but are arranged to bolt directly on to the engine clutch housing or into the truck chassis (see Chapter 4 for more detailed description of use). Machines are available for ratings up to about 1000 kW. In these machines load is usually controlled by an inlet control valve associated with a throttled outlet. By nature of their simplified design and lower mass, these machines are not capable of the same level of speed holding or torque measurement as the more conventional hydrokinetic designs.

Disk Dynamometers

These machines, not widely used, consist of one or more flat disks located between flat stator plates, with a fairly small clearance. Power is absorbed by intensive shearing of the water and torque is controlled as in variable-fill machines. Disk dynamometers have comparatively poor low-speed performance but may be built to run at very high speeds, making them suitable for loading gas turbines. A variation is the perforated disk machine, in which there are holes in the rotor and stators, giving greater power dissipation for a given size of machine.

Hydrostatic Dynamometers

Not very widely used now because of the arrival of high-powered AC dynamometers, these machines consist generally of a combination of a fixed-stroke and a variable-stroke positive displacement hydraulic pump/motor similar to that found in large off-road vehicle transmissions. The fixed-stroke machine forms the dynamometer. An advantage of this arrangement is that the dynamometer unit, unlike most other non-electrical machines, is capable of developing full torque down to zero speed and is also capable of acting as a source of power to "motor" the engine under test. Like AC and DC machines, power is regenerated into the electrical supply by way of the variable-stroke/constant-speed unit that runs synchronized with the mains and therefore, unlike AC and DC units, does not need a complex power control cabinet.

Electrical Motor-Based Dynamometers and Their Associated Control

The rapid development, since around 1985, of three core technologies has meant that the AC dynamometer based on an induction motor and its associated modern four-quadrant drive system has become the required device for any automotive test facility carrying out modern R&D work. The three technologies are:

1. The high power rated, insulated gate bipolar transistor or IGBT, allowing the synthesis of complex electrical supply waveforms with pulse-width modulation required for variable AC motor speed control in four quadrants.
2. Microprocessor-based controllers capable of multi-kHz calculation of torque and vector control calculations.
3. Optical absolute encoders capable of high-definition measurement of rotor speed and position.

The vector control microprocessor uses the feedback information from the encoder and drive to calculate the exact vector of voltage and frequency required to attain the set speed at the required torque. As the system is closed loop it constantly updates that v/f vector to achieve the required, and changing, torque/speed combination.

Thus, modern AC dynamometer systems covering the normal automotive engine range can have the following typical control part specification:

- Torque calculation with 4 kHz for air gap/1 kHz for shaft
- Processor generates dynamic set values for the dynamometer system at >3 kHz
- Capable of zero torque simulation
- Capable of simulation of gear-shifting oscillations in the drivetrain
- Torque response time better than 3 ms
- Current harmonics <4.5%
- Optical speed encoder with 1024 or 4096 pulses per resolution
- Power factor 1
- Speed control accuracy of ±1 rpm
- Overload capability 25% (1 min in 15 min).

Note that quoted "torque accuracy" figures may have to be disambiguated since the control system of the type of dynamometer described above will use, as its inner loop, an internal torque calculator system rather than the load cell or torque flange. In dynamometer specification documentation the torque accuracy figures should relate to calibrated accuracy of the whole torque measuring system[6] used by the test cell data acquisition system. The torque measuring accuracy will depend on the mechanical configuration of the dynamometer but the leading models quote accuracy ranges of between 0.3% and 0.1% of full scale (FS) when the dynamometer is absorbing engine power.

Together with the development of very low inertia and high torque permanent magnet motors, the continuing development of the technologies listed above, partially driven by the requirements of electric motor propulsion in vehicles, will allow increasingly dynamic simulations of real-life powertrain functions in the test cell.

Recirculating electrical power within powertrain test rigs is possible by using a combined power control system for two or more AC dynamometers; it is the modern electrical equivalent of the mechanical technique of recirculating torque in "back-to-back" rigs (Figure 10.15).

When planning a facility layout the designer should remember that AC drive cabinets are large and heavy and have to be transported in their upright position,[7] and positioned after the building work has been completed. The position of the drives should normally be within 15 meters of cable length from the dynamometer but this should be minimized so far as is practical to reduce

---

6. Beware of accuracy figures that relate only to the load-cell or torque flange rather than the installed system.

7. Finished doorway height is a frequent problem met when installing this type of equipment; often they are over 2.2 meters plus the height of a pedestal and any trolleys used to move them.

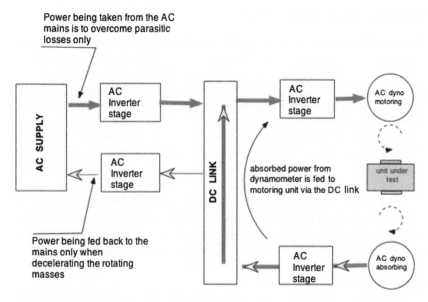

FIGURE 10.15 Diagram showing recirculation of electrical power in a transmission test rig comprising two matched AC dynamometer systems.

the high cost of the connecting power cables. A description of the requirements for the electrical installation of these machines is found in Chapter 5.

Direct Current (DC) Dynamometers

These machines, used as engine dynamometers, have a long pedigree in the USA and in the rest of the world when incorporated into chassis dynamometers. They are robust, easily controlled, and capable of full four-quadrant operation. Their torque-measuring systems are the same as those used on AC dynamometers, although the development of torque flanges postdate the production of many units. Speed measurement is usually via a tacho-generator or toothed wheel, or both.

The disadvantages of DC dynamometers include limited maximum speed and high inertia, which can present problems of torsional vibration (see Chapter 11), and limited rates of speed change. Because they contain a commutator the normal maintenance of DC machines will be higher than those based on induction motors. Control is almost universally by means of a thyristor-based power converter. Harmonic distortion of the regenerated AC mains waveform may be a problem if isolating transformers are not used (see Chapter 5).

A modern use of DC dynamometers is in chassis dynamometers within EMC test facilities where their absence, or low level, of radiofrequency emissions is an essential feature.

Eddy-Current Dynamometers

These machines make use of the principle of induction to develop torque and dissipate power. In the most common type, a toothed rotor of high magnetic permeability steel rotates, with a fine clearance, between water-cooled steel loss plates. A magnetic field parallel to the machine axis is generated by one or two annular coils (depending on the configuration) and motion of the rotor causes changes in the distribution of magnetic flux in the loss plates. This in turn gives rise to circulating eddy currents and the dissipation of power in the form of electrical resistive losses. Energy is transferred in the form of heat to cooling water circulating through passages in the loss plates, while some cooling is achieved by the radial flow of air in the gaps between rotor and plates.

Power is controlled by varying the current supplied to the annular exciting coils, and rapid load changes are possible. Eddy-current machines are simple and robust, providing adequate cooling water flow is maintained; the control system is simple and they are capable of developing substantial braking torque at quite low speeds. Unlike AC or DC dynamometers, however, they are not able to develop motoring torque.

There are two common forms of machine, both having air circulating in the gap between rotor and loss (cooling) plates, hence "dry gap":

1. Dry gap machines fitted with one or more tooth disk rotors (Figure 10.16). These machines have lower inertia than the drum machines and a very large installed user base, particularly in Europe; however, the inherent design features of their loss plates place certain operational restrictions on their use. It is absolutely critical to maintain the required water flow through the machines at all times; even a very short loss of cooling will cause the loss plates to distort, leading to the rotor/plate gap closing, with disastrous results. These machines must be fitted with flow detection devices interlocked with the cell control system; pressure switches should not be used since in a closed water system it is possible to have pressure without flow.
2. Dry gap machines fitted with a drum rotor. These machines usually have a higher inertia than the equivalent disk machine but may be less sensitive to cooling water conditions.

Although no longer so widely used in engine testing, an alternative form of eddy-current machine is also available. These "wet gap" machines are liable to corrosion if left static for any length of time, have higher inertia, and have a high level of minimum torque, arising from drag of the cooling water in the gap.

Powder Brake Dynamometers

These are a variant of eddy-current designs, the main feature of which is that they are able to absorb full-rated torque at zero rpm. There is a fine magnetic powder in the "air" gap between the rotor and the coil plates that restrains

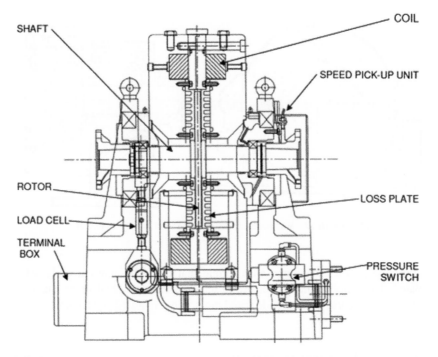

FIGURE 10.16 Section through a Froude AG trunnion bearing-mounted, dry-gap, disk-type, eddy-current dynamometer.

rotation in proportion to the coil excitation. Due to the difficulty in dissipating the heat generated, they are restricted in maximum power rating and rotational speed, typically <50 kW and 2000 rpm; therefore, such machines are very rarely seen in automotive test facilities, but this is changing since they are suited to power testing of small electrical motors.

Friction Dynamometers

These machines (Figure 10.17), in direct line of succession from the original "rope brake", consist essentially of water-cooled multidisk friction brakes. They are useful for low-speed applications, for example for measuring the power output of a large, off-road vehicle transmission at the wheels, and have the advantage of developing full torque down to zero speed for indefinite periods of time.

Air Brake Dynamometers

These devices, of which the Walker fan brake was the best-known example, are now largely obsolete. They consisted of a simple arrangement of radially

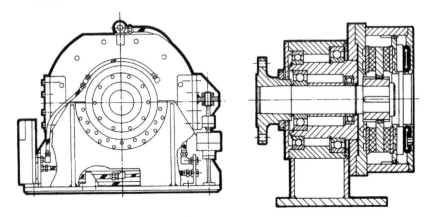

FIGURE 10.17 Water-cooled friction brake used as a dynamometer.

adjustable paddles that imposed a torque that could be approximately esti-mated. They survive mainly for use in the field testing of helicopter engines, where high accuracy over a wide speed range is not required and the noise is no disadvantage.

Hybrid and Tandem Dynamometers

For completeness, mention should be made of both a combined design that is occasionally adopted for cost reasons and the use of two dynamometers in line for special test configurations.

The DC or AC electrical dynamometer is capable of generating a motoring torque almost equal to its braking torque. However, the motoring torque required in engine testing seldom exceeds 30% of the engine power output. Since, for equal power absorption, AC and DC machines are more expensive than other types, it is sometimes worthwhile running an electrical motor or dynamometer in tandem with, for example, a variable-fill hydraulic machine. Control of these hybrid machines can be a more complex problem and the need to provide duplicate services, both electrical power and cooling water, is a further disad-vantage. The solution may, however, on occasion be cost-effective.

Tandem machines are used when the torque/speed envelope of the prime mover cannot be covered by a standard dynamometer; usually this is found in gas turbine testing when the rotational speed is too high for a machine fitted with a rotor capable of absorbing full-rated torque. The first machine in line has to have a shaft system capable of transmitting the combined torques.

Tandem machines are also used when the prime mover is producing power through two contrarotating shafts, as with some aero and marine applications; in these cases the first machine in line is of a special design with a hollow rotor shaft to allow the quill shaft connecting the second machine to run through it.

Rigging the Engine and Shaft Selection

Engine Testing. DOI: 10.1016/B978-0-08-096949-7.00011-X

PART 1. RIGGING OF ELECTRICAL AND FLUID SERVICES PLUS ENGINE MOUNTINGS

Installing a modern, electronically controlled, engine in a test cell and getting it to run correctly can be difficult and time consuming, even for test engineers working for the original equipment manufacturer (OEM); the problem is even more difficult for those working in organizations external to the OEM.

In addition to the physical connections required, such as the connecting shaft, fluid connectors, transducer plugs, engine supports, etc., it will be necessary, when dealing with modern automotive engines, to connect into or simulate signals from vehicle and powertrain control units. Control devices (engine control unit (ECU), transmission control unit (TCU), etc.), in their production form, will require signals from transducers and actuators that may not be present in the test cell.

The problems thus caused are legion in our industry and two examples met by the author are: an engine, that has never been in a vehicle, running only in "limp home mode" because the trunk lid was not closed, and another that was unable to be started because the ECU detected that the handbrake was not engaged. Such problems can be deeply frustrating and waste weeks of test time. The problem is not made easier by the lack of harmonization in the automotive industry, the complex control logic trees resulting from interacting control devices, and the level of secrecy surrounding access to propriety interface details or error codes. The ISO 15765-4:2011 standard is part of the worldwide harmonized on-board diagnostics (WWH-OBD) regulations intended to reduce these problems.

Besides the problems met in getting the ECU-controlled engines to run correctly, some test engineers have the need to get engines to run *incorrectly*, such as when there is a need to accelerate the collection of soot particles in exhaust filters; such requirements can create even greater problems and, naturally enough, no official advice will be forthcoming from the OEM or supplier chain.

In production testing, or any test where the same model of engine is repeatedly run, a modified ECU, specifically configured for the test regime, would be used. However, unless the test is part of a full hardware-in-the-loop (HIL) experiment, the physical presence of the dedicated ECU and some transducers and actuators with its connecting loom is vital to the function of a modern engine outside the vehicle. Typically the list will include:

- The lambda sensor mounted in the first section of exhaust ducting
- Idle running, throttle bypass valve
- The correct flywheel with crank position transducers
- The coolant temperature transducer
- The throttle position transducer
- The inlet manifold pressure transducer.

There are three general strategies available in resolving the problem of getting ECU-controlled engines to run in the test cell:

1. The missing devices have to be present as part of the rigging loom or they have to be simulated. It is not uncommon to see rigged engines in a cell with their mounting frame festooned with a significant part of a vehicle loom, including the ignition key module.
2. A special version of the ECU or VCU has to be obtained or created so that the need for signals from non-existent vehicle transducers or actuator feedback loops are switched off. There is a specialized market for programmable ECUs. These devices are most often used by motor sports vehicle tuners. Such devices can be adapted to be used to run a limited range of engine tests in cells, including tests on the mechanical integrity of ancillary devices, sealants, and seals. Note that the commercially available units that "piggy-back" the OEM control unit and modify the fueling and ignition signals in order to change the shape of the power curve and fuel consumption do not solve the problem of engine test cell running.
3. Hardware and software tools created and supported by specialist companies such as Engineering Tools Application and Services (ETAS; founded as part of the Bosch Group, www.etas.com/) and A&D Technology Inc. (http://www.aanddtech.com/) that allow their licensees various, discrete, permitted levels of access to the ECU software code and engine maps. This is an industrial standard methodology used worldwide in the calibration laboratories of all OEM, Tier 1 suppliers and research institutions; it requires full professionally trained specialists and a high level of powertrain test instrumentation; even so, access to the lowest logic and code levels is kept commercially secret by all major ECU developers.

Strategy 2 requires forward planning and some access to control devices' internal code. The OEMs are usually able to prepare special, nonproduction, versions of the ECU to allow out-of-vehicle Tier 1 and Tier 2 testing and their own calibration work to proceed. Any organization doing work for the OEM should negotiate for the supply of such a device at the outset of a project or face the cost and time consequences of strategy 3. Alternatively, and only in the limited circumstances where no emission calibration is required, a fully programmable after-market ECU may to be purchased that contains a basic cold start and normal running map for the chosen engine (if available); the wiring loom may have to be modified to suit.

Whatever ECU rigging strategy is used there are still some ex-vehicular hardware modules that are recommended always to be physically included within the test cell rigging rather than substituted with difficulty. Two common examples are:

- The throttle pedal controller, which is nowadays part of a complex, closed loop, "drive-by-wire" transducer and actuator system. This means that the test cell has to control the engine through a throttle actuator device linked to the throttle pedal arm. While this may be physically complex it has proved, in most cases, to be far more cost-effective than developing special simulation models of the required command/feedback signals.
- The ignition key system, which may use a key device that wirelessly communicates with its matched ECU as part of the powertrain shutdown and a vehicle security system.

Supply of Electrical Services to Engine Systems

Modern automotive test beds are now frequently supplied with an "engine services box" mounted close to the engine from which stabilized DC power is supplied for the ECU and switched supplies and earth connections for other ancillary devices. Starter-motor current outputs and switching for cranking are also included, negating the need for 12- or 24-volt batteries in the cell.

On-Board Diagnostic Systems (OBDs)

Since 1996 all cars and light trucks used on the public road in the USA have had to have ECUs that operate an on-board diagnostic system that is compliant with the SAE standard OBD-11. The EU made a version of OBD-11 called EOBD mandatory for gasoline vehicles in 2001 and for diesel vehicles of model year 2003.

So whatever mix of strategies used to rig and run engines is used, the test cell operators will need a device to read the OBD-11 to diagnose operational and rigging (missing signal) problems. All commercially available readers connect through a common type of 16-pin D-type connector; however, there are (in early 2011) five OBD-11 protocol types in use: J1850 PWM, J1850 VPW, ISO 9141-2, ISO 14230 KWP2000, and ISO 15765 CAN.

Each of these protocols differ electrically (pin designation although 12 V and GND are common) and by communication format. The code reader or scan tool used must be compatible with the specific protocol of the unit under test (UUT).

Rigging Engine Circuits and Auxiliaries

The connection between cell and engine systems must not be allowed to constitute a weak point in the physical or electrical integrity of the combined system; therefore, it is important to use connectors that are as least as robust as the matching engine component. In the case of high-volume testing the cell's rigging plugs and connectors must be of a significantly higher standard than the engine components, which are not designed to sustain many cycles of connection and disconnection. In both Europe and the USA there exist specialist companies whose catalogs list most of the items commonly needed for engine rigging [1].

Air Filter

Ex-vehicular units now tend to be large plastic moldings that require supplementary support if used in the test cell; some contain a thermally operated flow device designed to speed up engine warm-up, but which will only work correctly when installed in the correct under-bonnet (hood) location. When the engine is fed with treated combustion air the size and shape of the air filter may create problems and it should be replaced with a smaller, exposed element unit having the same resistance to air flow.

Auxiliary Drives

Some engines will have directly mounted, usually belt-driven, drives for auxiliary circuits such as air-conditioning, power steering hydraulics, etc. The rigging of these devices rather depends on the design of the experiment and test sequences being performed. Even in tests where these devices, or the loads they impose, are of no importance they often have to remain in place due to the configuration of the drive-belt system; in this case they may need to be either connected to a (dummy) circuit to prevent seizure, or modified to only consist of the drive wheel (pulley) supported in bearings.

Note that harmonized exhaust emission legislation specifies adjustments that have to be made in the recorded power output of some engine types when auxiliary drives have been removed (see Chapter 16).

Charge Air Coolers (Intercoolers)

The charge air coolers mounted in turbocharged diesel vehicles tend to be air/air type; these present some problems in rigging in the test cell. If the vehicle cooler is used as part of the engine rigging a spot-fan will be required to supply the cooling air flow, but this is impossible to control with much accuracy and puts a considerable amount of heat into the cell air, which may create problems when operating during high ambient temperatures.

Some vehicles use water/charge air coolers that have advantages for use in the test cell: they are physically smaller and easy to rig; they are also easier to control via a closed loop within the test controller; finally, they have a larger thermal sink rate during increasing engine power. Charge coolers of the water/air type suitable for test cell use can be bought in the automotive after-market; they need to be of high quality, pressure tested, and not overpressurized, since the outcome of any fluid leak into the charge air is not beneficial to the engine.

Coolant Circuit

Whatever coolant control module is used (see Chapter 7), it is important to locate it as close as possible to the engine. Normally the connection points used are those that in the vehicle would go to the radiator. Self-sealing couplings are not recommended due to their high friction head and it is important not to allow the weight of overlong pipes to put undue strain on clamped pipe connections. When filling the coolant system it is most important to bleed air out of the system at both the engine and control module high points; squeezing the pipes is simple and effective.

Oil Cooler

Engines, usually in the higher performance automotive range, may be fitted in the vehicle with auxiliary oil coolers, of which there are three major types:

- Oil coolers integrated into the engine body and circulated with coolant fluid. These normally need no rigging in the test cell.
- Coolant/oil coolers that are fitted as an external auxiliary to the engine; these can be rigged to run using cell primary cooling water or remain as part of the engine system.
- Oil coolers of the air/oil type that are, in a vehicle, a second and smaller version of the coolant circuit radiator. In most cases these are better replaced in the cell by a water/oil cooler circuit forming part of the cell's fluid control module (see Figures 7.3 and 7.4). Some testers use the vehicle type of cooler and direct a spot cooling fan through the matrix but this gives, at best, limited temperature regulation and puts extra heat into the cell.

Engine Handling and Support Systems

The common causes of destructive levels of engine vibration in the test cell are discussed in some detail in Chapter 9, but there are some general points of good practice that should be followed. It is common practice to rig diesels for test by supporting them at the drive shaft end with a profiled steel plate bolted to the face of the flywheel housing resting on damped mounts fixed to the mounting frame at either side of the engine (shown in Figure 11.1 and diagrammatically in Figure 9.3). The horizontal plane running through the mounting feet should generally be at or near that of the engine's crankshaft; this ensures that the fixings are in simple compression or tension and the connecting shaft is not subjected to lateral vibrations as the engine tries to turn about the crankshaft's centerline. For

FIGURE 11.1 Diesel engine in the process of being rigged. Note the profiled mounting frame bolted to the flywheel housing, which is supported by resilient mountings on the crankshaft centerline. A cast-iron tee-slotted rail runs either side of the engine, supported by screw-jack support columns bolted to a cast-iron bedplate solidly bolted to a concrete floor. *(Photo courtesy of University of Bradford.)*

short-duration hot and cold production tests the engine is often clamped rigidly to a common frame with the drive motor (dynamometer) using any convenient flat-machined surface, or even the sump fixing bolt-heads, but the whole frame is mounted on some form of soft mounts such as those shown in Figure 9.9.

Most gasoline engines will be rigged for test supported from the vehicle mounting points as is the case of the engine in Figure 11.2, but as warned in Chapter 9, this may not be satisfactory because of the differences between stiffnesses of the mounting frames and effects of other engine restraints. Failures of exhaust connections near the engine and fatigue failure of the support castings of ancillary units, such as the alternator, are common indicators of engine mounting problems.

FIGURE 11.2 A small automotive gasoline engine mounted in an adaptable, sectionalized, engine support frame fitted to a combined engine–dynamometer bedframe sitting on flexible mounts fitted to a concrete floor. *(Photo courtesy of University of Bradford.)*

Clearly it is not cost-effective to have a valuable test cell lying idle for days while problems related to engine rigging are being solved; therefore, it is common and sensible practice to rig engines on some sort of pallet or trolley system and carry out as much as possible of the rigging external to the test cell. In some large facilities engines are both pre-rigged on pallets, then started and run without load in a safe area, equipped with basic services, so as to check base function before being taken to the test cell.

Even with pre-rigging a considerable number of connections must be made to any engine in the cell before testing can proceed. These include the coupling shaft, fuel, cooling water, exhaust, and transducers.

The degree of complexity of the system adopted for transporting, installing, and removing the engine within the facility naturally depends on the frequency with which the engine is changed. In some research or lubricant test cells the engine is more or less a permanent fixture, but at the other extreme the test duration for each engine in a production cell will be measured in minutes, and the time taken to change engines must be cut to an absolute minimum.

There is a corresponding variation in the handling systems:

- Simple arrangements when engine changes are comparatively infrequent. The engine is fitted with test transducers then either craned into the cell and rigged in situ or mounted on a suitable stand such as that shown in Figure 11.2. All engine-to-cell connections are made subsequently by skilled staff with workshop backup before the engine is filled with liquids.
- The pre-rigged engine is mounted on a wheeled trolley or eletric-truck-maneuvered pallet carrying various transducers and service connections that are coupled to the engine before it is moved into the cell. An engine rigging workshop needs to be fitted with a dummy test cell station that presents datum connection points identical in position and detail to those in the cells, particularly the dynamometer shaft, in order for the engine to be pre-aligned. Clearly, to gain maximum benefit, each cell in the facility needs to be built with critical fixed interface items in identical positions from a common datum. These positions should be repeated by dummy interface points in any pre-rigging stand, in the workshop.
- For production test beds it is usual to make all engine-to-pallet connections before the combined assembly entering the cell. An automatic docking system permits all connections (including in some cases the driving shaft) to be made in seconds and the engine to be filled with liquids.

Workshop support in the provision of suitable fittings and adaptors always needs be made available. All pipe connections from the cell to the engine should be sufficiently flexible to allow for thermal and engine movement. Exhaust connection sections can be particularly short-lived; they should be treated as consumable items.

Making up of the engine to cell exhaust systems often requires a welding bay. *Electrical arc welding must not be allowed within the test cell because of the danger of damage to computers, instrumentation, and powertrain electronics by stray currents through a common earth.*

The workshop area used for de-rigging should be designed to deal with the inevitable spilled fluids and engine wash activities. Floor drains should run into an oil intercept unit.

Note Concerning Rigging Aero-Piston Engines Including Radial Air-Cooled Engines

Aero-piston engines of both modern and historic types usually present fewer electrical or control rigging problems than their automotive equivalents, because most are fueled by carburettors and fitted with a form of magneto ignition. Very few engines in light aircraft have ECUs of the automotive type; this is due in part to the smaller volume market and the, justifiably stringent, certification process requiring such electronic devices to be approved to the standards of full authority digital engine controls known as FADEC.

Historic aircraft engines may have to be run after major rebuilds to retain airworthiness certification. Such units are normally mounted using their airframe mounts but the connection to a dynamometer may require a special splined coupling that fits in place of the propeller (airscrew) hub.

Each propeller has its characteristic rising load with rising speed or "propeller law" curve that needs to be simulated in the dynamometer control system. Many analog control systems of the 1970s to 1980s era came with a propeller law circuit as standard, as it was considered the safe default mode when an operator was allowed into the cell and was able to manually open the throttle; over-revving was thus inhibited.

Alternatively, for testing light aircraft engines outside their airframe, mounting them on a torque-reacting test frame fitted with its matched propeller is possible. For air-cooled engines and radial units where the cooling airflow from the propeller is required, this is a standard, albeit noisy, test method that is usually carried out in an open-ended aircraft hanger building.

PART 2. SELECTION OF A SUITABLE DYNAMOMETER CONNECTING SHAFT

The selection of suitable couplings and shaft for the connection of the engine or transmission to the dynamometer is not a simple matter and the commonly used "suck it and see" method can sometimes lead to spectacular failures.

Incorrect choice or faulty system design may create a number of problems:

- Torsional oscillations
- Vibration of engine or dynamometer
- Whirling of coupling shaft
- Damage to engine or dynamometer bearings
- Excessive wear of shaft-line components
- Catastrophic failure of coupling shafts
- Engine starting problems.

The whole subject, the coupling of engine and dynamometer, can cause more trouble than any other routine aspect of engine testing and the problem of torsional vibration is not confined to the engine/dynamometer system. Automotive engineers may also have to deal with the problem in the engine/generator system of the new generation of hybrid vehicles; therefore, a clear understanding of the many factors involved is highly desirable. The following text covers in some detail all the main factors, but in some areas more extensive analysis may be necessary and appropriate references are given.

The Nature of the Problem

The special feature of the problem is that it must be considered afresh each time an engine not previously encountered is installed. It must also be recognized that unsatisfactory torsional behavior is associated with the whole system, engine, coupling shaft, and dynamometer, as shown diagrammatically in Figure 11.3, rather than with the individual components, all of which may be quite satisfactory in themselves or within an alternative system of the same nominal power.

Problems arise partly because the dynamometer is seldom equivalent dynamically to the system driven by the engine in service. In the case of a vehicle with rear axle drive, the driveline consists of a clutch, which may itself act as a torsional damper, followed by a gearbox, the characteristics of which are low inertia and some damping capacity. This is followed by a drive shaft and differential, itself having appreciable damping, two half-shafts and two wheels, both with substantial damping capacity and running at much slower speed than the engine, thus reducing their effective inertia. When coupled to a dynamometer this system, with its built-in damping and moderate inertia, is replaced by a single drive shaft connected to a single rotating mass, the dynamometer, running at the same speed as the engine. The clutch may be, but is usually not, retained.

Particular care is necessary where the moment of inertia of the dynamometer is more than about twice that of the engine. A further consideration that must be taken seriously concerns the effect of the difference between the engine mounting arrangements in the vehicle and on the test bed. This can lead to vibrations of the whole engine that can have consequential effects on the drive shaft.

Overhung Mass on Engine and Dynamometer Bearings

Care must be taken when designing and assembling a shaft system that the loads imposed by the mass and unbalanced forces do not exceed the overhung weight limits of the engine bearing at one end and the dynamometer at the other. Steel adaptor plates on either end of the shaft, required to adapt the bolt holes of the shaft to the dynamometer flange or engine flywheel, can increase

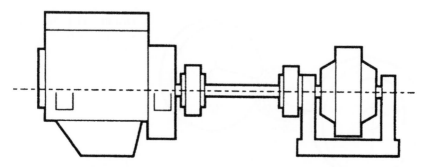

FIGURE 11.3 Simple form of dynamometer–engine driveline.

the load on bearings significantly and compromise the operation of the system, in which case the use of high-grade aluminum is recommended. Dynamometer manufacturers produce tables showing the maximum permissible mass at a given distance from the coupling face of their machines; the equivalent details for most engines are much more difficult to obtain, but the danger of overload should be kept in mind by all concerned.

Background Reading

The mathematics of the subject is complex and not readily accessible. Den Hartog [2] gives what is possibly the clearest exposition of fundamentals. Ker Wilson's classical treatment in five volumes is probably still the best source of comprehensive information; his abbreviated version [3] is sufficient for most purposes, but was out of current print in late 2010. MEP have published a useful practical handbook [4], while Lloyd's Register [5] gives rules for the design of marine drives that are also useful in the present context.

A listing of the notation used is to be found at the end of this chapter.

Torsional Oscillations and Critical Speeds

In its simplest form, the engine–dynamometer system may be regarded as equivalent to two rotating masses connected by a flexible shaft (Figure 11.4).

Such a system has an inherent tendency to develop torsional oscillations. The two masses can vibrate 180° out of phase about a node located at some point along the shaft between them. The oscillatory movement is superimposed on any steady rotation of the shaft. The resonant or critical frequency of torsional oscillation of this system is given by:

$$n_c = \frac{60}{2\pi} \sqrt{\frac{C_c(I_e + I_b)}{I_e I_b}} \tag{11.1}$$

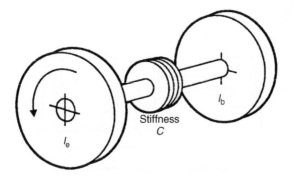

FIGURE 11.4 Two-mass system (compare with Figure 11.3).

If an undamped system of this kind is subjected to an exciting torque of constant amplitude T_{ex} and frequency n, the relation between the amplitude of the induced oscillation θ and the ratio n/n_c is as shown in Figure 11.5.

At low frequencies the combined amplitude of the two masses is equal to the static deflection of the shaft under the influence of the exciting torque, $\theta_0 = T_{ex}/C_s$. As the frequency increases the amplitude rises and at $n = n_c$ it becomes theoretically infinite: the shaft may fracture or nonlinearities and internal damping may prevent actual failure. With further increases in frequency the amplitude falls and at $n - \sqrt{2n_c}$ it is down to the level of the static deflection. Amplitude continues to fall with increasing frequency.

The shaft connecting engine and dynamometer must be designed with a suitable stiffness C_s to ensure that the critical frequency lies outside the normal operating range of the engine, and with a suitable degree of damping to ensure that the unit may be run through the critical speed without the development of a dangerous level of torsional oscillation. Figure 11.5 also shows the behavior of a damped system. The ratio θ/θ_0 is known as the dynamic magnifier M. Of particular importance is the value of the dynamic magnifier at the critical frequency, M_c. The curve of Figure 11.5 corresponds to a value $M_c = 5$.

Torsional oscillations are excited by the variations in engine torque associated with the pressure cycles in the individual cylinders (also, though usually of less importance, by the variations associated with the movement of the reciprocating components).

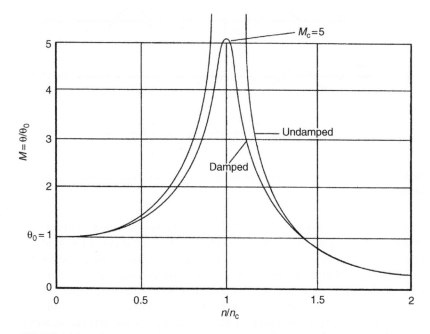

FIGURE 11.5 Relationship between frequency ratio, amplitude, and dynamic amplifier M.

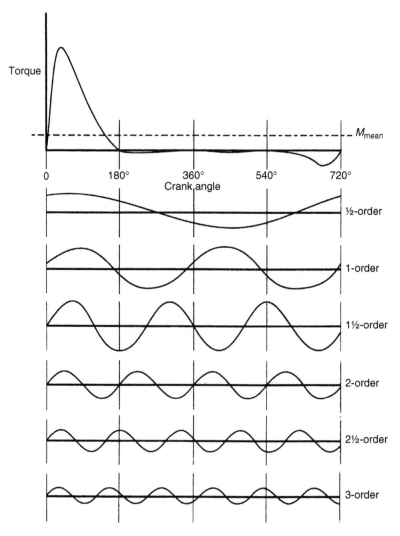

FIGURE 11.6 Harmonic components of turning moment, single-cylinder four-stoke gasoline engine.

Figure 11.6 shows the variation in the case of a typical single-cylinder four-stroke gasoline engine. It is well known that any periodic curve of this kind may be synthesized from a series of *harmonic components*, each a pure sine wave of a different amplitude having a frequency corresponding to a multiple or submultiple of the engine speed, and Figure 11.6 shows the first six components.

The *order* of the harmonic defines this multiple. Thus, a component of order $N_0 = \frac{1}{2}$ occupies two revolutions of the engine, $N_0 = 1$ one revolution, and so on. In the case of a four-cylinder four-stroke engine, there are two firing strokes per revolution of the crankshaft and the turning moment curve of Figure 11.6 is repeated at intervals of 180°.

In a multicylinder engine the harmonic components of a given order for the individual cylinders are combined by representing each component by a vector, in the manner illustrated in Chapter 9, for the inertia forces. A complete treatment of this process is beyond the scope of this book, but the most significant results may be summarized as follows.

The first major critical speed for a multicylinder in-line engine is of order:

$$N_0 = N_{cyl}/2 \quad \text{for a four-stroke engine} \qquad (11.2a)$$

$$N_0 = N_{cyl} \quad \text{for a two-stroke engine} \qquad (11.2b)$$

Thus, in the case of a four-cylinder four-stroke engine, the major critical speeds are of order 2, 4, 6, etc. In the case of a six-cylinder engine, they are of order 3, 6, 9, etc.

The distinction between a major and a minor critical speed is that in the case of an engine having an infinitely rigid crankshaft it is only at the major critical speeds that torsional oscillations can be induced. This, however, by no means implies that in large engines having a large number of cylinders the minor critical speeds may be ignored.

At the major critical speeds the exciting torques T_{ex} of all the individual cylinders in one line act in phase and are thus additive (special rules apply governing the calculation of the combined excitation torques for Vee engines).

The first harmonic is generally of most significance in the excitation of torsional oscillations, and for engines of moderate size, such as passenger vehicle engines, it is generally sufficient to calculate the critical frequency from equation (11.1), then to calculate the corresponding engine speed from:

$$N_c = n_c/N_0 \qquad (11.3)$$

The stiffness of the connecting shaft between engine and dynamometer should be chosen to that this speed does not lie within the range in which the engine is required to develop power.

In the case of large multicylinder engines the "wind up" of the crankshaft as a result of torsional oscillations can be very significant and the two-mass approximation is inadequate; in particular, the critical speed may be over-estimated by as much as 20% and more elaborate analysis is necessary. The subject is dealt with in several different ways in the literature; perhaps the easiest to follow is that of Den Hartog. The starting point is the value of the mean turning moment developed by the cylinder, M_{mean} (Figure 11.6).

Values are given for a so-called "p factor", by which M_{mean} is multiplied to give the amplitude of the various harmonic excitation forces. Table 11.1, reproduced from Den Hartog, shows typical figures for a four-stroke medium-speed diesel engine.

Exciting torque is:

$$T_{ex} = pM_{mean} \qquad (11.4)$$

TABLE 11.1 p Factors

Order	$^1/_2$	1	$1^1/_2$	2	$2^1/_2$	3...	8
p factor	2.16	2.32	2.23	1.91	1.57	1.28	0.08

The relation between M_{mean} and i.m.e.p. (*indicated mean effective pressure*) is given by:

$$\text{For a four-stroke engine} \quad M_{mean} = p_i \frac{B^2 S}{16} \times 10^{-4} \quad (11.5a)$$

$$\text{For a two-stroke engine} \quad M_{mean} = p_i \frac{B^2 S}{8} \times 10^{-4} \quad (11.5b)$$

Lloyd's Rulebook, the main source of data on this subject, expresses the amplitude of the harmonic components rather differently, in terms of a "component of tangential effort", T_m. This is a pressure that is assumed to act upon the piston at the crank radius $S/2$. Then, exciting torque per cylinder:

$$T_{ex} = T_m \frac{\pi B^2}{4} \frac{S}{2} \times 10^{-4} \quad (11.6)$$

Lloyd's give curves of T_m in terms of the indicated mean effective pressure p_i and it may be shown that the values so obtained agree closely with those derived from Table 11.1.

The amplitude of the vibratory torque T_v induced in the connecting shaft by the vector sum of the exciting torques for all the cylinders, ΣT_{ex}, is given by:

$$T_v = \frac{\sum T_{ex} M_c}{(1 + I_e/I_b)} \quad (11.7)$$

The complete analysis of the torsional behavior of a multicylinder engine is a substantial task, though computer programs are available that reduce the effort required. As a typical example, Figure 11.7 shows the "normal elastic curves" for the first and second modes of torsional oscillation of a 16-cylinder Vee engine coupled to a hydraulic dynamometer. These curves show the amplitude of the torsional oscillations of the various components relative to that at the dynamometer, which is taken as unity. The natural frequencies are respectively $n_c = 4820$ c.p.m. and 6320 c.p.m. The curves form the basis for further calculations of the energy input giving rise to the oscillation. In the case of the engine under consideration, these showed a very severe fourth-order oscillation, $N_0 = 4$, in the first mode. (For an engine having eight cylinders in line the first major critical speed, from equation (11.2a), is of order $N_0 = 4$.)

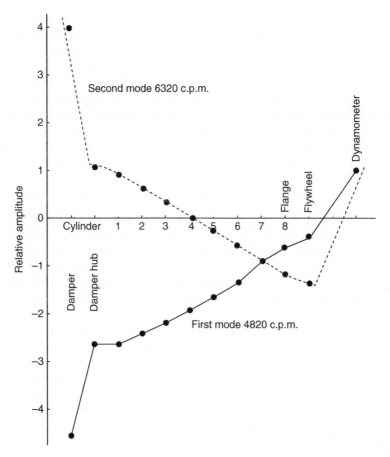

FIGURE 11.7 Normal elastic curves for a particular 16-cylinder Vee engine coupled to a hydraulic dynamometer. *(Taken from an actual investigation)*

The engine speed corresponding to the critical frequency of torsional oscillation is given by:

$$N_c = n_c/N_0 \qquad (11.8)$$

giving, in the present case, $N_c = 1205$ rev/min, well within the operating speed range of the engine. Further calculations showed a large input of oscillatory energy at $N_0 = 4\frac{1}{2}$, a minor critical speed, in the second mode, corresponding to a critical engine speed of $6320/4\frac{1}{4} = 1404$ rev/min, again within the operating range. Several failures of the shaft connecting engine and dynamometer occurred before a safe solution was arrived at.

This example illustrates the need for caution and for full investigation in setting up large engines on the test bed.

It is not always possible to avoid running close to or at critical speeds, and this situation is usually dealt with by the provision of torsional vibration dampers, in which the energy fed into the system by the exciting forces is absorbed by viscous shearing. Such dampers are commonly fitted at the nonflywheel end of engine crankshafts. In some cases it may also be necessary to consider their use as a component of engine test cell drivelines, when they are located either as close as possible to the engine flywheel or at the dynamometer. The damper must be "tuned" to be most effective at the critical frequency and the selection of a suitable damper involves equating the energy fed into the system per cycle with the energy absorbed by viscous shear in the damper. This leads to an estimate of the magnitude of the oscillatory stresses at the critical speed. For a clear treatment of the theory, see Den Hartog [2].

Points to remember concerning torsional vibration problem avoidance are:

- As a general rule, it is good practice to avoid running the engine under power at speeds between 0.8 and 1.2 times critical speed. If it is necessary to take the engine through the critical speed, this should be done off-load and as quickly as possible. With high inertia dynamometers the transient vibratory torque may well exceed the mechanical capacity of the driveline and the margin of safety of the driveline components may need to be increased.
- Problems frequently arise when the inertia of the dynamometer much exceeds that of the engine: a detailed torsional analysis is desirable when this factor exceeds 2. This situation usually occurs when it is found necessary to run an engine of much smaller output than the rated capacity of the dynamometer.
- The simple "two-mass" approximation of the engine–dynamometer system is inadequate for large engines and may lead to overestimation of the critical speed.

Design of Coupling Shafts

The maximum shear stress induced in a shaft, diameter D, by a torque T Nm is given by:

$$\tau = \frac{16T}{\pi D^3} \text{ Pa} \tag{11.9a}$$

In the case of a tubular shaft, bore diameter d, this becomes:

$$\tau = \frac{16TD}{\pi(D^4 - d^4)} \text{ Pa} \tag{11.9b}$$

For steels, the shear yield stress is usually taken as equal to 0.75 × yield stress in tension [6]. A typical choice of material would be a nickel–chromium–molybdenum steel, to specification BS 817M40 (previously EN 24), heat treated to the "T" condition.

The various stress levels for this steel are roughly as follows:

Ultimate tensile strength	not less than 850 MPa (55 t.s.i.)
0.1% proof stress in tension	550 MPa
Ultimate shear strength	500 MPa
0.1% proof stress in shear	300 MPa
Shear fatigue limit in reversed stress	±200 MPa.

It is clear that the permissible level of stress in the shaft will be a small fraction of the ultimate tensile strength of the material.

Stress Concentrations, Keyways, and Keyless Hub Connection

For a full treatment of the very important subject of stress concentration, see Ref. [7]. There are two particularly critical locations to be considered:

- At a shoulder fillet, such as is provided at the junction with an integral flange. For a ratio fillet radius/shaft diameter = 0.1, the stress concentration factor is about 1.4, falling to 1.2 for $r/d = 0.2$.
- At the semicircular end of a typical rectangular keyway, the stress concentration factor reaches a maximum of about 2.5 × nominal shear stress at an angle of about 50° from the shaft axis. The author has seen a number of shaft failures at this location and angle.

Cyclic stresses associated with torsional oscillations are an important consideration and as, even in the most carefully designed installation involving an internal combustion engine, some torsional oscillation will be present, it is wise to select a conservative value for the nominal (steady-state) shear stress in the shaft.

In view of the stress concentration inherent in shaft keyways and the backlash present that can develop in splined hubs, keyless hub connection systems of the type produced by the Ringfeder Corporation® are now widely used. These devices are supported by comprehensive design documentation; however, the actual installation process must be exactly followed for the design performance to be ensured.

Stress concentration factors apply to the cyclic stresses rather than to the steady-state stresses. Figure 11.8 shows diagrammatically the Goodman diagram for a steel having the properties specified above. This diagram indicates approximately the relation between the steady shear stress and the permissible oscillatory stress. The example shown indicates that, for a steady torsional stress of 200 MPa, the accompanying oscillatory stress (calculated after taking into account any stress concentration factors) should not exceed ±120 MPa. In the absence of detailed design data, it is good practice to design shafts for use in engine test beds very conservatively, since the consequences of shaft failure can be so serious. A shear stress calculated in

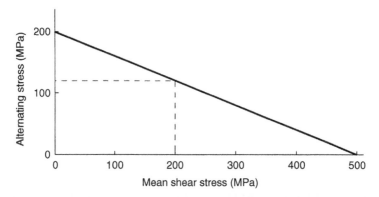

FIGURE 11.8 Goodman diagram, steel shaft in shear.

accordance with equation (11.9) of about 100 MPa for a steel with the properties listed should be safe under all but the most unfavorable conditions. To put this in perspective, a shaft of 100 mm diameter designed on this basis would imply a torque of 19,600 Nm, or a power of 3100 kW at 1500 rev/min.

The torsional stiffness of a solid shaft of diameter D and length L is given by:

$$C_s = \frac{\pi D^4 G}{32L} \tag{11.10a}$$

For a tubular shaft, bore d:

$$C_s = \frac{\pi (D^4 - d^4)}{32L} \tag{11.10b}$$

Shaft Whirl

Shaft whirl is sometimes called descriptively "skipping rope" vibration; it may be caused by a bent shaft tube or occur within plain shaft bearings, but in the engine–dynamometer system the coupling shaft is usually supported at each end by a universal joint or flexible coupling. Such a shaft will "whirl" at a rotational speed N_w (also at certain higher speeds in the ratio 2^2N_w, 3^2N_w, etc.).

The whirling speed of a solid shaft of length L is given by:

$$N_w = \frac{30\pi}{L^2} \sqrt{\frac{E\pi D^4}{64W_s}} \tag{11.11}$$

It is desirable to limit the maximum engine speed to about $0.8N_w$.

When using rubber flexible couplings it is essential to allow for the radial flexibility of these couplings, since this can drastically reduce the whirling speed. It is sometimes the practice to fit self-aligning rigid steady bearings at the center of flexible couplings in high-speed applications, but these are liable to give fretting problems and are not universally favored.

The whirling speed of a shaft is identical to its natural frequency of transverse oscillation. To allow for the effect of transverse coupling flexibility, the simplest procedure is to calculate the transverse critical frequency of the shaft plus two half-couplings from the equation:

$$N_t = \frac{30}{\pi} \sqrt{\frac{k}{W}} \qquad (11.12a)$$

where $W =$ mass of shaft $+$ half-couplings and $k =$ combined radial stiffness of the two couplings.

Then whirling speed N taking this effect into account will be given by:

$$\left(\frac{1}{N}\right)^2 = \left(\frac{1}{N_w}\right)^2 + \left(\frac{1}{N_t}\right)^2 \qquad (11.12b)$$

Couplings

The choice of the appropriate coupling for a given application is not easy: the majority of driveline problems probably have their origin in an incorrect choice of components for the driveline, and are often cured by changes in this region. A complete discussion would much exceed the scope of this book, but the reader concerned with driveline design should obtain a copy of Ref. [4], which gives a comprehensive treatment together with a valuable procedure for selecting the best type of coupling for a given application. A very brief summary of the main types of coupling follows.

Quill Shaft with Integral Flanges and Rigid Couplings

This type of connection is best suited to the situation where a driven machine is permanently coupled to the source of power, when it can prove to be a simple and reliable solution. It is not well suited to test bed use, since it is intolerant of relative vibration and misalignment.

Quill Shaft with Toothed or Gear Type Couplings

Splined, solid quill shafts have been successfully used as "mechanical fuses" in motor sport test-bed layouts between the engine flywheel and a bearing-mounted coupling assembly in a dummy gearbox housing. Such installations experience frequent severe power variations that such quills are designed to absorb, and protect more valuable engine or torque transducer components that would otherwise be damaged.

In applications such as World Rally Car powertrain testing, a 25 mm shank diameter quill shaft made from a 2.5% nickel–chromium–molybdenum high-carbon steel (S155) hardened and shot peened has been used successfully in extremely arduous torsional regimes. In such a layout the coupling of the dummy gearbox is connected to a conventional cardan shaft (see below).

Hollow shaft gear couplings are very suitable for high powers and speeds and are used in gas-turbine test facilities; they can deal with vibration and some degree of misalignment, but this must be very carefully controlled to avoid problems of wear. Boundary lubrication of the coupling teeth is particularly important, as once local tooth-to-tooth seizure takes place deterioration, even of nitrided surfaces, may be rapid and catastrophic. Such shafts are inherently stiff in torsion.

Conventional "Cardan Shaft" with Universal Joints

There are a number of design variations of this type of commonly used shaft, shown in Figure 11.9, some having a torsionally soft rubber joint at one end. They are readily available from a number of suppliers, and are probably the preferred solution in the majority of automotive and up to medium-speed engine test assemblies. However, the standard ex-vehicular shafts may not be suitable as they will not be designed to run at engine flywheel speeds in excess of those met in vehicle applications. A small amount of misalignment between the two ends is usually recommended by manufacturers, particularly in endurance applications, to avoid fretting of the needle rollers within the joints. The shaft should be installed so that an amount of longitudinal movement in the center section is possible in either direction.

Elastomeric Element Couplings

There is a vast number of different designs on the market and selection is not easy. Reference [8] may be found helpful. The great advantage of these couplings is that their torsional stiffness may be varied widely by changing the elastic elements and problems associated with torsional vibrations and critical speeds dealt with (see the next section).

FIGURE 11.9 A cardan shaft with universal ("Hooks type") joints at both ends and splined center, allowing some length adjustment.

Damping: The Role of the Flexible Coupling

The earlier discussion leads to two main conclusions: The engine–dynamometer system is susceptible to torsional oscillations and the internal combustion engine is a powerful source of forces calculated to excite such oscillations. The magnitude of these undesirable disturbances in any given system is a function of the damping capacities of the various elements: the shaft, the couplings, the dynamometer, and the engine itself. The couplings are the only element of the system whose damping capacity may readily be changed, and for our purposes the damping capacity of the remainder of the system may be neglected.

The dynamic magnifier M (Figure 11.5) has already been mentioned as a measure of the susceptibility of the engine–dynamometer system to torsional oscillation. Now, referring to Figure 11.4, let us assume that there are two identical flexible couplings, of stiffness C_c, one at each end of the shaft, and that these are the only sources of damping. Figure 11.10 shows a typical torsionally resilient coupling in which torque is transmitted by way of a number of shaped rubber blocks or bushes that provide torsional flexibility, damping, and a capacity to take up misalignment. The torsional characteristics of such a coupling are shown in Figure 11.11.

These differ in three important respects from those of, say, a steel shaft in torsion:

1. The coupling does not obey Hooke's law: the stiffness or coupling rate $C_c = \Delta T/\Delta\theta$ increases with torque. This is partly an inherent property of the rubber and partly a consequence of the way it is constrained.

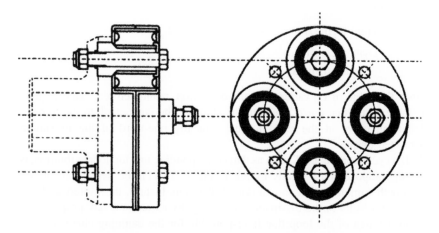

FIGURE 11.10   Rubber bush-type torsionally resilient coupling.

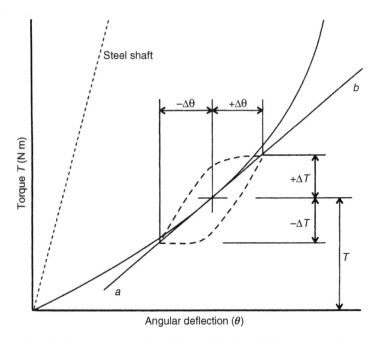

FIGURE 11.11 Dynamic torsional characteristic of multiple-bush-type coupling.

2. The shape of the torque–deflection curve is not independent of frequency. Dynamic torsional characteristics are usually given for a cyclic frequency of 10 Hz. If the load is applied slowly, the stiffness is found to be substantially less. The following values of the ratio dynamic stiffness (at 10 Hz) to static stiffness of natural rubber of varying hardness are taken from Ref. [4].

Shore (IHRD) hardness	40	50	60	70
$\dfrac{\text{Dynamic stiffness}}{\text{Static stiffness}}$	1.5	1.8	2.1	2.4

Since the value of C_c varies with the deflection, manufacturers usually quote a single figure that corresponds to the slope of the tangent ab to the torque–deflection curve at the rated torque, typically one-third of the maximum permitted torque.

3. If a cyclic torque $\pm \Delta T$, such as that corresponding to a torsional vibration, is superimposed on a steady torque T (Figure 11.11), the deflection follows a path similar to that shown dotted. It is this feature, the hysteresis loop, which results in the dissipation of energy, by an amount ΔW proportional to the area of the loop that is responsible for the damping characteristics of the coupling.

Damping energy dissipated in this way appears as heat in the rubber and can, under adverse circumstances, lead to overheating and rapid destruction of the elements. The appearance of rubber dust inside coupling guards is a warning sign.

The damping capacity of a component such as a rubber coupling is described by the *damping energy ratio*:

$$\psi = \frac{\Delta W}{W}$$

This may be regarded as the ratio of the energy dissipated by hysteresis in a single cycle to the elastic energy corresponding to the wind-up of the coupling at mean deflection:

$$W = \frac{1}{2} T\theta = \frac{1}{2} T^2/C_c$$

The damping energy ratio is a property of the rubber. Some typical values are given in Table 11.2. The dynamic magnifier is a function of the damping energy ratio; as would be expected, a high damping energy ratio corresponds to a low dynamic magnifier. Some authorities give the relation:

$$M = 2\pi/\psi$$

However, it is pointed out in Ref. [9] that for damping energy ratios typical of rubber the exact relation:

$$\psi = (1 - e^{-2\pi/M})$$

is preferable. This leads to the values of M shown in Table 11.3, which correspond to the values of ψ given in Table 11.2.

It should be noted that when several components, e.g. two identical rubber couplings, are used in series the dynamic magnifier of the combination is given by:

$$\left(\frac{1}{M}\right)^2 = \left(\frac{1}{M_1}\right)^2 + \left(\frac{1}{M_2}\right)^2 + \left(\frac{1}{M_3}\right)^2 + \dots \qquad (11.13)$$

(this is an empirical rule, recommended in Ref. [5]).

TABLE 11.2 Damping Energy Ratio ψ

Shore (IHRD) Hardness	50/55	60/65	70/75	75/80
Natural rubber	0.45	0.52	0.70	0.90
Neoprene		0.79		
Styrene–butadiene (SBR)		0.90		

TABLE 11.3 Dynamic Magnifier *M*

Shore (IHRD) Hardness	50/55	60/65	70/75	75/80
Natural rubber	10.5	8.6	5.2	2.7
Neoprene		4.0		
Styrene–butadiene (SBR)		2.7		

An Example of Drive Shaft Design

The application of these principles is best illustrated by a worked example. Figure 11.2 represents an engine coupled by way of twin multiple-bush-type rubber couplings and an intermediate steel shaft to an eddy-current dynamometer, with dynamometer starting.

Engine specification is as follows:

Four-cylinder four-stroke gasoline engine, swept volume 2.0 liter, bore 86 mm, stroke 86 mm

Maximum torque	110 Nm at 4000 rev/min
Maximum speed	6000 rev/min
Maximum power output	65 kW
Maximum b.m.e.p.	10.5 bar
Moments of inertia	$l_e = 0.25$ kgm^2
	$l_d = 0.30$ kgm^2

Table 11.4 indicates a service factor of 4.8, giving a design torque of $110 \times 4.8 = 530$ Nm.

It is proposed to connect the two couplings by a steel shaft of the following dimensions:

Diameter	$D = 40$ mm
Length	$L = 500$ mm
Modulus of rigidity	$G = 80 \times 10^9$ Pa

From equation (11.9a), torsional stress $\tau = 42$ MPa (very conservative).
From equation (11.10a):

$$C_s = \frac{\pi \times 0.04^4 \times 80 \times 10^9}{32 \times 0.5} = 40,200 \text{ Nm/rad}$$

Consider first the case when rigid couplings are employed:

$$n_c = \frac{60}{2\pi} \sqrt{\frac{40,200 \times 0.55}{0.25 \times 0.30}} = 5185 \text{ c.p.m.}$$

TABLE 11.4 Service Factors for Dynamometer/Engine Combinations

Dynamometer Type	Number of Cylinders									
	Diesel					Gasoline				
	1/2	3/4/5	6	8	10+	1/2	3/4/5	6	8	10+
Hydraulic	4.5	4.0	3.7	3.3	3.0	3.7	3.3	3.0	2.7	2.4
Hyd. + dyno. start	6.0	5.0	4.3	3.7	3.0	5.2	4.3	3.6	3.1	2.4
Eddy current (EC)	5.0	4.5	4.0	3.5	3.0	4.2	3.8	3.3	2.9	2.4
EC + dyno. start	6.5	5.5	4.5	4.0	3.0	5.7	4.8	3.8	3.4	2.4
DC + dyno. start	8.0	6.5	5.0	4.0	3.0	7.2	5.8	4.3	3.4	2.4

Reproduced by courtesy of Twiflex Ltd.

For a four-cylinder four-stroke we have seen that the first major critical occurs at order $N_0 = 2$, corresponding to an engine speed of 2592 rev/min. This falls right in the middle of the engine speed range and is clearly unacceptable. This is a typical result to be expected if an attempt is made to dispense with flexible couplings.

The resonant speed needs to be reduced and it is a common practice to arrange for this to lie between either the cranking and idling speeds or between the idling and minimum full-load speeds. In the present case these latter speeds are 500 and 1000 rev/min respectively. This suggests a critical speed N_c of 750 rev/min and a corresponding resonant frequency $n_c = 1500$ cycles/min.

This calls for a reduction in the torsional stiffness in the ratio:

$$\left(\frac{1500}{5185}\right)^2$$

i.e. to 3364 Nm/rad.

The combined torsional stiffness of several elements in series is given by:

$$\frac{1}{C} = \frac{1}{C_1} + \frac{1}{C_2} + \frac{1}{C_3} + \cdots \tag{11.14}$$

This equation indicates that the desired stiffness could be achieved with two flexible couplings, each of stiffness 7480 Nm/rad. A manufacturer's catalog shows a multibush coupling having the following characteristics:

Maximum torque	814 Nm (adequate)
Rated torque	170 Nm
Maximum continuous vibratory torque	±136 Nm
Shore (IHRD) hardness	50/55
Dynamic torsional stiffness	8400 Nm/rad.

Substituting this value in equation (11.14) indicates a combined stiffness of 3800 Nm/rad. Substituting in equation (11.1) gives $n_c = 1573$, corresponding to an engine speed of 786 rev/min, which is acceptable.

It remains to check on the probable amplitude of any torsional oscillation at the critical speed. Under no-load conditions the i.m.e.p. of the engine is likely to be around 2 bar, indicating, from equation (11.5a), a mean turning moment $M_{mean} = 8$ Nm.

From Table 11.1, p factor $= 1.91$, giving $T_{ex} = 15$ Nm per cylinder:

$$\sum T_{ex} = 4 \times 15 = 60 \, \text{Nm}$$

Table 11.4 indicates a dynamic magnifier $M = 10.5$ and combined dynamic magnifier from equation (11.13) $= 7.4$.

The corresponding value of the vibratory torque, from equation (11.7), is then:

$$T_v = \frac{60 \times 7.4}{(1 + 0.25/0.30)} = \pm 242 \, \text{Nm}$$

This is outside the coupling continuous rating of ± 136 Nm, but multiple-bush couplings are tolerant of brief periods of quite severe overload and this solution should be acceptable provided the engine is run fairly quickly through the critical speed. An alternative would be to choose a coupling of similar stiffness using SBR bushes of 60/65 hardness. Table 11.4 shows that the dynamic magnifier is reduced from 10.5 to 2.7, with a corresponding reduction in T_v.

If in place of an eddy-current dynamometer we were to employ a DC machine, the inertia I_b would be of the order of 1 kgm^2, four times greater. This has two adverse effects:

- Service factor, from Table 11.4, increased from 4.8 to 5.8.
- The denominator in equation (11.7) is reduced from $(1 + 0.25/0.30) = 1.83$ to $(1 + 0.25/1.0) = 1.25$, corresponding to an increase in the vibratory torque for a given exciting torque of nearly 50%.

This is a general rule: The greater the inertia of the dynamometer, the more severe the torsional stresses generated by a given exciting torque.

An application of equation (11.1) shows that for the same critical frequency the combined stiffness must be increased from 3364 to 5400 Nm/rad. We can meet this requirement by changing the bushes from Shore hardness 50/55 to Shore hardness 60/65, increasing the dynamic torsional stiffness of each coupling from 8400 to 14,000 Nm/rad (in general, the usual range of hardness numbers, from 50/55 to 75/80, corresponds to a stiffness range of about 3:1, a useful degree of flexibility for the designer).

Equation (11.1) shows that with this revised coupling, stiffness n_c changes from 1573 to 1614 cycles/min, and this should be acceptable. The oscillatory torque generated at the critical speed is increased by the two factors mentioned above, but reduced to some extent by the lower dynamic magnifier for the harder rubber, $M = 8.6$ against $M = 10.5$. As before, prolonged running at the critical speed should be avoided.

For completeness, we should check the whirling speed from equation (11.11). The mass of the shaft per unit length is $W_s = 9.80$ kg/m:

$$N_w = \frac{30\pi}{0.50^2} \sqrt{\frac{200 \times 10^9 \times \pi \times 0.04^4}{64 \times 9.80}} = 19,100 \text{ rpm}$$

The mass of the shaft + half-couplings is found to be 12 kg and the combined radial stiffness 33.6 MN/m. From equation (11.12a):

$$N_t = \frac{30}{\pi} \sqrt{\frac{33.6 \times 10^6}{12}} = 16,000 \text{ rpm}$$

then from equation (11.12b), whirling speed = 12,300 rev/min, which is satisfactory.

Note, however, that if shaft length were increased from 500 to 750 mm, whirling speed would be reduced to about 7300 rev/min, which is barely acceptable. This is a common problem, usually dealt with by the use of tubular shafts, which have much greater transverse stiffness for a given mass.

There is no safe alternative, when confronted with an engine of which the characteristics differ significantly from any other run previously on a given test bed, to following through this design procedure.

An Alternative Solution

The above worked example makes use of two multiple-bush-type rubber couplings with a solid intermediate shaft. An alternative is to make use of a conventional propeller shaft with two universal joints with the addition of a coupling incorporating an annular rubber element in shear to give the necessary torsional flexibility. These couplings, shown in Figure 11.12, are generally softer than the multiple-bush type for a given torque capacity, but are less tolerant of operation near a critical speed. If it is decided to use a conventional universal joint shaft, the supplier should be informed of the maximum speed at which it is to run. This will probably be much higher than is usual in the vehicle and may call for tighter than usual limits on balance of the shaft.

Shock Loading of Couplings Due to Cranking, Irregular Running, and Torque Reversal

Systems for starting and cranking engines are described later in this chapter, where it is emphasized that during engine starting severe transient torques can arise. These have been known to result in the failure of flexible couplings of apparently adequate torque capacity. The maximum torque that may be necessary to get

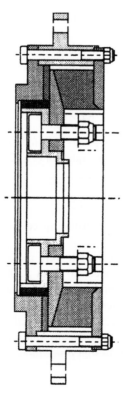

FIGURE 11.12   Annular-type rubber coupling.

a green engine over top dead center (TDC) or that can be generated at first firing should be estimated and checked against maximum coupling capacity.

Irregular running or imbalance between the powers generated by individual cylinders can create exciting torque harmonics of lower order than expected in a multicylinder engine and should be remembered as a possible source of rough running. Finally, there is the possibility of momentary torque reversal when the engine comes to rest on shutdown.

However, the most serious problems associated with the starting process arise when the engine first fires. Particularly when, as is common practice, the engine is motored to prime the fuel injection pump, the first firing impulses can cause severe shocks. Annular-type rubber couplings (Figure 11.12) can fail by shearing under these conditions: in some cases it is necessary to fit a torque limiter or slipping clutch to deal with this problem.

Axial Shock Loading

Engine test systems that incorporate automatic shaft docking systems have to provide for the axial loads on both the engine and dynamometer imposed by such a system. In some high-volume production facilities an intermediate pedestal bearing isolates the dynamometer from both the axial loads of normal docking operation and for cases when the docking splines jam "nose to nose"; in these cases the docking control system should be programmed to back off the engine, spin the dynamometer, and retry the docking.

Selection of Coupling Torque Capacity

Initial selection is based on the maximum rated torque with consideration given to the type of engine and number of cylinders, dynamometer characteristics, and inertia. Table 11.4 shows recommended service factors for a range of engine and dynamometer combinations. The rated torque multiplied by the service factor must not exceed the permitted maximum torque of the coupling.

Other manufacturers may adopt different rating conventions, but Table 11.4 gives valuable guidance as to the degree of severity of operation for different situations. Thus, for example, a single-cylinder diesel engine coupled to a DC machine with dynamometer start calls for a margin of capacity three times as great as an eight-cylinder gasoline engine coupled to a hydraulic dynamometer.

Figure 11.13 shows the approximate range of torsional stiffness associated with three types of flexible coupling: the annular type as illustrated in Figure 11.12, the multiple-bush design of Figure 11.10, and a development of

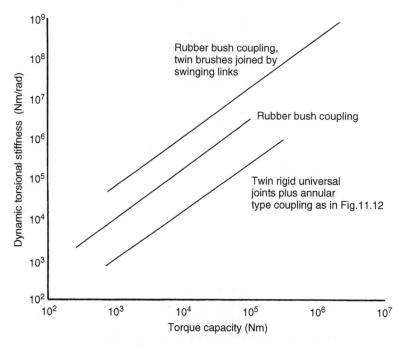

FIGURE 11.13 Ranges of torsional stiffness for different types of rubber coupling.

the multiple-bush design which permits a greater degree of misalignment and makes use of double-ended bushed links between the two halves of the coupling. The stiffness values in Figure 11.13 refer to a single coupling.

The Role of the Engine Clutch

Vehicle engines in service are invariably fitted with a flywheel-mounted clutch and this may or may not be retained on the test bed. The advantage of retaining the clutch is that it acts as a torque limiter under shock or torsional vibration conditions. The disadvantages are that it may creep, particularly when torsional vibration is present, leading to ambiguities in power measurement. When the clutch is retained a dummy (no internal gearing) gearbox is usually used to provide an outboard bearing. Clutch disk springs may have limited life under test-bed conditions.

Balancing of Driveline Components

This is a matter that is often not taken sufficiently seriously and can lead to a range of troubles, including damage (which can be very puzzling) to bearings, unsatisfactory performance of such items as torque transducers, transmitted vibration to unexpected locations, and serious driveline failures. It has sometimes been found necessary to carry out in situ balancing of a composite engine driveline where the sum of the out-of-balance forces in a particular radial relationship causes unacceptable vibration; specialist companies with mobile plant exist to provide this service.

As already stated, conventional universal joint-type cardan shafts are often required to run at higher speeds in test-bed applications than is usual in vehicles; when ordering, the maximum speed should be specified and, possibly, a more precise level of balancing than standard specified.

BS 5265 [10] gives a valuable discussion of the subject and specifies 11 balance quality grades. Driveline components should generally be balanced to grade G 6.3 or, for high speeds, to grade G 2.5. The Standard gives a logarithmic plot of the permissible displacement of the center of gravity of the rotating assembly from the geometrical axis against speed. To give an idea of the magnitudes involved, G 6.3 corresponds to a displacement of 0.06 mm at 1000 rev/min, falling to 0.01 mm at 5000 rev/min.

Alignment of Engine and Dynamometer

This appears to be simple but is a fairly complex and quite important matter. For a full treatment and description of alignment techniques, see Ref. [4]. Differential thermal growth, and the movement of the engine on its flexible mountings when on load, should be taken into account and if possible the mean position should be determined. The laser-based alignment systems now available greatly reduce the effort and skill required to achieve satisfactory levels of accuracy. In particular, they are able to bridge large gaps between flanges without any compensation for droop and deflection of arms carrying dial indicators, a considerable problem with conventional alignment methods.

There are essentially three types of shaft having different alignment requirements to be considered:

- Rubber bush and flexible disk couplings should be aligned as accurately as possible as any misalignment encourages heating of the elements and fatigue damage.
- Gear-type couplings require a minimum misalignment of about 0.04° to encourage the maintenance of an adequate lubricant film between the teeth.
- Most manufacturers of universal joint propeller shafts recommend a small degree of misalignment to prevent brinelling of the universal joint needle rollers. Note that it is essential, in order to avoid induced torsional oscillations, that the two yokes of the intermediate shaft joints should lie in the same plane.

Unless the shaft used has a splined center section allowing length change the distance between end flanges will be critical, as incorrect positioning can lead to the imposition of axial loads on bearings of engine or dynamometer.

Guarding of Coupling Shafts

Not only is the failure of a dynamometer/UUT coupling shaft potentially dangerous, as very large amounts of energy can be released, it is not particularly uncommon.

Under the latest machinery directives the shaft guard is considered as a secondary protection device, with the cell doors being the primary protection.

The ideal shaft guard will contain the debris resulting from a failure of any part of the driveline and prevent accidental or casual contact with rotating parts, while being sufficiently light and adaptable not to interfere with engine rigging and alignment.

A guard system that is very inconvenient to use will eventually be misused or neglected.

A substantial guard, preferably made from a steel tube not less than 6 mm thick, split and hinged in the horizontal plane for shaft access, is an essential precaution. The hinged "lid" should be interlocked with the emergency stop circuit to prevent engine cranking or running while it is open. Many guard designs include shaft restraint devices, made of wood or a nonmetallic composite, that loosely fit around the tubular portion, and are intended to prevent a failing shaft from whirling. These restraints should not be so close to the shaft as to be touched by it during normal starting or running, otherwise they will be the cause of failure rather than a prevention of damage.

Engine-to-Dynamometer Coupling: Summary of Design Procedure

1. Establish speed range and torque characteristic of the engine to be tested. Is it proposed to run the engine on load throughout this range?
2. Make a preliminary selection of a suitable drive shaft. Check that maximum permitted speed is adequate. Check drive shaft stresses and specify material. Look into possible stress raisers.
3. Check manufacturer's recommendations regarding load factor and other limitations.
4. Establish rotational inertias of engine and dynamometer, and stiffness of proposed shaft and coupling assembly. Make a preliminary calculation of torsional critical speed from equation (11.1) (in the case of large multicylinder engines, consider making a complete analysis of torsional behavior).
5. Modify specification of shaft components as necessary to position torsional critical speeds suitably. If necessary, consider use of viscous torsional dampers.
6. Calculate vibratory torques at critical speeds and check against capacity of shaft components. If necessary, specify "no go" areas for speed and load.
7. Check whirling speeds.
8. Specify alignment requirements.
9. Design shaft guards.

PART 3. FLYWHEELS AND ENGINE STARTING SYSTEMS

No treatment of the engine–dynamometer driveline would be complete without mention of flywheels, which may form a discrete part of the shaft system.

A flywheel is a device that stores kinetic energy. The energy stored may be precisely calculated and is instantly available. The storage capacity is directly related to the mass moment of inertia, which is calculated by:

$$I = k \times M \times R^2$$

where I = moment of inertia (kgm^2), k = inertia constant (dependent upon shape), m = mass (kg), and R = radius of flywheel mass.

In the case of a flywheel taking the form of a uniform disk, which is the common form found within dynamometer cells and chassis dynamometer designs:

$$I = \frac{1}{2}MR^2$$

The engine or vehicle test engineer may expect to deal with flywheels in the following roles:

1. As part of the UUT, as in the common case of an engine flywheel, where it forms part of the engine–dynamometer shaft system and contributes to

the system's inertial masses taken into account during a torsional analysis.

2. As part of a kinetic energy capture and recovery system (KERS), wherein the braking energy of the vehicle is used to store and then release energy within an encased flywheel system as in the Flybrid® CFT KERS.

3. As part of the test equipment where one or more flywheels may be used to provide actual inertia that would, in "real life", be that of the vehicle or some part of the engine-driven system.

Flywheel Safety Issues

The uncontrolled discharge of energy from any storage device is hazardous. The classic case of a flywheel failing by bursting is now exceptionally rare and invariably due to incompetent modification rather than the nineteenth century problems of poor materials, poor design, or overspeeding.

In the case of engine flywheels, the potential source of danger in the test cell is that the shaft fixed on the flywheel may be quite different in mass and fixing detail from its final application connection; this can and has caused overload leading to failure. Cases are on record where shock loading caused by connecting shafts touching the guard system due to excessive engine movement has created shock loads that have led to the cast engine flywheel fracturing, with severe consequential damage.

The most common hazard of test-rig-mounted flywheels is caused by bearing or clutch failure, where consequential damage is exacerbated by the considerable energy that is available to fracture connected devices, and because of the length of time that the flywheel and connected devices will rotate before the stored energy is dissipated and movement is stopped.

It is vital that flywheels are guarded in such a manner as to prevent absolutely accidental entrainment of clothing or cables, etc.

A common and easy to comprehend use of flywheels is as part of a vehicle brake testing rig. In these machines flywheels supply the energy that has to be absorbed and dissipated by the brake system under test; the rig motor is only used to accelerate the chosen flywheel combinations up to the rotational speed required to simulate the vehicle axle speed at the chosen vehicle speed. Flywheel brake rigs have been made up to the size that can provide the same kinetic energy as fully loaded high-speed trains.

Flywheels are also used on rigs used to test all types of transmission units and automatic automotive gearboxes.

Test-rig flywheel sets need to be rigidly and securely mounted and balanced to the highest practical standard. Multiples of flywheels forming a common system that can be engaged in different combinations and in any radial relationship require particular care in the design of both their base frame and individual bearing supports. Such systems can produce virtually infinite combinations of shaft balance and require each individual mass to be as well balanced and aligned on as rigid a base as possible.

Simulation of Inertia[1] Versus Iron Inertia

Modern AC dynamometer systems and control software have significantly replaced the use of flywheels in chassis and engine dynamometer systems in the automotive industry. Any real shortcoming in the speed of response or accuracy of the simulation is usually considered, particularly in chassis dynamometers, to be of less concern than the mechanical simplicity of the electric dynamometer system and the reduction in required cell space. Transmission test rigs still tend to benefit from the inclusion of flywheels at the nondrive end of their wheel dynamometers where they can be easily changed or removed.

Finally, it should be remembered that, unless engine rig flywheels are able to be engaged through a clutch onto a rotating shaft, the engine starting/cranking system will have to be capable of accelerating engine, dynamometer, and flywheel mass up to engine start speed.

Engine Cranking and Starting in a Test Cell

Starting an engine when it is connected to a dynamometer may present the cell operator with a number of rigging problems, and is a factor to be remembered when selecting the dynamometer. If the engine is fitted with a starter motor the cell system must provide the high current DC supply and associated switching; in the absence of an engine-mounted starter a complete system to start and crank the engine must be available that compromises neither the torsional characteristics nor the torque measurement accuracy.

Engine Cranking: No Starter Motor

The cell cranking system must be capable of accelerating the engine to its normal starting speed and of disengaging when the engine fires. An AC or DC (four-quadrant) dynamometer, suitably controlled, will be capable of starting the engine directly.

The power available from any four-quadrant machine will always be greater than that required; therefore, excessive starting torque must be avoided by the engine controller's alarm system, otherwise an engine locked by seizure or fluid in a cylinder will cause severe damage to the driveline.

The preferred method of providing other types of dynamometer with a starting system is to mount an electric motor at the non-engine end of the dynamometer shaft, driving through an over-running or remotely engaged

---

1. Some readers may object to the phrase "simulation of inertia" since one is simulating the effects rather than the attribute, but the concept has wide industrial acceptance.

clutch, and generally through a speed-reducing belt drive. The clutch half containing the mechanism should be on the input side, otherwise it will be affected by the torsional vibrations usually experienced by dynamometer shafts. The motor may be mounted above, below, or alongside the dynamometer to save cell length.

The sizing of the motor must take into account the maximum breakaway torque expected, usually estimated as twice the average cranking torque, while the normal running speed of the motor should correspond to the desired cranking speed. The choice of motor and associated starter must take into account the maximum number of starts per hour that may be required, both in normal use and when dealing with a faulty engine. The running regime of the starter motor can be demanding, involving repeated bursts at overload, with the intervening time at rest, and an independent cooling fan may be necessary.

Some modern diesel engines, when green,[2] require cranking at more than the normal starting speed, sometimes as high as 1200 rev/min, in order to prime the fuel system. In such cases a two-speed or fully variable speed starter motor may be necessary.

The system must be designed to impose the minimum parasitic torque when disengaged, since this torque will not be sensed by the dynamometer measuring system.

In some cases, to avoid this source of inaccuracy, the motor may be mounted directly on the dynamometer carcase and permanently coupled to the dynamometer shaft by a belt drive. This imposes an additional load on the trunnion bearings, which may lead to brinelling, and it also increases the effective moment of inertia of the dynamometer. However, this design has worked well in practice and has the advantage that, providing the combined machine is set up correctly with no residual load on the load cell, motoring and starting torque may be measured by the dynamometer system.

An alternative solution is to use a standard vehicle engine starter motor in conjunction with a gear ring carried by a "dummy flywheel" incorporated in the driveline, but this may have the disadvantage of complicating the torsional behavior of the system.

Engine-Mounted Starter Systems

It must be remembered that the standard engine-fitted starter system may be inadequate to accelerate the whole dynamometer system and the engine up to starting speed, particularly when connected to a "green" engine that may exhibit a very high breakaway torque and require prolonged cranking at high speed to prime the fuel system before it fires.

---

2. A green engine is one that has never been run. The rubbing surfaces may be dry; therefore, the breakaway torque will be higher than for a previously run engine, the fuel system may need priming, and there is always the possibility that it is faulty and incapable of starting.

However, if the engine is rigged with an adequate starter motor, all that has to be provided is the necessary 12 V or 24 V supply and an electrical relay system operated from the control desk. The traditional approach has been to locate a suitable battery as close as possible to the starter motor, with a suitable battery charger supply. This system is not ideal, as the battery needs to be in a suitably ventilated box to avoid the risk of accidental shorting, and will take up valuable space. Batteries can be heavy and handling them in confined cell spaces may be considered as an H&S risk not worth taking.

Special transformer/rectifier units designed to replace batteries for this duty are on the market. They will include an "electrical services box" to provide additional power for ignition systems and diesel glow plugs. In large integrated systems there may be a 12- or 24-volt bus bar system for the DC supplies.

Non-Electrical Starting Systems

Diesel engines, larger than the automotive range, are usually started by means of compressed air, which is admitted to the cylinders by way of starting valves. In some cases it is necessary to move the crankshaft to the correct starting position by opening cylinder valves to prevent compression then either by barring using holes in the flywheel or using an engine-mounted inching motor. The test facility should include a compressor and a receiver of capacity at least as large as that recommended for the engine in service.

Compressed air or hydraulic motors are sometimes used instead of electric motors to provide cranking power, but have no obvious advantages over a DC electric motor, apart from a marginally reduced fire risk in the case of a compressed air system, provided the supply is shut off automatically in the case of fire.

NOTATION

Frequency of torsional oscillation	n (cycles/min)
Critical frequency of torsional oscillation	n_c (cycles/min)
Stiffness of coupling shaft	C_s (Nm/rad)
Rotational inertia of engine	I_e (kgm^2)
Rotational inertia of dynamometer	I_b (kgm^2)
Amplitude of exciting torque	T_x (Nm)
Amplitude of torsional oscillation	θ (rad)
Static deflection of shaft	θ_0 (rad)
Dynamic magnifier	M
Dynamic magnifier at critical frequency	M_c
Order of harmonic component	N_0
Number of cylinders	N_{cyl}
Mean turning moment	M_{mean} (Nm)
Indicated mean effective pressure	p_i (bar)
Cylinder bore	B (mm)
Stroke	S (mm)

Component of tangential effort	T_m (Nm)
Amplitude of vibratory torque	T_v (Nm)
Engine speed corresponding to n_c	N_c (rev/min)
Maximum shear stress in shaft	τ (N/m^2)
Whirling speed of shaft	N_w (rev/min)
Transverse critical frequency	N_t (cycles/min)
Dynamic torsional stiffness of coupling	C_c (Nm/rad)
Damping energy ratio	ψ
Modulus of elasticity of shaft material	E (Pa)
Modulus of rigidity of shaft material	G (Pa)

(for steel, $E = 200 \times 10^9$ Pa, $G = 80 \times 10^9$ Pa)

REFERENCES

[1] UK rigging parts catalog: Demon Tweeks, http://www.demon-tweeks.co.uk; RS Components, http://uk.rs-online.com

[2] J.P. Den Hartog, Mechanical Vibrations, McGraw-Hill, Maidenhead, UK, 1956. ISBN-13: 978-0486647852.

[3] W. Ker Wilson, Vibration Engineering, Griffin, London, 1959. ISBN-13: 978-0852640234.

[4] M.J. Neale, P. Needham, R. Horrell, Couplings and Shaft Alignment, Mechanical Engineering Publications, London, 1998. ISBN-13: 978-1860581700.

[5] Rulebook, Chapter 8. Shaft vibration and alignment. Lloyd's Register of Shipping, London.

[6] W.C. Young, Roark's Formulas for Stress and Strain, McGraw-Hill, New York, 1989. ISBN-13: 978-0070725423.

[7] W.D. Pilkey, Peterson's Stress Concentration Factors, Wiley, New York, 1997. ISBN-13: 978-0470048245.

[8] BS 6613. Methods for Specifying Characteristics of Resilient Shaft Couplings.

[9] W. Ker Wilson, Practical Solution to Torsional Vibration Problems (5 vols), Chapman & Hall, London, 1963. ISBN-13: 978-0412076800.

[10] BS 5265 Parts 2 and 3. Mechanical Balancing of Rotating Bodies.

FURTHER READING

BS 4675 Parts 1 and 2. Mechanical Vibration in Rotating Machinery.

BS 6716. Guide to Properties and Types of Rubber.

BS 6861 Part 1. Method for Determination of Permissible Residual Unbalance.

E.J. Nestorides, A Handbook of Torsional Vibration, Cambridge University Press, 1958. ISBN-13: 978–0521203524.

Test Cell Safety, Control, and Data Acquisition

Engine Testing. DOI: 10.1016/B978-0-08-096949-7.00012-1

PART 1. MACHINERY AND CONTROL SYSTEM SAFETY: STOPPING, STARTING, AND CONTROL IN THE TEST CELL

The safe collection, verification, manipulation, display, storage, and transmission of data is the prime consideration in the design and operation of any test facility. The almost universal adoption of software programmable devices and systems in all industries means there is a commensurate rapid development in worldwide safety regulations covering their use. These create, in 2011, a complex subject, involving overlapping standards and directives, encoded in multiple acronyms.

There are some fundamental concepts common to almost all of the latest safety legislation, since all attempt to be relevant to the age of digital, programmable controls. These concepts are included here because they are highly relevant to anyone integrating complex control systems with programmable devices. This chapter should be read following study of Chapter 5, which covers the electrical power distribution and installation aspects of test facilities.

The following features apply to powertrain test cell systems when they are taken together to form "machinery" under the wide meaning of that word used in legislation.

Essential features of machinery design, integration, and most modern safety legislation are:

• Machinery manufacturers must provide products and systems of inherently safe design and construction.
• Machinery must not start unexpectedly (e.g. key lock switches to protect remote areas).

- Manufacturers and users must take the necessary preventative measures in relation to risks that cannot be eliminated (e.g. interlocked shaft guards).
- Manufacturers and system integrators must inform users of any residual risks (e.g. discharge time of a battery simulator DC bus); therefore, good documentation is vital.
- Particular to systems controlled by programmable devices, hazardous situations must not arise as the result of:
  - reasonably foreseeable human error (misuse)
  - errors in control logic, software, or hardware.
- Hazards, their effects, and the probability of their occurrence have to be quantified by the use of factors such as:
  - component's mean time to failure (MTTF)
  - identification of common cause (of) failure (CCF)
  - severity of injury that might result from the hazard
  - frequency of exposure to the hazard
  - (ease of) possibility of avoiding the hazard.

While the manufacturer is responsible for delivering the safety components and the system integrator is responsible for the system, it must be the responsibility of the user to select the safe operating parameters when integrating the unit under test (UUT). The degree to which the user can be "reasonably" foreseen to misuse a system, before rendering it hazardous, will probably be decided in the future by lawyers equipped with perfect hindsight.

European Machinery Safety Regulations

There are two major pieces of safety legislation covering the European Economic Area:

- The Machinery Directive 2006/42/EC, which covers many qualitative aspects of safety, some appropriate to specific industries, such as ergonomics, hygiene, and environment.
- The "New Machinery Directive" (EN ISO 13849-1), which tends to concentrate on the quantitative categorization of hazards, with particular concentration on machinery under the control of programmable control systems. Depending on the performance levels (PL) identified, the directive leads to the appropriate choice of control system components and architecture.

At the time of writing, there is still a great deal of discussion about the detailed implementation of EN ISO 13849-1, which from the start of 2011 has replaced the machinery directive EN 954-1 under which test cells had been designed and built up to and during 2010. There is also some current confusion concerning the status, under legislation, of levels of modifications carried out to test

TABLE 12.1 Legislative Safety Levels to the Two Different European Standards

Performance Level (PL) ISO 13849-1	Average Probability of a Failure to a Danger Level per Hour ISO 13849-1	EN/IEC 62061 Safety Integration Level (SIL)
a	$\geq 10^{-5}$ to $<10^{-4}$	
b	$\geq 3 \times 10^{-6}$ to $< 10^{-5}$	1
c	$\geq 10^{-6}$ to $< 3 \times 10^{-6}$	1
d	$\geq 10^{-7}$ to 10^{-6}	2
e	$\geq 10^{-8}$ to 10^{-7}	3

facilities that were built under older legislation, that have been running for some years and under established procedures.

Much of the detail of the directive affects the test plant system manufacturer, supplier, and system integrator rather than the test engineers that uses the plant, although the latter bear the heavy responsibility of understanding the safe operation of the system and are advised not to make unauthorized modifications to it that materially affect its (safe) operation.

The five performance level ratings (PLr) a, b, c, d, and e, defined in ISO 13849, together with the equivalent system integration levels (SIL) defined in IEC 62061, are shown in Table 12.1. The major test plant suppliers in Europe are generally working to PL(d) and SIL 2, but it is the responsibility of the systems integrator to ensure that the control system(s) are compliant with the European Harmonized EN ISO 13849-1 safety standard.

The new legislation is tending to move designers towards the use of standardized system and subsystem architecture using components that meet or exceed the PL levels required. Component catalogs are now listing such features as mean time to dangerous failure (MTTFd) for items used in automated (test) systems in a form such as:

[Device name]: MTTFd: 50 years, DC: medium, Cat 3, PLe.

As a final note it should be remembered that, according to recent reports, 50% of industrial accidents are caused by operator error or "human behavior"; therefore, *training* in the operation and maintenance of powertrain test equipment is vital, even after safety legislation has rendered it as safely designed and constructed as is reasonably possible.

US Machinery Safety Regulations

In the USA the Occupational Safety and Health Administration (OSHA) is the federal organization that oversees industrial safety. However, any test facility project manager will also need to understand the local requirements of the National Fire Protection Association (NFPA) standards and most major automotive original equipment manufacturers (OEMs) have their own regulations specific to powertrain testing.

The OSHA publishes regulations in "Title 29 of the code of Federal Regulation" (29 CFR)[1] and standards covering industrial machinery are in Part 1910. Standards such as 29CFR 1910.147, Subpart J, which covers the rules of locking out machines during maintenance, or Subpart O, which covers guarding, are mandatory. Many OSHA standards have equivalent ANSI standards and while some standards are voluntary, or advisory, local fire chiefs or responsible officials may insist that they be met. Once again, early contact with local officialdom is the best way of avoiding problems later in a modification or build project.

Emergency Stop Function

Don't Start Something You Can't Stop.

The emergency stop (EM stop) function is, in older hard-wired[2] systems, sometimes quite separate from any automatic alarm monitoring system embedded into the cell control system.

Most emergency stop systems may be visualized as a chain of switches and relays in series, all of which need to be closed before engine ignition or some other significant process can be initiated. When the chain is opened, normally by use of manually operated press buttons or relays linked to an oil pressure switch, the process is shut down and the operated button is latched open until manually reset. This feature allows staff working on any part of the test cell complex to use a safety procedure based on using an EM stop station as a self-protection device by pressing in an EM stop button and hanging a "Do Not Start" notice on it. The notice has to bear the time of application and name of the user; in normal circumstances only the person named may remove it.

In modern systems built to the New Machinery Directive, the EM stop chain will probably be operated through a specially constructed safety PLC connected to double-pole switches and relays all built to Cat 3, PLe standards, and the wiring scheme will follow an architecture that meets the same performance level.

EN ISO 13849-1 requires that all field-mounted devices in the EM stop chain, including push buttons, shaft guard limit switches, cell door limit switches, etc., need to be double-pole (Cat 3) devices.

---

1. Standards may be downloaded from www.osha.gov.

2. The term "hard wired" in this context means that the instrument or device is connected, via dedicated cables directly with switches, relays or instruments in its discrete circuit.

In general, the EM stop system is initiated by humans rather than devices; the exceptions are safety critical systems such as the fire alarm system, which often have a hard-wired safety relay connection in the EM stop circuit. This particular "fire alarm" feature is not always fitted or desirable because, provided the cell is fully staffed, the damage to a valuable prototype engine, through the rapid sequence of events that would result from the automatic, and possibly spurious, triggering of a fire alarm-initiated shutdown, can be avoided by the more appropriate actions of trained operators.

The traditional result of triggering the emergency stop circuit was to shut off electrical power from all devices directly associated with the running of the test cell. The major perceived risk was seen as electrical shock. In this model dynamometers are switched to a "de-energized" condition, fuel supply systems are shut off, ventilation fans are shut down, but services shared with other cells are not interrupted and fire dampers are not closed.

Modern practice, particularly for cells fitted with electrical four-quadrant dynamometers, is for the EM stop function to shut off engine fueling and ignition; shaft rotation is then stopped through the controlled application of load, using a pre-programmed "fast stop" sequence. Power to the dynamometer is cut when the zero-speed signal is recorded.

BS EN 60204-1: 2006 (Safety of Machinery) recognizes the following categories of emergency stop:

- *Uncontrolled stop (removal of power): Category 0.* This is the traditional function commonly used in engine cells fitted with manual control systems and two quadrant dynamometers.
- *Controlled shutdown (fast stop with power removed at rotation stop or after timed period): Category 1.* This is the normal procedure for engine cells fitted with four-quadrant dynamometers where, under some circumstances, shutting down the engine fueling may not stop rotation (see "runaway" engines below).
- *Controlled stop (fast stop with power retained at rotation stop): Category 2.* Normally operated as a "fast stop" button rather than an "EM stop" button in modern test cell practice (see below).

The control desk area may be fitted with the following "emergency action" buttons:

1. Large red emergency stop button (latched). One will be desk mounted, but there may be at least two repeater buttons mounted on the cell wall and on any auxiliary control box.
2. Fast stop button (hinged cover to prevent accidental operation) mounted only on the control desk.
3. Fire suppression release button. Wall mounted and possibly break-glass protected.

It should be clear to all readers that operational training is required to enable operators to distinguish between situations and to use the most appropriate emergency control device.

A rare and sometimes confusing emergency that can occur when testing diesel engines is when the engine "runs away". In this uncontrolled state the engine will not respond to shutting of the fuel rack as it is being fueled from its own lubrication oil. This can happen if the sump has been overfilled or seals in the turbocharger fail; engines with worn bores and those built in a horizontal configuration can be particularly susceptible. When dealing with a runaway engine, hitting a Category 0 EM stop is completely the wrong action; cutting the fuel supply will have little effect and removing load from the dynamometer will allow the engine to accelerate to a destructive overspeed. The engine must be stalled by applying load with an automated fast stop function or under manual control of the dynamometer. To recognize the risk of such events and avoiding them means that the training of staff, particularly those running with worn or partially rebuilt diesel engines, is important.

It is important to use the *safety interaction matrix*, introduced in Chapter 5, to record what equipment will be switched, to what state, in the event of any of the many alarm or shutdown states being monitored. There can be no one rule that fits all cases; the risk assessment process that determines the safety actions matrix for a test facility must differentiate between the situations of, for example: a remotely monitored transmission endurance cell and an F1 powertrain cell manned by several highly trained engineers.

Everything possible should be done to avoid spurious or casual operation of the EM stop system or any other system that initiates cutting off power to cell systems. Careful consideration has to be given before allowing the shutting down of plant that requires long stabilization times such as humidity control for combustion air or oil temperature control heaters. It is also important not to exceed the numbers of starts per hour of motors driving large fans, etc. that would cause overheating of windings and the operation of thermal protection devices.

Hard-Wired Instrumentation in the Test Facility

Hard-wired instrumentation and safety systems, once widely required, are becoming increasingly rare with the availability of digital devices that meet the requirement of safety legislation. Only devices that are part of gas detection and fire suppression systems are now required to be hard-wired, in the latter case using cables compliant with BS 5839.

Power and signal cabling to the several motors built into a test facility need to be marshaled within a motor control panel (MCP) that, for multi-cell facilities, will be large and complex; therefore, it is usually positioned on the services floor with ventilation equipment, whose motors, amongst others, are connected to it. The status of the services operated from the MCP (motorized dampers, fan motors, cooling water pumps, etc.) needs to be visible to the test cell operator(s) either by a simple indicator lamp board or through a mimic panel being part of a building management system (BMS).

Computerized Monitoring of Test Cell and UUT Alarm Signals

All computer-based, test cell control systems allow the safe operating envelope for both test cell equipment and the UUT to be defined and protected by user-defined alarms. Alarms can be programmed in many ways. Their setting and modification may be protected by a hierarchy of passwords, but the default form is usually based on the following alarm levels:

- High-level warning—no automatically initiated action
- High-level test shutdown or other automatic remedial action
- Low-level warning—no automatically initiated action
- Low-level test shutdown or other automatic remedial action.

Where alarm channels are inappropriate to a measurement (e.g. low coolant temperature = shutdown), these alarm levels can be switched off or assigned values outside the operating range.

Primary alarms, often high-level shutdowns, that are protected by the highest password level access, should refer to limits of performance of the installed test plant, rather than the UUT, whose performance limits should be lower. These test plant limits, if exceeded, would endanger the integrity of facility equipment and therefore personnel.

Levels set to protect the UUT may be global, meaning they are active throughout the whole test or test stage specific. The global alarms usually protect the integrity of the whole test system while the stage alarms may protect either the UUT or the validity of the test data. Thus, it might be necessary during a particular test sequence to hold the engine's coolant temperature between close limits while a series of secondary measurements are made; by setting the high and low warning alarm limits to those limits the validity of the test conditions is audited.

Some test sequence editors will allow alarm actions to be set by Boolean logic statements; they may also allow the sensitivity (time-lag) to be changed to avoid false alarms caused by short-duration peaks.

Most engine test software contains a rolling buffer that allows examination of the values of all monitored channels for a selectable time before any alarm shutdown. Known as "historic" or "dying seconds" logging, this buffer will also record the order in which the alarm channels have triggered, information of help in tracing the location of the prime cause of a failure.

Security of Electrical Supply

The possible consequences of a failure of the electricity supply, or of such events as individual subcircuit fuse failures, must be carefully thought out and recorded in the safety matrix. The safe shutdown of rotating equipment in the case of a mains electrical supply failure needs to be considered carefully, since the dynamometers will have lost the ability to absorb torque and will have effectively become flywheels.

Equally, an automatic or unsafe restart after power is restored following a mains supply failure is particularly dangerous and must be inhibited by system design, the supply switching and by operator training.

It is usual to design safety systems as normally de-energized (energize for operation), but all aspects of cell operation must be considered. Best design practice of cell control logic and now legislation absolutely requires that equipment, such as solenoid-operated fuel valves at the cell entry, must be reset after mains failure through an orderly, operator initiated, restart procedure.

The installation of an uninterrupted power supply (UPS) for critical control and data acquisition devices is to be recommended. This relatively inexpensive piece of equipment needs to provide power for the time required to save or transfer computer data.

Building Management Systems (BMS) and Services Status Displays

Test cell function is usually totally dependent on the operation of a number of building and site services, remote and hidden from the cell operator. A failure of one of these services may directly or indirectly (through alarm state caused by secondary effects) result in a cell shutdown. The common and time-wasting alternative to the display of the service's running status is for operators to tour the facility, checking the status of individual pieces of plant at start-up, after a fault condition and shutdown. As stated elsewhere, a centralized control and indication panel for services is of great value; however, such system status displays must indicate the actual rather than the control demand status.[3]

A modern BMS can have an interactive mimic display showing not only the running state of services, but also the status of fire dampers, valves, and the temperature and pressure of fluids. These PLC-based[4] systems may have the

---

3. The Three Mile Island nuclear power station accident, initiated by a failure of the main feed water pumps, was greatly exacerbated by incorrect actions taken by operations staff whose only indications showed a critical valve as being closed, but in fact this only represented that the signal to close the valve was sent; in fact, it was open. A "positive feedback" lamp in the control room indicating the true position of the valve had been eliminated in original construction to save time and money.

4. Programmable logic controller (PLC) devices now have control capabilities indistinguishable from those of many powerful PCs, so differentiation is difficult. In this volume the term is reserved for devices that run fixed control routines and are normally programmed with ladder-type logic diagrams.

complete start, stop, and alarm state sequence logic programmed into them, but cell designers must carefully consider the interaction of the BMS and the cell controller. All alarm states triggered by one system must have the correct (safe) response within the other. It is particularly important to avoid test shut-downs being triggered by trivial or unrelated fault conditions detected by the BMS. It is equally important that the building services local to a cell are controlled appropriately to any fault in the cell; for example, a high hydrocarbon alarm has to switch the affected cell ventilation to maximum purge, rather than shutting ventilation down.

Test Cell Start-Up and Shutdown Procedures

Failure of any test cell service, or failure to bring the services on stream in the correct order, can have serious consequences and it is essential to devise and operate a logical system for start-up and shutdown. In all but highly automated production tests a visual inspection of the UUT must be made before each test run. Long-established test procedures can create complacency and lead to accidents. When investigating cell accidents, the author has noted the frequency with which the phrase "I assumed that …" arises, as in "I assumed that the night shift had reconnected the drive shaft".

Visual inspection beats assumption every time.

Checks Before Start-Up

On the basis that routine calibration procedures have been followed, and that instrument calibrations are correct and valid, then the following type of pre-run checks should be made or status confirmed:

- No "work in progress" or "lockout" labels are attached to any system switches.
- Engine/dynamometer alignment within set limits and each shaft bolt has been tightened to the correct torque and the fact recorded and signed for in the control desk diary or appropriate record.
- Shaft guard in place and centered so that no contact with shaft is possible (if appropriate, it is a good practice to rock the engine on its mounts by hand to see that the rigging system, including exhaust tubing, is secure and flexes correctly).
- All loose tools, bolts, etc., removed from the test bed.
- Engine support system tightened down.
- Fuel system connected and leak proof.
- Engine oil at the correct level.
- The valves for all fluid services are in the correct position for running.
- Fire suppression system is indicating a "healthy" status.
- Ventilation system available and switched on to purge cell of flammable vapor before start-up of an engine.
- Check that any cell access doors, remote from the control desk, have warnings set to deter casual entry during test.

The engine start process is then initiated.

Checks Immediately After Start-Up

- Oil pressure is above trip setting. In the case of manually operated cells the alarm override is released when pressure is achieved.
- If stable idle is possible and access around the engine is safe, a quick in-cell inspection should be made. In particular, look for engine fuel or oil leaks, cables or pipes chafing or being blown against the exhaust system and listen for abnormal noises.
- In the case of a run scheduled for prolonged unmanned running, test the emergency shutdown system by operating any one of the stations.
- Restart and run the test.

Checks Immediately After Shutdown

- Allow a cooling period with services and ventilation left on.
- Shut off fuel system as dictated by site regulations (draining down of day-tanks, etc.).
- Carry out data saving, transmission, and backup procedures.

PART 2. OPEN- AND CLOSED-LOOP CONTROL OF ENGINE AND DYNAMOMETER

True open-loop control of a modern engine is an arguable concept because they start and run under the control of their ECU, which, for most of the time, is a closed-loop controller.

Open-loop control of the rotational speed and load applied to transmission test rigs is possible in manual control mode, during which primary alarms should still be active. If the automatic control of speed or load is poor, the best strategy in investigating instabilities is to eliminate as much of the control system as possible by switching to open-loop control where possible. In this way the source of the disturbing influence or the interaction of two systems may be identified.

The control software programmed in a modern engine ECU and electric dynamometers makes wide use of ROM-based look-up tables rather than (sometimes additional to) complex control algorithms. However, the controllers of many single- and two-quadrant dynamometers and auxiliary devices, such as coolant temperature controllers, use proportional-integral-derivative (PID) feedback loop-based control (Figure 12.1).

There are may web-based tools and books on the subject of PID tuning and no space here to go into detail; it is sufficient to record that the roles of each of the terms expressed mathematically in Figure 12.1 are as follows:

- Proportional gain effect—the controller output is proportional to the error or a change in measurement.

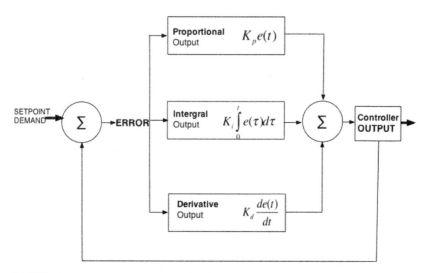

FIGURE 12.1 PID feedback control diagram. Tuning parameters are: K_d, derivative gain; K_i, integral gain; K_p, proportional gain; SP − CV, set point control variable minus the measured value; t, time; Σ, error; τ, dummy integration variable.

- Integral gain effect—the controller output is proportional to the amount of time the error is present. Integral action tends to eliminate offset from the set point.
- Derivative gain effect—the controller output is proportional to the rate of change of the measurement or error.

Up to the early 1980s, analog PID controllers were based on electronic hardware containing one miniaturized, variable-resistor "pot" per gain term. The tuning of these devices was somewhat of a "black art" because there was no visual or numerical indication of what value of gain was being used, other than the observed effects on the rotating machinery when the pots were adjusted (see Table 12.2). Nowadays PI and PID controllers are almost universally based on digital computers and PLCs that present to the operator a number, albeit a rather arbitrary number, relating to the channel gain and thus the size of changes in value being applied may be judged.

Problems in Achieving Control

Some engines are inherently difficult to control because of the shape of their torque–speed characteristics. Highly turbocharged engines may have abrupt changes in the slope of the power–speed curve, others may have sluggish response to step demand changes due, for example, to the time taken for speed changes in the turbocharger; such characteristics can make the optimization of a PID-based control system, which has to operate over the full performance envelope, very difficult.

TABLE 12.2 The Effects, on the Control of Steady-State and Step Demand Performance, of Increasing Any One PID Parameter Independently

Parameter	Stability	Overshoot	Rise Time	Settling Time	Steady-State Error
K_p	Degrades	Increases	Decreases	Little effect	Decreases
K_i	Degrades	Increases	Decreases	Increases	Large decrease
K_d	Improves	Small decrease	Small decrease	Small decrease	No effect

Inexperienced attempts to adjust a full three-term PID controller can lead to problems; it is well worth recording the settings before starting to make changes so that one can at least return to the starting point. It is good practice to make sure that elementary sources of trouble, such as mechanical backlash and "stick-slip" in control linkages, are eliminated before delving into control variables in software.

Sinusoidal speed signals caused by poor encoder installation or intermittent speed signals caused by the wrong pickup clearance at the engine flywheel are frequently the cause of poor engine/dynamometer control. A speed encoder must be accurately aligned and fitted with a coupling that does not impose lateral loads on the instrument's shaft.

While the development of complex, adaptive, and model-based controllers will continue, the fundamental requirements such as good mechanical and noise-free electrical connections still have to be observed.

The Test Sequence: Modes of Prime-Mover/Dynamometer Control

A test program for an engine coupled to a dynamometer is, primarily, a sequence of desired values of engine torque and speed. This sequence is achieved by manipulating only two controls: the engine's power output, usually by way of its throttle[5] system, and the dynamometer torque absorption setting.

For any given setting of the throttle, the engine has its own inherent torque–speed characteristic and similarly each dynamometer has its own torque–speed curve for a given control mode setting. The interaction of these two characteristics determines the inherent stability, or otherwise, of the engine–dynamometer combination.

---

5. It is convenient to refer to the throttle, even when it is an electric signal direct to the engine control unit or, in the case of a diesel engine, actuation of the fuel pump rack setting.

TABLE 12.3 Control Mode Nomenclatures

English (first term refers to Engine)	German (first term *may* refer to Dynamometer)
Speed	"n"
Position	"Alpha"
Torque	Torque

Note on Mode Nomenclature

In this book the naming convention of modes of control used in the UK and widely throughout the English-speaking world has been adopted. The alternative nomenclature and word order used in many major German-speaking and mainland European organizations is shown in Table 12.3.

The engine or throttle control may be manipulated in three different ways:

- To maintain a constant throttle opening (position mode)
- To maintain a constant speed (speed mode)
- To maintain a constant torque (torque mode).

The dynamometer control may be manipulated:

- To maintain a constant control setting (position mode)
- To maintain a constant speed (speed mode)
- To maintain a constant torque (torque mode)
- To reproduce a particular torque–speed characteristic (power law mode).

Various combined modes are possible, and in this book they are described in terms of engine-mode/dynamometer-mode.

Position/Position Mode

This describes the classic manually operated engine test. With the engine running the throttle is set to any fixed position from that giving idle to WOT, the dynamometer control similarly set in such a position to absorb the required torque, and the system settles down, hopefully in a stable state. There is no feedback: this is an open-loop system. Figure 12.2a shows a typical combination, in which the engine has a fairly flat torque–speed characteristic at fixed throttle opening, while the dynamometer torque rises with speed. The two characteristics meet at an angle approaching 90° and operation is quite stable.

Certain designs of variable-fill water-brake dynamometer, with simple outlet valve control, may become unstable at light loads; this can lead to hunting, or to the engine running away.

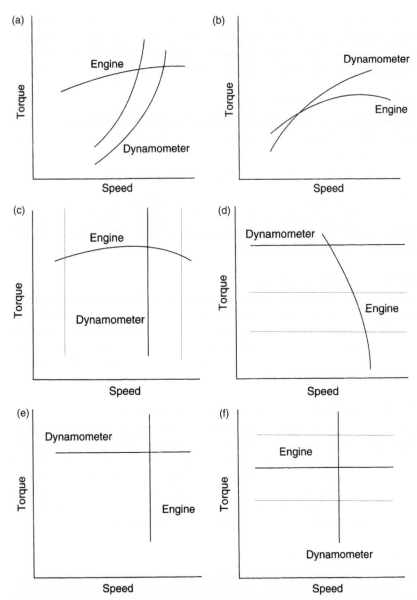

FIGURE 12.2 (a) Position/position mode with stable control of hydraulic or eddy-current dynamometer. (b) Position/position mode with unstable control of hydraulic dynamometer. (c) Position/speed mode point. (d) Position/torque mode testing of a speed governed engine (speed droop characteristic exaggerated). (e) Speed/torque mode. (f) Torque/speed mode. Speed is held constant by the dynamometer and engine throttle controls torque at that speed.

When the two characteristics meet at an acute angle, as shown in Figure 12.2b, good control will not be achieved. Position/position mode is useful in system fault-finding in that the inherent stability of the engine or dynamometer, independent of control system influence, may be checked.

Position and Power Law Mode

This is a control mode that can be visualized as the operator driving a boat; as the throttle is opened, the dynamometer torque rises with a curve speed characteristic of the form:

$$\text{Torque} = \text{Constant} \times \text{Speed}^n$$

When $n = 2$ this approximates to the torque characteristic of a marine propeller and the mode is thus useful when testing marine engines. It is also a safe mode, helping to prevent the engine from running away if the throttle is opened manually by an operator in the cell.

Position and Speed Mode

In this mode the throttle position is set manually but the dynamometer adjusts the torque absorbed to maintain the engine speed constant whatever the throttle position and power output, as shown in Figure 12.2c. As the speed setting line of the dynamometer moves back and forth, the intersection with the engine torque curve gives the operating point. This is a very stable mode, generally used for plotting engine power curves at full and part throttle openings.

Position and Torque Mode (Governed Engines)

Governed engines have a built in torque–speed characteristic, usually slightly "drooping" (speed falling up to 3% as torque increases). They are therefore not suited for coupling to a dynamometer in speed mode; they can, however, be run with a dynamometer in torque mode. In this mode the automatic controller on the dynamometer adjusts the torque absorbed by the machine to a desired value (Figure 12.2d). Control is quite stable. Care must be taken not to set the dynamometer controller to a torque that may stall the engine.

Speed and Torque Mode

This is a useful mode for running in a new engine, when it is essential not to apply too much load. As the internal friction of the engine decreases it tends to develop more power and since the torque is held constant by the dynamometer, the tendency is for the speed to increase; this is sensed by the engine speed controller, which acts to close the throttle, shown in Figure 12.2e.

Torque and Speed Mode

This is an approximate simulation of the performance of a vehicle engine when climbing a hill. The dynamometer holds the speed constant while the engine controller progressively opens the throttle to increase torque at the set speed (Figure 12.2f).

Precautions in Modes of the Four-Quadrant Dynamometer

Where the dynamometer is also capable of generating torque it will be necessary to take precautions in applying some of the above modes of control. Such precautions are built into all reputable commercial systems, but need to be understood by those assembling their own test plant or writing control software. When a four-quadrant dynamometer is in speed mode it is capable of maintaining the set speed even if the engine has failed in some way and has ceased to deliver power. Therefore, when in speed mode such dynamometers testing engines should be working in "absorb only" rather than "motor and absorb" control mode.

Throttle Actuation

Except in cases when an engine throttle lever is directly cable controlled, with a marine type of "tee handle" system at the control desk, engines require the test cell to be equipped with some form of remote and precise throttle, or throttle-pedal unit, actuation, for both manual or automatic control. The actuators may be based on rotary servo-motors, printed circuit motors or linear actuators, and fitted with some form of positional feedback to allow closed-loop control.

Most throttle actuators need be fitted with some form of stroke adjustment to allow for 0% (shut) to 100% (wide-open throttle = WOT) positions to be determined for different engines and linkage configurations. Setting up this precise relationship between the controller's datum of "closed" and WOT with the physical position of the engine's fueling device is an important, safety-critical, rigging task. It must be arranged in such a way that engine movement on its mountings does not induce throttle movement.

Some units will have a force transducer built in that can be used as a force-limiting device or a friction clutch can provide the same protection to the physical system. Typical specification ranges of proprietary devices are:

Constant force applied at Bowden cable connection	100–400 N
Maximum stroke length	100–160 mm
Cable connector speed of travel	0.5–1.5 m/s
Travel resolution and positional repeat accuracy	±0.05 mm
Maximum shifting force	20 N (push and pull).

*It must be remembered that the forces applied by the actuator to the engine lever will depend on the **mechanical leverage** within the engine rigging.*

The specification of positional repeat accuracy (above) requires that all cables and linkages used must be free from backlash or good closed-loop control will not be achieved. An important safety feature of a throttle actuator is that it should "fail safe"; that is, automatically move the engine control lever to the Stop position if there is a power failure.

While entirely possible, it is not common nowadays to drive the electronic throttle controller (ETC) directly as a discrete "analog out" voltage because the ECU requires a number of feedback signals from the ETC in order to function correctly, which is why the complete throttle pedal assembly is invariably used.

PART 3. TEST CONTROL SOFTWARE, CHOOSING SUPPLIER, AND SEQUENCE EDITING

The world's engine test industry is served by a relatively small number of companies whose flagship product is a control system and comprehensive suite of software designed specifically for automating engine, powertrain, transmission and vehicle tests, acquiring test data, and displaying results.

Several of these companies also make the major modules of instrumentation required for testing engines such as dynamometers, engine fluid temperature controllers, fuel meters, and exhaust emission equipment. However, many facilities are made up of modules of plant coming from several of these specialist suppliers, along with locally made building services, all of which have to be integrated into one optimally functioning whole.

The software suites sold by major test equipment suppliers are the outcome of many man-years of work; it is not cost-effective or very sensible for, even a highly specialized user, however experienced, to undertake the task of producing their own test control hardware and software from scratch.

Prospective buyers of automated test systems should, if possible, examine the competitive systems at the sites of other users doing similar work. As with any complex system, the following general questions should be considered, in no order of importance:

- Does the system support ASAM (Standardization of Automation and Measuring Systems) [1] standard interfaces for communication with third-party devices? For engine calibration and many research tasks the ASAP3 communication interface with an engine ECU is vital and for data exchange a modern cell control system must meet the requirements of the ASAM ODS standard, which deals with data modeling, interfaces, physical storage, and data exchange.
- How intuitive and supportive to the work does the software appear to be, or does the software provider understand the user's world in which the software has to function?
- Purchase cost is clearly important, but also consider the cost of ownership, including technical support, upgrades, and additional licenses.
- Can the post-processing and reporting required be carried out directly by the new software or can data be transferred to other packages in use on site?
- How good is the training and service support in the user's geographical location?

- In the case of international companies, does the software support screen displays of the languages required?
- Does the proposed system run on hardware and software platforms that are supported by the company's IT policy? This is a subject that can generate considerable internal heat in organizations. The author has witnessed in the past cases where compliance with a corporate IT diktat, based on policies for use of computers in business systems rather than industrial control, has compromised the choice and increased the price of test cell computer systems.
- Has the system been used successfully in the user's industrial sector?
- Can existing dynamometers, instrumentation, or new third-party instrumentation be integrated with the new system without performance degradation?[6]
- Is data security robust and can the post-processing, display, and archiving required be accommodated?
- Can the data generated by the new system be stored in the existing data format? This "heritage support" may be both important and difficult to achieve.

Editing and Control of Complex Engine Test Sequences

An engine or transmission test program can consist of a succession of stages of speed and torque settings, together with the demand values sent to other devices in the test cell system, the whole format looking rather like a spreadsheet. Modern, highly dynamic test sequences still have to contain the same fundamental speed, torque, analog and digital signals, both in and out, but they more resemble, or can actually be constructed from, a Simulink[7] block diagram.

Software suites designed to semi-automate engine calibration work are capable of calculating and generating the test set points within a test sequence designed to find boundary conditions, such as engine knock.

Most powertrain test profiles or sequences, with their precisely ramped changes in values, are beyond the capacity of a human operator to manually run accurately and repeatably, a fact borne out by the variability seen in the results from human-driven, vehicle emission, drive cycles.

Although the terminology used by test equipment suppliers may vary, the principles involved in building up test profiles will be based on the same essential component instructions.

Test sequence "editors" range from those that require the operator to have knowledge of software code, through those that present the operator with an

---

6. The English idiom "You can't make a silk purse out of a pig's ear" is a valid warning to those facility managers who believe that replacing an old control system with the latest digital controller will always radically improve the physical response of the existing electromechanical system of a previous design generation.

7. MATLAB and Simulink are industrial standard tools in complex mathematical modeling and are produced by MathWorks®.

interactive "form fill" screen, to those where blocks of subsequences are assembled in the required order of running.

The nature and content of the questions shown on the screen will depend on the underlying logic that determines how the answers are interpreted. It is important that the user should understand the test cell logic, otherwise there is a risk of calling up combinations of control that the particular system cannot run, though it may not be able to indicate exactly where the error lies. For example, if the control mode is set for throttle as "position" and dynamometer as "speed", the editor will require instructions in terms of these parameters; in the example it will not be able to accept a throttle control instruction in terms of speed because it requires a percentage figure.

Test Sequence Elements

In those test sequence editors that are based on a simple "form fill" layout, the following elements will be included:

- *Mode of control* (covered earlier in the chapter). As a historical note, most analog-based control systems required engine speed and load to be brought down to some minimum value between any change of test mode, while modern digital controllers are able to make a "bumpless transfer", a term still used.
- *Engine speed and torque or throttle position.* The way in which these parameters are set will depend on the control mode. A good sequence editor will present only viable options.
- *Ramp rate.* This is the acceleration rate required or the time specified for transition from one state to the next.
- *Duration or "end condition" of stage.* The duration of the stage may be defined in several ways:
  - at a fixed time after the beginning of the stage
  - on a chosen parameter reaching a specified value, e.g. coolant tempera-ture reaches 80 °C as in a "warm-up" stage
  - on reaching a specified logic condition, e.g. on completion of a fuel consumption measurement.
- *Choice of next stage.* At the completion of each stage (stage x), the editor will "choose" the next stage. Typical instructions governing the choice are:
  - run stage $x + 1$
  - rerun stage x a total of y times (possible combinations of "looped" stages may be quite complex and include "nested" loops)
  - choose the next stage on the basis of a particular analog or digital state being registered (conditional stage).
- *Events to take place during a stage.* Examples are the triggering of ancillary events such as fuel consumption measurement or smoke density

measurement. These may be programmed in the same way as duration or end condition above.

- *Nominated alarm table.* This is a set of alarm channels that are activated during the stage in which they are "called". The software and wired logic must prevent stage alarms from modifying and overriding primary "global" safety alarms.

PART 4. DATA ACQUISITION AND THE TRANSDUCER CHAIN

The electronically encoded information, defining the state of any measured parameter, travels through a cable from the transducer, then through signal conditioning, usually involving amplification and/or analog-to-digital conversion (ADC), to a storage device and one or more displays. During this journey it is subject to a number of possible sources of error and corruption, many of which are covered in more detail in Chapter 5.

Data Channel Names

It is *vital* for test data to have a consistent and coherent naming convention attached to each data channel. If data is to be shared between cells or sites, or is rerun at a later date, it is essential that an identical naming convention is used. The process of attaching a name to a measurement channel, from transducer through terminal number, to display and finally to the calibration certificate, is done at the time of system calibration and is sometimes called the system's parameterization. It should be obvious to the reader that without such discipline there is no realistic chance of anything but the most crude comparison of test results and all chances of true correlation are effectively lost.

Calibration

All professional engine test software suites will contain calibration routines for all commonly used transducers and instrument types and, in cases of industrial standard equipment, the exact models. Often these routines follow a "form fill" format that requires the operator to type in confirmation of test signals in the correct order, permitting zero, span, intermediate, and ambient values to be inserted in the calibration calculations. Such software should not only lead the operator through the calibration routines, but should also store and print out the results in such a way as to meet ISO 9001 quality control and certification procedures. The appropriate linearization and conversion procedures required to turn transducer signals into the correct engineering units will be built into the software.

For the most demanding calibration the complete chain from transducer to final data store and indication needs to be included, but for normal post-commissioning work it is not practical, nor cost/time-effective. Common practice for temperature channels is to use certified transducers and a calibrated instrument called a resistive decade box, which is injected into the measuring system at the point into which the temperature transducer is plugged. Several measuring points throughout the transducer's range are simulated, checked, and adjusted against the display. Similarly, pressure channels are calibrated by attaching a portable pressure calibration device, made by companies such as Druck or Fluke, to the transducer housed in a transducer box.

Transducer Boxes and Distributed I/O

To measure the various temperatures and pressures of the UUT (engine or vehicle), transducer probes have to be attached and their signal cables connected into the data acquisition system. In many cases the transducer probes will be intrusive to the UUT and fitted into test points sealed with compression fittings. Along with the single probe cables there will be cables from other instruments such as optical encoders requiring connection within the test cell in order for the signals, in raw form, to be transmitted to the signal conditioning device. This network of short transducer cables needs to be marshaled and collected at a transducer box near the engine.

Modern best practice is to reduce the vulnerability of signal corruption by conditioning the signal as near as possible to the transducer and to transmit them to the computer through a high-speed communication link such as a parallel SCSI or IEEE 1394 high-speed serial bus; such implementations have greatly reduced the complexity of test cell installation.

Older installations will have a loom of discrete cables in a loom running from the transducer box, through the cell wall to a connection strip in the control cabinet where signal conditioning takes place.

Transducer boxes will contain some or all of the following features:

- External, numbered or labeled, thermocouple or PRT sockets.
- Pressure transducers of ranges suitable for the UUT, possibly including two with special ranges: inlet manifold pressure and ambient barometric.
- External labeled sockets for dedicated channels or instruments.
- Regulated power supply sockets for engine-mounted devices such as an ECU.

Transducer boxes may have upwards of 50 cables or pipes connected to them from all points of the engine under test; this needs good housekeeping to avoid entanglement and contact with hot exhaust components.

The transducer box can be positioned:

1. On a swinging boom positioned above the engine, which gives short cable lengths and an easy path for cables between it and the control room. However, unless positioned with care it will be vulnerable to local

overheating by radiation or convection from the engine; in these cases, forced ventilation of the enclosed boom duct and transducer box should be provided.

2. On a pedestal alongside the engine, but this may give problems in running cable back to the control room.

3. On the test cell wall with transducer cables taken from the engine through a lightweight boom.

Choice of Instruments and Transducers

It is not the intention to attempt a critical study of the vast range of instrumentation that may be of service in engine testing. The purpose is rather to draw the attention of the reader to the range of choices available and to set down some of the factors that should be taken into account in making a choice.

Table 12.4 lists the various measurements. Within each category the methods are listed in approximately ascending order of cost.

Note on Wireless Transducer Systems

Wireless transducers, such as those used in tire pressure monitors, are now in common use in vehicles. It is not clear at the time of writing how the more demanding applications in powertrain testing can benefit or will use the technology. Presently there are several physical wireless interfaces used, including IEEE 802.11 (WLAN), IEEE 802.15.1 (Bluetooth), IEEE 802.16 (WiMax), IEEE 802.20, Ultra Wide Band (UWB), and IEEE 802.15.4, all involved in several development initiatives by major companies. It is dangerous to predict the pace of technology and there are already test cell installations making use of wireless technology, but concern over "packet loss" and data security will have to be dealt with before this technology takes the place of conventional wired systems in installations other than where transducers are embedded in rotating components (drive shafts, tires, motor rotors, etc.).

The subject of instrument accuracy is treated at more length in Chapter 19, but note that:

- Many of the simpler instruments, e.g. spring balances, manometers, Bourdon gauges, liquid-in-glass thermometers, cannot be integrated with data logging systems but may still have a place, particularly during commissioning or fault-finding of complex distributed systems.
- Accuracy costs money. Most transducer manufacturers supply instruments to several levels of accuracy. In a well-integrated system a common level is required; over- or under-specification of individual parts compared with the target standard either wastes money or compromises the overall performance.
- It is cheaper to buy a stock item rather than to specify a special measurement range.

TABLE 12.4 Common Instrumentation and Transducers for Frequently Required Measurements

Measurement	Principal Applications	Method or Transducer
Time interval	Rotational speed	Tachometer Single impulse trigger Starter ring gear and pick-up Optical encoder
Force, quasi-static	Dynamometer torque	Dead weights and spring gauge Hydraulic load cell Strain gauge load cell
Force, cyclic	Stress and bearing load investigations	Attached strain gauges plus wireless transmitters Strain gauge transducer Piezoelectric transducer
Pressure	Fluid flow systems, gas or liquid	Liquid manometer Bourdon gauge Strain gauge pressure transducer
Pressure cyclic	In-cylinder, inlet and exhaust events, fuel rail	Toughened strain gauge transducer Capacitative transducer Piezoelectric transducer
Position	Throttle and other actuators	Mechanical linkage Linear variable displacement transducer (LVDT) Rotary optical encoder
Displacement cyclic	Needle lift Crank top dead center (TDC)	Eddy-current transducer Inductive transducer Capacitive transducer
Acceleration	NVH investigations Shaft balancing	Strain gauge accelerometer Piezoelectric accelerometer
Temperature	Fluids, air and gases, mechanical components, in-cylinder	Liquid in glass Types of thermocouple PRT Thermistor Electrical resistance Optical pyrometer

- Some transducers, particularly force and pressure, can quickly be destroyed by overload. Ensure adequate capacity or overload protection.
- Always read the maker's catalog with care, taking particular note of accuracies, overload capacity, and fatigue life.

- Some transducers and instruments cannot be calibrated at site but have to be returned to the manufacturer; support contracts covering such devices may be time and cost efficient.

Measurement of Time Intervals and Speed

Time can be measured to great accuracy. However, for linked events, which occur very close together in time and are sensed in different ways, with different instruments, it can be very difficult to establish the exact order of events or to detect simultaneity. This is of prime importance in combustion analysis and engine calibration when highly transient events such as the pressure pulses in fuel rails need to be taken into account.

A single impulse trigger, such as a hole in the rim of the flywheel, is satisfactory for providing a position and timing datum for counting purposes, but is not ideal for locating top dead center (TDC; see Chapter 15 for detailed coverage of TDC measurement). Multiple impulses picked up from a starter ring gear or flywheel drillings used by the engine's ECU may be acceptable.

For precise indications of instantaneous speed and speed crank angle down to tenths of a degree, an optical encoder is necessary. However, mounting arrangements of an encoder on the engine may be difficult and if, as is usually the case, the drive is taken from the nonflywheel end of the crankshaft, torsional effects in the crankshaft may lead to errors that may differ from cylinder to cylinder. When encoders are fitted to the free ends of both dynamometer and the UUT, particularly an engine, there will be a lack of observed synchronization that is due to torsional wind-up and release in the shaft system, so care must be exercised when deciding the source of signals for different purposes.

Force, Quasi-Static

The strain gauge transducer has become the almost universal method of measuring forces and pressures. The technology is very familiar and there are many sources of supply. There are advantages in having the associated amplifier integral with the transducer as the transmitted signals are less liable to corruption, but the operating temperature may be limited.

Hydraulic load cells are simple and robust devices still fairly widely used in portable dynamometers.

Measurement of Cyclic Force

The strain gauge transducer has a limited fatigue life and this renders it unsuitable for the measurement of forces that have a high degree of cyclic variation, though in general there will be no difficulty in coping with moderate variations such as result from torsional oscillations in drive systems.

Piezoelectric transducers are immune from fatigue effects but suffer from the limitation that the piezoelectric crystal, that forms the sensing element, produces an electrical charge that is proportional to pressure change and thus requires integration by a charge amplifier. Such transducers are unsuitable for steady-state measurements but are the universal choice for in-cylinder and fuel injection pressure measurements. Piezoelectric transducers and signal conditioning instrumentation are in general more expensive than the corresponding strain gauge devices.

Measurement of Pressure

To avoid ambiguity, it is important to specify the mode of measurement when referring to a pressure value. The three modes are:

- Absolute
- Relative
- Differential.

Pressures measured on a scale that uses a complete vacuum (zero pressure, as in a theoretical vessel containing no molecules) as the reference point are said to be *absolute* pressures.

Atmospheric pressure at sea level varies,[8] but is approximately 10^5 Pa (1000 mbar) and is an *absolute* pressure because it is expressed with respect to zero pressure. Most engine test instrumentation is designed to measure pressure values that are expressed with respect to atmospheric pressure; thus, they indicate zero when their measurement is recording atmospheric pressure. In common speech relative pressures are often referred to as *gauge* measurements. Because of the difference between an absolute pressure value and a gauge pressure value, the actual value of atmospheric pressure in powertrain test facilities needs to be recorded daily as a reference datum for all calculations.

Note that the barometric pressure measured in an engine test cell, and experienced by the engine, will often be some 50 Pa lower than the value taken elsewhere in the facility due to the effect of a pressure-balanced ventilation system. Engine inlet manifold pressure will vary between pressures above and (generally) *below* the "reference" barometric pressure; this is another form of relative pressure, sometimes called a "negative gauge pressure".

In applications such as the measurement of flow through an orifice, when it is necessary to measure the difference in pressure between two places, the reference pressure may not necessarily be either zero or atmospheric pressure

---

8. For reference concerning the range of ambient barometric pressures, to date the highest recorded atmospheric pressure, 108.6 kPa (1086 mbar), occurred in Mongolia, 19 December 2001. The lowest recorded atmospheric pressure at sea level, 87.0 kPa (870 mbar), occurred in the Western Pacific during Typhoon Tip on 12 October 1979.

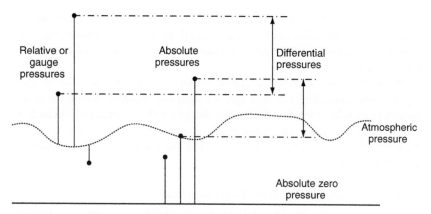

FIGURE 12.3   Diagram showing relationship between absolute, relative, and differential pressures.

but one of the measured values; such measurements are of *differential* pressures (see Figure 12.3).

Electronic Pressure Transducers

There are large ranges of strain gauge and piezoelectric pressure transducers made for use in engine testing, most having the following characteristics:

- Pressure transducers normally consist of a metal cylinder made of a stainless steel, with a treaded connection at the sensing end and a signal cable attached at the other.
- The transducer body will contain some form of device able to convert the pressure sensed at the device inlet into an electrical output. The signal may be analog millivolt, 4–20 mA, or a digital output connected to a serial communications interface.
- The output cable and connective circuitry will depend on the transducer type, which may require four-, three-, or two-wire connection, one of which will be a stabilized electrical supply.
- Each transducer will have an operating range that has to be carefully chosen for the individual pressure channel; they will often be destroyed by over-pressurization and be ineffective at measuring pressures below that range.
- Each transducer will be capable of dealing with specific pressurized media; therefore, use with fuels or special gases must be checked at the time of specification.

Capacitative transducers, in which the deflection of a diaphragm under pressure is sensed as change in capacity of an electrical condenser, may be the appropriate choice for low pressures, where a strain gauge or piezoelectric sensor may not be sufficiently sensitive.

Calibration of the complete pressure channel may be achieved by use of certified, portable calibrators. Some special pressure transducers such as barometric sensors and differential pressure transducers may need annual off-site certification.

Pressure sensors used in the engine testing environment, because of their expense and vulnerability to damage, are normally mounted within the transducer box with flexible tubes connecting them by self-sealing couplings to their measuring point. Their vulnerability to incorrect pressure or fluid makes it imperative that the connectors are clearly marked at their connection points on the transducer box.

Other Methods of Pressure Measurement

The liquid manometer has a good deal to recommend it as a device for indicating low pressures. It is cheap, effectively self-calibrating, and can give a good indication of the degree of unsteadiness arising from such factors as turbulence in a gas flow. It is recommended that manometers are mounted in the test cell with the indicating column fixed against the side of the control room window.

The traditional Bourdon gauge with analog indicator is the automatic choice for many fluid service sensing points such as compressed air receivers.

Displacement

Inductive transducers are used for "noncontact" situations, in which displacement is measured as a function of the variable impedance between a sensor coil and a moving conductive target.

The Hall effect transducer is another noncontact device in which the movement of a permanent magnet creates an induced voltage in a sensor made from gold foil. It has the advantage of very small size and is used particularly as an injector needle lift indicator.

Capacitive transducers are useful for measuring liquid levels, very small clearances, or changes due to wear.

Acceleration/Vibration

There are three basic types of transducer for measuring vibration:

- *Accelerometers*. These are based on piezoelectric sensors that generally have the widest frequency range and typically an output of 100 mV/g. They are extremely sensitive to the method of attachment and it is vital to follow the manufacturer's instructions. Accelerometers are normally used for high-frequency vibration readings and the output is integrated electronically to velocity (mm/s). Piezoelectric sensors should not be used for frequencies of less than about 3 Hz but tend to be more robust than strain gauge units.

- *Velocity transducers.* These usually work by the physical movement of an internal part within a magnetic field; they are therefore more prone to damage than accelerometers and they have a lower frequency range.
- *Proximity transducers.* Sometimes called "eddy probes", these are used as noncontract vibration transducers. They tend to be rather delicate until correctly installed but can measure a very wide range of frequencies and are particularly good at frequencies below the range of the other types of transducer. The normal air gap between transducer face and the vibration surface is usually between 2.5 and 5 mm.

Temperature Measurement

Thermocouples

Most of the temperatures measured during engine testing do not vary significantly at a high rate of change, nor is the highest degree of accuracy cost-effective. The types of thermocouple commonly used in engine testing are listed below.

For the majority of temperature channels the most commonly used transducer is the type K thermocouple (see Table 12.5), having a stainless steel grounded probe and fitted with its own length of special thermocouple cable and standard plug. To produce a calibrated temperature reading, all thermocouples need to be fitted within a system having a "cold junction" of known temperature with which their own output is compared. Thermocouples of any type may be bought with a variety of probe and cable specifications depending on the nature of their installation; they should be considered as consumable items and spares kept in stock.

PRTs

When accuracy and long-term consistency in temperature measurement higher than that of thermocouples, or for measuring temperatures below 0 °C, is required then platinum resistance thermometers (PRTs) are to be recommended. The PRT works on the principle of resistance through a fine platinum

---

TABLE 12.5 Commonly Used Thermocouple Types

Type	Internal Materials	Temperature Range (°C)
Type T	Cu–Ni/Cu	0 to +350
Type J	Fe–Ni/Cu	0 to +800
Type K	Ni/Cr–Ni/Al/Mn	−40 to +1200

wire as a function of temperature. PRTs are used over the temperature range
−200 to 750 °C.

Because of their greater durability or accuracy over time, it is common
practice to use PRTs in preference to thermocouples for within-device control
circuits. They require different signal conditioning from thermocouples and, for
engine-rigged channels, are fitted with dedicated plugs and sockets within the
transducer box.

Thermistors

Thermistors (thermally sensitive resistors) have the characteristic that,
dependent on the nature of their sensing "bead", they exhibit a large change in
resistance, either increasing or decreasing, over a narrow range of temperature
increase. Because they are small and can be incorporated within motor
windings they are particularly useful as safety devices. Thermistors are
generally available in bead or surface mounting form and in the range from
−50 to 200 °C.

Pyrometers

Optical or radiation pyrometers, of which various types exist, are noncontact
temperature measuring devices used for such purposes as flame temperature
measurement and for specialized research purposes such as the measurement
of piston-ring surface temperatures by sighting through a hole in the cylinder
wall. They are effectively the only means of measuring very high
temperatures.

Finally, suction pyrometers, which usually incorporate a thermocouple as
temperature-sensing device, are the most accurate available means of
measuring exhaust gas temperatures.

Other Temperature-Sensing Devices

Once a standard tool in engine testing, liquid-in-glass thermometers are now
very rarely seen; however, they are cheap, simple, and easily portable. On the
other hand, they are fragile, not easy to read, have a relatively large heat
capacity, and a slow response rate. They retain some practical use during
commissioning of process plant distributed over a wide location. The interface
between the body, solid, liquid, or gas, of which the temperature is to be
measured and the thermometer bulb needs careful consideration.

Vapor pressure and liquid-in-steel thermometers present the same
interface problems and are not as accurate as high-grade liquid-in-glass
instruments, but are more suitable for the test cell environment and are more
easily read.

Mobile, hand-held thermal imaging devices are more useful for fault-
finding and safety work than for collection of test data. They are used for on-
site commissioning work on large stationary engine systems.

Smart[9] Instrumentation and Communication Links

The test cell control and data acquisition computer will, along with taking in individual transducer signals, also have to switch on and off and acquire data from complex modules of instrumentation, a list of which includes:

- Fuel consumption devices—gravimetric or mass flow
- Oil consumption
- Engine "blow-by"
- Exhaust gas emission analysis equipment
- Combustion analysis (indicating) instrument.

The integration of these devices within the control system will require special software drivers to allow for communication covering basic control, data acquisition, and calibration routines.

The interface between device and test cell computer may be by way of conventional analog I/O, 0–10 V, digital I/O, or by serial interface such as RS232.

Combustion analyzers and other complex instruments are increasingly equipped with high-speed outputs based on CAN bus (ISO 11898 compatible) with baud rates, which are dependent on cable length, of about 1 Mb/s; 15 m is usually the supplied length, with 25 m as the recommended maximum.

Control for Endurance Testing and "Unmanned" Running

Techniques such as key life testing, aimed at accelerating the attrition experienced by powertrain components during real life, attempt to reduce the duration and therefore the cost of endurance tests. However, there are several, fairly standard, engine and transmission test sequences that are of 400 hours or more duration, which are very expensive to run, and call for a level of performance and reliability in the test equipment that is an order of magnitude higher than that of the UUT. For such work, duplication of transducers and careful running of cables to prevent deterioration by wear or heat over long periods are good practices.

Cells running endurance tests should not be vulnerable to shutdowns caused by faults within indirectly connected systems. For the same reason shared systems should be minimized so far as is possible. The amount of data generated over 1000 hours of cyclic running can be enormous, so data filtering and compression are usually required.

Unmanned running, in the context of powertrain testing, means that no operator is present outside set working hours and during that time the test cell runs

---

9. The term "smart" is used, without intentional irony, to cover instrumentation containing embedded programmable devices or computers and carrying out their own specialized data processing.

under automatic control. Strict procedures have to be in place to cover the event of a shutdown alarm occurring; these include safe automatic shutdown, remote alarm monitoring, competent technician support, and backup being "on call".

Control for Dynamic Engine Drive Cycles

In the late 1990s there was much debate on the exact definition of dynamic or transient testing; much depended on the type of UUT and the technology being used. What was considered a highly dynamic test sequence in one industrial sector was considered positively pedestrian in another; therefore, the authors in the early edition of this book, and others in the industry, classified the titles being used in terms of four characteristic time scales, where the times quoted represented the time in which the control instruction is sent and the change in force at an AC dynamometer air gap is sensed. These were:

- <2 ms = high dynamic
- <10 ms = dynamic
- <50 ms = transient
- >100 ms is in the region where steady-state conditions begin to exist, in automotive work.

It is perhaps worth pointing out that before the arrival of modern dynamometer controllers during what may be described as "classical" steady-state testing, the transition ramp rates between successive test conditions were highly variable and were rarely used to trigger data collection. In modern drive cycles, the transient states are the test sequence stages.

Exhaust emission legislation continues to be based on engine and vehicle testing incorporating transient drive cycles but, because it is necessary for the majority of major test laboratories worldwide to achieve these cycles, they can be considered to be quite benign with low acceleration rates and maximum speeds.

At the stage we are at (in early 2012) in the "drive cycle test legislation versus test technology for driving cycles" development cycle, the technology of controller, drives, and high-power, low-inertia AC dynamometers can be said to be ahead of legislation. However, the technology required to measure gaseous and particulate exhaust emissions during these drive cycles is not in the same position and development is still being forced by legislation, as is discussed briefly in Chapter 16.

In the transient test frequency band range of 0.2 seconds, we are concerned essentially with torsional vibrations in the engine–transmission–road wheel complex, ranging from two-mass engine–vehicle judder, with a frequency typically in the range 5–10 Hz, to much higher frequency oscillations involving the various components of the powertrain.

This kind of powertrain investigation, however, involves the engine to a secondary degree; in fact, for such work the engine may now be replaced by a lower inertia, permanent magnet, electric motor-based dynamometer such as the AVL "Dynoprime".

One of this range, a water-cooled machine with a 370 kW rating and an inertia of 0.11 kgm^2, is matched with a drive and control that can generate sinusoidal oscillations into the shaft system at between 12 and 300 Hz at full power.

In the transient test time scale, work is characteristic of gear shifts; correct behavior in this range is critical in a system called upon to simulate these events. This is a particularly demanding area, as the profile of gearshifts is extremely variable. At one extreme we have the fast electrohydraulic shifting of a race car gearbox, at the other gear changes in the older commercial vehicle, which may take more than 2 seconds, and in between the whole range of individual driver characteristics, from aggressive to timid.

Whatever the required profile, the dynamometer must impose zero torque on the engine during the period of clutch disengagement and this requires precise following of the "free" engine speed. The accelerating/decelerating torque required is proportional to dynamometer inertia and the rate of change of speed:

$$T = I\mathrm{d}\omega/\mathrm{d}t$$

where T = torque required (Nm), I = dynamometer inertia (kgm^2), and dω/dt = rate of speed change in rad/s.

A gear change involves throttle closing, an engine speed change, up or down, of perhaps 2000 rev/min, and the reapplication of power. Ideally, at every instant during this process the rate of supply of fuel and air to the engine cylinders and the injection and ignition timings should be identical with those corresponding to the optimized steady-state values for the engine load and speed at that instant; it will be clear that this optimizing process makes immense demands on the mapping of the engine management system.

A special problem of engine control in this area concerns the response of a turbocharged engine to a sudden demand for more power. An analogous problem has been familiar for many years to process control engineers dealing with the management of boilers. If there is an increase in demand the control system increases the air supply before increasing the fuel input; on a fall in demand, the fuel input is reduced before the air. The purpose of this "air first up, fuel first down" system is to ensure that there is never a deficiency of air for proper combustion. In the case of a turbocharged engine only specialized means, such as variable turbine geometry, can increase the air flow in advance of the fuel flow. A reduction in value from the (optimized) steady-state value, with consequently increased emissions, is an inevitable concomitant to an increase in demand. The control of maximum fuel supply during acceleration is a difficult compromise between performance and drivability on the one hand and a clean exhaust on the other.

Data Display

The manner in which data is displayed on the operators' screens is usually highly configurable by the user, perhaps dangerously so. Individual operators, if allowed, will adapt the data display according to their own abilities in using the software and to suit their own immediate requirements and their esthetic tastes; the result may not produce a clear representation of data according to its operational importance, or be readily understandable by colleagues. Most large test facilities impose the use of one or more, task-appropriate, data display templates on test cell operatives as part of their processes under ISO 9001.

These comments and the example shown in Figure 12.4 only cover the main "engine driver's" screen, the main function of which is to ensure that those parameters of major significance to the test and those covered by primary alarms, such as engine oil pressure, engine coolant, etc., are able to be monitored within a clear uncluttered display. There may be several other screens displaying data from other cell instrumentation and the cell services, all of which can be considered of secondary importance to the safe running of the in-cell machinery and the UUT.

The appropriate use of analog or digital representation is briefly discussed in Chapter 19, but it will be found informative to visit facilities testing widely different types of prime movers in order to see the best practices found to be appropriate to each specialization. As an example, in those test cells testing gas

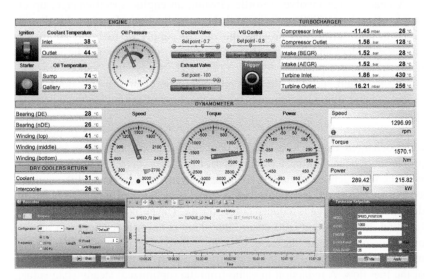

FIGURE 12.4 An example of a test cell operator's display screen. The analog representation of speed, torque, power, and oil pressure follows good practice. The test sequence was concerned with turbocharger operation; therefore, there is a display section covering that item. The bottom graph represents a scrolling 60-second log of speed, torque, and other chosen parameters. *(Courtesy of University of Bradford.)*

turbines and some medium-speed diesel engines, it is found that there is a wide use of multiple vertical bar displays; this is because it is the equal balance of multiple temperatures or pressure that has to be immediately apparent to the operator, rather than the exact measurement.

One great advantage of the digital display of an analog (dial) format, as shown in the engine speed display in Figure 12.4, is that the dial can be easily scaled appropriately, from zero to maximum, to the limits of operation of the UUT.

The advent of "touch-screen" technology and screen-embedded (virtual) buttons creates a new set of screen configuration problems for cell users, a subject worthy of a thesis in its own right.

REFERENCE

[1] Association of Standardization of Automation and Measuring Systems (ASAM), www. asam.net/.

FURTHER READING

R.P. Benedict, Fundamentals of Temperature, Pressure and Flow Measurements, John Wiley, 1984. ISBN-13: 978-0471893837.
BS 6174. Specification for Differential Pressure Transmitters with Electrical Outputs.
National Physical Laboratory, Good Practice Guides. http://www.npl.co.uk/.
N.S. Nise, Control Systems Engineering, John Wiley, 2011. ISBN-13: 978-0470646120.

Data Handling, the Use of Modeling, and Post-Test Processing

PART 1. DATA COLLECTION AND TRANSMISSION

It is perhaps the ever-increasing speed of the recording, processing, transmission and storage of data, and the availability of virtual models of vehicle components that are responsible for the most revolutionary changes in the practice of vehicle and powertrain testing in the last 15 years.

Engine Testing. DOI: 10.1016/B978-0-08-096949-7.00013-3

It should be appreciated that, if given lack of care in their collection and storage, computerized data collection and processing systems are inherently no more accurate, nor traceable, than paper systems. All data should have an "audit trail", back to the calibration standards of the instruments that produced them, to prove that the information is accurate, to the level required by the user. It is a prime responsibility of the test cell manager to ensure that the installation, calibration, recording, and post-processing methods associated with the various transducers and instruments are in accordance with the maker's instructions and the user's requirements.

It is necessary to repeat a statement made in the previous chapter:

It is vital *for test data to have a consistent and coherent naming convention attached to each data channel; if data is to be shared between cells and sites it is vital that an identical naming convention is shared.*

Bad data are still bad data, however rapidly recorded and skillfully presented.

Traditional Paper-Based Test Data Collection

It should not be assumed that traditional methods of data collection should be rejected; they may still represent the best method of monitoring a device or system under test. This is particularly true in cases where it is necessary to ensure an operator regularly patrols the test field, as is the case with very large marine installation or industrial process plant. It is also the most cost-effective solution in certain cases where budgets and facilities are very limited.

Figure 13.1 shows a traditional "test sheet". The example records the necessary information for plotting curves of specific fuel consumption, b.m.e.p., power output, and air/fuel ratio of a gasoline engine at full throttle and constant speed, with a total of 13 test points. All the necessary information to permit a complete and unambiguous understanding of the test, long after the test engineer has forgotten all about it, is included.

Such simple test procedures as the production of a power–speed curve for a rebuilt diesel engine can perfectly well be carried out with recording of the test data by hand on a pro-forma test sheet. In spite of the low cost of computer hardware there may be little justification for the development and installation of a computerized control and data acquisition system in the engine rebuild shop of, for example, a small city bus company, handling perhaps less than 20 engines per month. If an impressive "birth certificate" for the rebuilt engine is required, the test sheet may easily be transferred to a computer spreadsheet and accompanied by a computer-generated performance curve.

In the data flow shown in Figure 13.2, the only post-processing of the test data is the conversion of test data into an engine "birth certificate", which may be simply a reduced and tidy copy of the original data. The archive may be legally required and be available for use for technical and commercial statistical analysis.

TEST SHEET—I.C. ENGINES

UNIT NO.

DATE	CUSTOMER		WORKS ORDER NO.

DYNAMOMETER TORQUE ARM 265mm

ENGINE VARIABLE COMP	BORE 85mm	STROKE 825mm	CYLINDERS 1	SWEPT VOL. 468 cc	FUEL SHELL 4-STAR	OIL SHELL 20W30
BAROMETER 755mmHg	AIR TEMP. 21°C	AIR BOX SIZE	150 liter	ORIFICE DIA. 18.14 mm	FLOWMETER NO.	FUEL GAUGE 'O'

POWER kW = $\dfrac{F_N n}{36,000}$ b.m.e.p = $7123\,F_N\ kN/m^2$ FUEL liter/Hour = $\dfrac{180}{1}$ NOTE: COMPRESSION RATIO 7.1

TACHO rev/min	COUNTER rev (2x)	COUNTER sec	rev/min (n)	BRAKE LOAD (F_N)	POWER kW	b.m.e.p kN/m^2	1sec 50/cc	FUEL liter/hour	FUEL liter/kW hr	EXH °C	HEAD cmH_2O (h_o)	TEMP °C (T_A)	VOL F/R l/s (V_a)	EFFY η VOL	MASS F/R g/s (m_a)	a/l RATIO	REMARKS
RICH 2000	780		2012	50.5	2.022	360	47.0	3.830	1.357	620	6.15	21	4.02	0.827	6.89	7.38	
	834		2002	67.0	3.726	477	50.0	3.600	0.966		6.15		4.92	0.630	5.89	7.85	
	891		1022	68.0	3.630	484	52.5	3.429	0.945	500	6.13		4.91	0.655	5.88	8.23	
	962		2025	71.0	3.993	506	57.0	3.158	0.791		6.10		4.90	0.620	5.87	8.92	
	1044		1989	74.0	4.089	527	63.0	2.857	0.699		6.05	21.5	4.88	0.629	5.84	9.81	
	1156		2010	74.5	4.160	530	69.0	2.609	0.627		6.00		4.86	0.620	5.82	10.71	
	1304		2019	74.5	4.178	530	77.5	2.323	0.556		5.97		4.85	0.616	5.80	11.99	
	1355		1995	74.5	4.129	530	81.5	2.209	0.535	675	5.93		4.83	0.621	5.79	12.58	
	1560		2035	69.0	3.900	491	92.0	1.957	0.502	685	5.93	19	4.83	0.609	5.79	14.20	
	1615		1988	65.2	3.601	464	97.5	1.846	0.513		5.87		4.81	0.620	5.76	14.98	
	1698		2008	60.0	3.347	427	101.5	1.773	0.530	673	5.95		4.84	0.618	5.80	15.70	
	1792		2010	52.0	2.903	370	107.0	1.682	0.580		6.03		4.88	0.623	5.84	16.66	
WEAK	1908		2000	35.0	1.944	249	114.5	1.572	0.809	682	6.03	19	4.88	0.626	5.84	17.83	

DENSITY OF FUEL 0.75 kg/liter

FIGURE 13.1 A traditional test sheet.

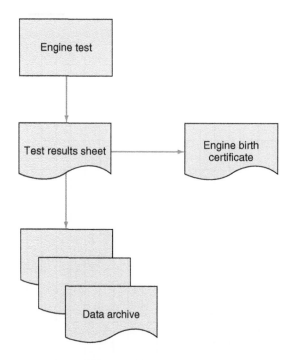

FIGURE 13.2 Data flow based on pre-computerized systems.

Chart Recorder Format of Data

Chart recorders represent a primitive stage in the process of automated data acquisition. Multi-pen recorders, having up to 12 channels giving continuous analog records, usually of temperatures, in various colors, still have a useful role, particularly in the testing of dispersed systems such as installed marine installations or of large engines "in the field" where robustness and the ease with which the record may be annotated are of value.

It is the manner in which the data is displayed as a colored graph of values against time, now emulated by computerized displays, which is the real value to test engineers who need the clear and immediate indication of trends, rates of change, and the interaction of data channels, in a paper form that can be notated contemporaneously with events.

Computerized Data Recording with Manual Control

At all levels of test complexity the appropriate degree of sophistication in the management of the test and the recording of the data calls for careful consideration.

As the number of channels of information increases and there is a need to calculate and record derived quantities such as power and fuel consumption, the task of recording data by hand becomes tedious and prone to error. There are a number of simple PC-based data acquisition packages on the market, particularly in the USA, that do not have any control function and are designed for the after-market engine tuning and testing facility. These systems, at their most basic, take time-stamped "snapshots" of all connected data channels on the key command from the operator.

Fully Computerized Test Cell Control and Data Handling Systems

Most of the following text will be based on these fully computerized systems now found in good technical universities as frequently as they are in the facilities of original equipment manufacturers (OEMs). A small danger for students having their first, perhaps only, experience of engine testing on these modern systems is that the majority of that experience is vicarious, through watching a screen displaying highly processed numbers.

Note on Analog-to-Digital Conversion (ADC) Accuracy

Whatever the level of computerization in the test facility, the signals from all transducers having analog outputs have to be digitized, conditioned, and linearized in the computer, both for display and storing in the chosen format. The converted values are usually stored electronically in binary form, so the resolution is usually expressed in bits, and proprietary systems are usually available with ADC of 8-, 12-, and 16-bit resolution. These should be selected as appropriate to the accuracy and the significance of the signals to be processed.

It tends to give a false illusion of accuracy (see Chapter 19) to use 16-bit or higher resolution of ADC for measurement channels having an inherent inaccuracy an order of magnitude greater. For example, a thermocouple measuring exhaust gas temperature may have a range of 0–1000 °C and an accuracy of $>\pm3$ °C. The 16-bit resolution would be to about 0.016 °C, while the entirely adequate, perhaps cheaper, 12-bit resolution would be 0.25 °C.

Data Transfer Systems and Protocols

The electronically controlled powertrains of today, installed within electronically controlled vehicles, are tested within electronically monitored and controlled test facilities, so to cope with the two-way flow of this data traffic the development of types of high-speed data transmission systems has had to take place. A computer "bus" is a subsystem that allows the orderly and prioritized transfer of data between components inside a computer and between computers and connected equipment; the test engineer will meet a number of different types in interfacing with both test objects and test equipment.

Computer buses can use both parallel and bit-serial connections, and can be wired in either a multidrop or daisy chain topology, or connected by switched hubs, as in the case of USB.

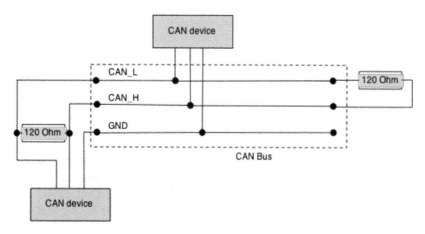

FIGURE 13.3   The common physical media of the CAN bus (ISO 11898-2), terminated at both ends with 120 ohms when running at high speeds to prevent "reflections" and to unload the open collector transceiver drivers.

The *controller area networking, or CAN bus*, is a multi-master, broadcast[1] serial bus that has become a standard in automotive engineering and is one of the protocols used in OBD-11 vehicle diagnostics. The CAN bus runs over either a shielded twisted pair (STP), unshielded twisted pair (UTP), or ribbon cable, and each node uses a male nine-pin D connector. The most common physical media for a CAN bus is a twisted pair, 5 V differential signal, which allows operation in the high-noise environments of some automotive installations (see Figure 13.3). The nodal devices are typically sensors, actuators, and other vehicle control devices such as the ABS (antiskid braking) system; these devices are connected through a CAN controller.

To ensure that CAN bus devices from different suppliers interoperate and that a good standard of EMC and RFI tolerance is maintained, the ISO 11898-2 (cars) and SAE J1939 (heavy vehicles) families of standards have been developed.

The high level of inbuilt error checking also gives the protocol the high level of reliability required in some of the safety-critical automotive systems.

Pressure on the data transfer rates beyond the capabilities of the original CAN bus have meant that several other protocols and standards (some optically based) have been developed; many are manufacturer or manufacturer consortium specific, and it is not clear at the time of writing which will become a future standard. Examples are:

---

1. Broadcasting means transmitting a data packet that will be received by every device on the network.

- *FlexRay*, which is designed to be faster and more reliable than CAN bus.
- *TTP (Time-Triggered Protocol)*, which is an open and modular control technology that has been used for aeronautical FADEC (Full Authority Digital Engine Control) systems.

The main test-bed computer may be physically linked, through CAN bus or another standard communication bus, to both the UUT and major modules of microprocessor-controlled instrumentation. A common example of such an interface is that of exhaust emission analyzing equipment. Linking the test-bed control computer with such an instrument is rarely a trivial task and it is essential, when choosing a main control system, to check on the availability of the correct version of software, device drivers, and compatibility of data transfer protocols.

Powertrain Mapping and Calibration

In this usage the term "calibration" is used by test engineers to describe the process by which the operational and control characteristics of a vehicle and its powertrain system is optimized, through manipulation of all its control mechanisms, to suit a specific geographical/vehicle/driver combination, while remaining within the limits of environmental and safety legislation.

As powertrains increase in the complexity of every area of technology, in order to meet legislative standards, the task of powertrain calibration is being made more difficult by the increase in the number of parameters with which it has to deal.

This "curse of dimensionality" is driving the calibration process beyond the capabilities of individuals and their available time scales and is driving the development of special tools, techniques and software, often through the collaboration of industry and academia, worldwide.

The optimized control parameters, when found, are stored within the "map", essentially a multi-axis look-up table or set of tables, held in the memory of the engine control unit (ECU), the powertrain, and vehicle control units. Thus, "mapping" an engine, outside its intended vehicle, requires that it be run on a test bed capable of simulating not only the vehicle's road-load characteristics, but also the full range of the vehicle's performance envelope, including the dynamic transitions between states.

To reduce the testing time required to produce viable ECU maps, and to keep the complexity of the operator's role in the mapping process to sensible levels, software tools designed to semi-automate the tasks are used.

During the mapping process, in order to discover boundary conditions, such as "knock", the engine will be run through test sequences that seek and confirm, by variation of control parameters of which ignition timing and fueling are only two, the map position of these boundaries at a wide range of vehicle operating conditions. While running these tests to find the limits of safe operation, conditions of excessive temperatures, cylinder pressure, and "knock" require fast detection and correct remedial action that may be beyond the capabilities of a human operator. Mapping tools such as CAMEO®, produced by AVL, contain test sequence control algorithms that allow some degree of automation for the location of these boundary conditions.

A typical mapping sequence is that called "spark sweep" wherein, at a given speed/torque operating point, the spark timing is varied until a knock condition is approached, the timing is then backed off and the knock point approached with finer increments to confirm its map position. The test sequence then sweeps the ignition timing so that a fault condition, such as high exhaust temperature, is located at the other end of the sweep. Given sufficient data the software can then calculate and extrapolate the boundary conditions for a wide operating envelope without having to run all of it.

Advanced design of experiment (DoE) tools such as those embedded in products like AVL CAMEO allow the test engineer to check the plausibility of data, the generation of data models, and obtain visualization of system behavior while the test is in progress.

The ultimate target of the calibration engineer is to be able to set the operating profile of the test engine, then set the sequence running, only to return when the task of producing a viable engine map has been automatically completed; perhaps this will be achieved within the lifetime of this edition.

Virtual Models in Powertrain Calibration and "Hardware-in-Loop" Testing

All models are wrong but some are useful.

(George E. P. Box)

George Box was not a test engineer but an economist, and we are all aware of the veracity of some of their models. The pressure on "time to market" and the development programs associated with engines and powertrain units do not allow the testing of embedded control systems to wait for a vehicle or engine prototype to be available. All new development schedules require that hardware-in-loop (HIL) simulation will be used in parallel with the development of the mechanical and electrical modules, and such simulation requires virtual models of the missing modules.

While modeling of the mechanical and electric functions and interactions is becoming demonstrably ever more useful, feedback from the vehicle aftermarket reveals that some of the models used in powertrain optimization are not yet fully reflecting real-life use or the expectations of drivers. The problem may be seen more as a marketing problem than a calibration engineer's problem because the nonspecialist driver's expectation of the essential parts of driving experience has remained regionally static[2] while the vehicle's technology, and therefore function, continues to change significantly. A modern vehicle with a continuously variable transmission (CVT) feels and sounds disturbingly different from, for example, the five-speed manual transmission-equipped car that any European driver with 30 years experience has as a mental model for a vehicle's "normal" behavior. The author had occasion to "rescue" a terrified driver from a powerful car whose DSG powertrain and vehicle stability system (VSC) became "undrivable" in moderately deep snow. The driver had no idea how to switch the gearbox to manual or to switch off the VSC and was determined to sell the car at the first opportunity. Perhaps a control system and driver display model that was more sympathetic to the average, technically unaware, driver could have been used in the powertrain development?

This raises the questions of how drivability is defined and how it is included in the model. Presently the calibration engineer's proxy for drivability is the rate of change of actuators; no doubt this area of modeling will develop.

Since HIL test techniques are only as good as the models on which they are based, a heavy responsibility rests on the quality control, verification, and correct application of models. However, as the total store of performance data, including previously unmodeled fault combination events, builds during their life, the models should improve. Model improvement requires that the in-service feedback loop works efficiently and quickly, rather than working through the motoring press and, that modern indicator that HIL has failed, the model recall process for a control software upgrade.

Currently (2011) most HIL testing, shown diagrammatically in Figure 13.4, is used by engineers to carry out two key tasks of ECU calibration:

- Vehicle performance optimization during the creation of control parameter maps.
- Testing of the diagnostic performance of the ECU under fault conditions as required by OBD legislation and proprietary service support.

A key benefit of HIL testing is that the ECU can be pre-tested in its full range of operation and in all primary fault conditions without endangering the hardware it will subsequently control. The multitude of vehicle and engine system faults, with their intended, and unintended, interactions can be examined in this virtual test environment, to an extent that would be simply impossible in the traditional engine or vehicle test bed. Along with the practicality of testing operation and fault conditions, the HIL test sequences

---

2. "Regionally" because drivers in different geographic locations learn to drive in vehicles and locations that are significantly different; thus, the average lifetime resident driver of the greater Detroit area (flat, grid pattern road layout, freeways, and automatic transmissions) has a totally different mental model of car behavior from that of a similarly aged resident of Glasgow (hills, bends, roundabout junctions, and manual transmissions).

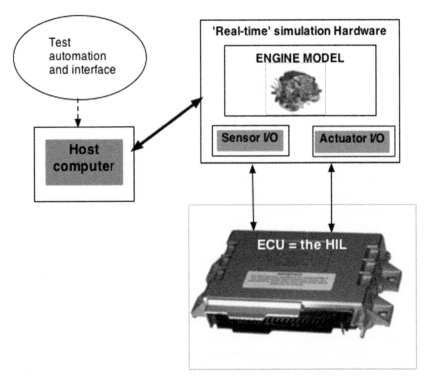

FIGURE 13.4 The engine or complete powertrain is simulated in a model programmed into high-speed electronic hardware together with models of sensors and actuators. The model simulation has to be an order of magnitude faster in operation than the ECU with which it is connected. The ECU is real, it is the hardware in the (simulation) loop. The host computer provides the test automation and the GUI interface.

give the great benefit of repeatability and the ability to rerun transient conditions to determine cause and effect.

As experience builds the models used become more verifiable but test engineers have to decide on the operational constraints that need to be imposed on the system. These constraints are determined both by the boundary conditions required to ensure that, within all "normal" real-life conditions:

- The engine runs in safe conditions of combustion (knock avoidance, etc.)
- Emissions in defined test conditions are within legislative limits
- Good drivability within the limits of the final vehicle model.

While boundary conditions and abnormal operation conditions can be safely and "cheaply" explored during HIL testing and it is therefore ideal in system development and university teaching, it can remove the tester from the reality of their test object; this decoupling of an increasingly important part of the test process needs to be guarded against. Driving the famous Nordschleife of the Nürburgring on a computer gaming machine may be convincingly realistic to those used to screen-based simulations but it does not adequately prepare one for the real thing.

Transmission Testing and a Role for HIL

Gearboxes and individual modules of transmission systems are increasingly controlled by embedded electronic controllers whose optimization requires HIL testing using chassis, engine, and driver models.

Much transmission development testing is designed to improve unit durability and efficiency, the latter by reducing internal friction and oil churn. Accuracy in measurement of power losses in transmissions is critically dependent upon good control of system temperature. Since efficiency will increase from its initial figure as gear surfaces bed in and, in multishaft units, it will differ with different meshing states and operating ratios, a well-designed run-in period is required before critical testing is carried out.

Control and actuation strategy will affect losses but recent developments such as dual-clutch, direct-shift gearboxes (DSG) tend only to "shift" the losses from gear-driven hydraulic pumps and actuators to electrical actuators, powered from the engine alternator.

When testing the latest generation of transmissions such as continuously variable (CVT) and automated manual (AMT) units, the rig has to have either the ECU and transmission control unit (TCU) in circuit or some form of specially designed HIL test rig controller has to simulate those units. The new, low-inertia AC dynamometer/motors are the most important actuator in such an HIL set-up, since the engine's operating model can now be run electrically while the vehicle's road load and final drive can be modeled through the absorption dynamometer(s).

The gear change, based on an empirical driver model, can be produced by a robot, as described in Chapter 17. Such tests will require careful installation and commissioning, and it may take considerable time before useful work can be carried out; even then they have not replaced driver testing on test tracks with pre-production vehicles. Again, the curse of dimensionality appears, this time in the variability of driver operation, road configuration, and control system interactions, which make simulation within a test cell difficult to assess. However, critical driver maneuvers such as prolonged "holding" on a hill by spinning the clutch, or a sudden change from braking to rapid acceleration (joining a traffic stream), can be and are being brought into test sequences. Nevertheless, visitors to powertrain test facilities may still observe early prototype vehicles being driven around the site transporting (transmission) development engineers armed with laptops.

PART 2. MANAGEMENT OF DATA: SOME GENERAL PRINCIPLES

Modern, high-speed data acquisition systems are capable of pulling in vast amounts of data that may be totally irrelevant to the purposes of the test or stored in a way that makes it virtually impossible to correlate with other tests.

The first basic rule is to keep a consistent naming scheme for all test channels, without which it will not be possible to carry out post-processing or comparison of multiple tests. Since, for practical purposes, all engine and vehicle test work is relative, rather than absolute, this ability of comparison and statistical analysis of results is vital.

Mere acquisition of the data is the "easy" part of the operation. The real skill is required in the post-processing of the information, the distillation of the significant results, and the presentation of these at the right time, to the right people, and in a form they will understand. This calls for an adequate management system for acceptance or rejection, archiving, statistical analysis, and presentation of the information.

Data is as valuable as it is accessible by the people that need to use it. The IT manager, in charge of the storage and distribution network of the data, is important in the test facility team and should be local, accessible, and involved. One vital part of the management job is to devise and administer a disaster recovery procedure and backup strategy.

Normally the IT manager will have under his or her direct control the host computer system into which, and out of which, all the test-bed level and post-processing stations are networked. Host computers can play an important role in a test facility, providing they are properly integrated. The relationship between the host computer, the subordinate computers in post-processing offices, and the test cells must be carefully planned and disciplined. The division of work between cell computer and host may vary widely. Individual cells should be able to function and store data locally if the host computer is "offline", but the amount of post-processing done at test-bed level varies according to the organizational structures and staff profiles.

In production testing the host may play a direct supervisory role, in which case it should be able to repeat on its screen the display from any of the cells connected to it and to display a summary of the status of all cells.

With modern communications it is possible for the host computer to be geographically remote; some suppliers will offer modem linking as part of a "site support" contract. This ability to communicate with a remote computer raises problems of security of data.

Post-Acquisition Data Processing, Statistics, and Data Mining

Much of this book is directly and indirectly concerned with the ways in which the accuracy of raw data is as high as practically possible and corruption of the raw values is avoided during its primary processing and storage.

The possibilities of data corruption and misrepresentation by post-processing are endless and the audit trail may be tortuous; therefore, a pre-requisite of any test data handling system is to maintain an inviolate copy of the original data, proof against accidental or malicious alteration.

The first level of post-processing, whereby individual tests are converted into a report for the use of a wider group than those conducting the test, such as production or QA, is very formulaic and optimized for its specific purpose and statistical analysis.

In research and development work the test data may be subjected to later analysis not directly related to the initial purpose of the test and used statistically in a wider investigation.

The processes whereby data becomes statistics contain terms and concepts that are frequently misused. It is vital that the basic, mathematically correct, nomenclature of data analysis is understood by both those at its source and its end use. In particular there is common confusion in the use of the terms "data" and "statistic", as in data analysis and statistical analysis. An illustrative example follows:

> The torque produced by a particular engine at 3000 rpm is not a statistic, it is data, a fact (accurate within the restraints of the experiment). To a statistician it is a *specimen*.
>
> The torque produced by 10% of the same engine type at 3000 rpm is also data. To a statistician it is a *sample*.
>
> The torque produced by all the engines of the same type at 3000 rpm is also data. To a statistician it is a *population*.
>
> The average of the population's (all the engines) torque at 3000 rpm is not a statistic, it is data, but the average of the torque value produced by taking the average of the sample is a statistic.

A statistic is an estimate of a parameter based on data collected from a sample of a population.

In the case of averaging, commonly used in simple processing of powertrain data, qualification of the simple average has to be made because we need to know if the spread of torque figures was over a narrow or wide range of values, and whether there were numerically more engines above or below the average value. The first qualification is *standard deviation* and the second is called the *skew* of the population. The processing of such data sets can be carried out by any technical spreadsheet program; the presentation of the results requires understanding of the underlying data and the test processes.

Since test acquired data is expensive in time and money any business will need to make the widest possible use of it.

A very small powertrain or vehicle test operation usually stores test data in a serially dated, flat form, database. Most importantly they also store the data as the developing and applied experience of their operative(s). Thus, when test-related problems or phenomena are met more than once the system relies on human memory to find previous occurrences so that resolution can be based on previous data or methodology. Large organizations, having vast quantities of data and a turnover of many employees, have to develop the same learning ability by structuring data storage so it can be efficiently mined in the future and so that it adds to the corporate wisdom. Data mining [1] is a practice carried out by staff trained in the subject and usually far removed from the test cells; it, like the design and use of a relational database, are highly complex subjects and are served by their own specialist literature.

All this means that the test engineer has to be provided with tools containing proven data handling and presentation templates that use audited formulae. Additionally, security of data needs to be built into any tool so that accidental or deliberate changes to data can be seen or prevented.

The difference between statistical analysis and data mining is the subject of specialist debate. For readers interested in data produced from testing engines and transmissions the following has been found useful:

Statistical analysis uses a model to characterize a pattern in the data; data mining uses the pattern in the data to build a model.

In other words, during statistical analysis you know what pattern in your data you are looking for, while in data mining (which is usually carried out on very large data sets) you are looking for patterns but you may, before starting, have little idea of their form or use.

Two recommendations to test and development engineers, based on experience and observation of data analysis, can be summed up by:

• Restrain the impulse to extrapolate beyond available data.
• Be warned by the exclamation attributed to an analyst of geological data, "If I hadn't believed it, I would never have seen it."

The presentation of data in graphical format has rather less literature than statistics, but is increasingly misused with the aid of powerful graphic software.[3] This is one more reason for the test engineer to be provided with proven (data graphing) tools containing appropriate presentation templates that use audited formulae.

Data Analysis Tools for the Test Engineer

Proprietary data analysis and presentation tools tend to be specific to areas of research such as population trends or medical research, but a few have been adapted and designed for the engineering test industry.

A tool widely used by high-end research laboratories is a software suite produced by AVL List called Concerto®, which is specifically designed to deal with the different types of data from test bed, combustion analysis and emission

---

3. The use of test result graphs seems to the author to suffer, on occasion, from two generic faults: the first is the trend towards small, multicolored, three-dimensional graphs from which only a very subjective impression can be made, rather than data points read. The second is the almost complete absence of error bars used in line graphs of post-processed data. All data points do not have the same degree of certainty and it is useful to know, and remind oneself of, the relative degrees of uncertainty.

equipment, and provides most of the mathematical, statistical, and graphical tools useful to automotive development engineers. If the "upstream" users of the test facility, i.e. the engineers requesting test work to be carried out, are not aware of the capabilities of the data acquisition system they will not make optimum use of it. It is essential that they should be given proper training in the capabilities of the facility and its data in order to use it to maximum advantage. This statement will appear to be a truism yet, in the experience of the author, training in the capabilities of modern test equipment and data analysis is often underfunded by purchasers after the initial investment, which leads to the facility's skills and use leveling out well below best possible practice.

Physical Security of Data

The proliferation and increased reliability of devices used for the storage and processing of electronic data has made the risk of its security more of a problem of misappropriation than of accidental loss. The privately owned computer system on which this book is being written automatically and wirelessly backs up all changed files every 30 minutes to a separate, secure storage device. The danger of catastrophic loss through hardware failure is greatly reduced from the level of that same risk that existed when the first edition was written in 1992, but the danger of human data management failings, of misfiling or overwriting, are undiminished; it's all about management.

In the twenty-first century, data security relies on management procedures: the appropriate distribution, access to and archiving of data, and the use of only authorized post-processing tools and templates.

The control system software will have its own security system to guard against unauthorized copying and use. It will probably be sold with a user license that will, through embedded code capable of shutting down the system, define and restrict the licensee to use the program on a single, identified machine.

Security of data systems from malicious programs is a priority of the IT department and the imposition of rules concerning internet access on the company computer system. Some test facilities, such as those involved with military or motor sport development, can be under constant and real threat to the confidentiality of their data. Physical restriction of access of personnel is built into the design of such test facilities. However, modern miniaturized data storage devices and phone cameras seem to conspire to make security of data increasingly difficult; it is a management role to balance the threat with sensible working practices.

Meanwhile, differential access to computer network systems has to be imposed by password or personal "swipe card".

In most large engine test facilities there are three levels of password-enforced security:

- *Operator.* Typically at this level tests may be started, stopped, and run only when the "header information", which includes engine number, operator name, and other key information, has been entered. Operator code may be verified. No alarms can be altered and no test schedules edited, but there will be a text "notepad" for operator comments, to be stored with the other data.

- *Supervisor.* On entry of password this level will allow alteration of alarm levels and test schedules, also activation of calibration routines, in each case by way of a form fill routine and menu entry. It will also give access to fault-finding routines such as the display of signal state tables.
- *Engineer.* On entry of password this level is allowed access to calibration and debugging routines.

There will certainly be a fourth and lowest level of access available only to the system supplier, which can be called "commissioning"; this allows system set-up and will not be available to the user.

REFERENCE AND FURTHER READING

[1] R. Nisbet, J. Elder IV, G. Miller, Handbook of Statistical Analysis and Data Mining Applications, Academic Press, 2009. ISBN-13: 978-0123747655.

SAE J1939. Standards collection on the web at: http://www.sae.org/products/j1939a.htm

Chapter 14

Measurement of Fuel, Combustion Air, and Oil Consumption

Engine Testing. DOI: 10.1016/B978-0-08-096949-7.00014-5

INTRODUCTION

Until the arrival of instruments capable of measuring "instant" air and liquid flows, the standard test methods recorded cumulative consumption over a period of time, or measured the number of revolutions over which a known mass of liquid fuel was consumed. While these methods and instruments used are still viable, increasingly test engineers are required to record the actual, transient, consumption during test sequences that simulate "real-life" operation.

Modern engine designs incorporate a number of fuel spillback or fuel return strategies that make the design of test cell consumption systems more complex than before.

In simple terms it is necessary now to measure the amount of fuel entering the metering and conditioning system, rather than leaving it on its way to the engine. The systems frequently have to control the pressure of the fuel return line and remove vapor bubbles and the heat energy picked up during the passage through the engine's fuel rail and pressure regulator circuit. Therefore, liquid fuel conditioning and consumption measurement is no longer a matter of choosing the appropriately sized instrument, but rather of designing a complex pipe circuit incorporating instruments, heat exchangers, and the engine.

The engine test engineer is likely to meet the measurement of air flow in two quite separate guises, both of which are covered in this chapter:

- The measurement of the air consumption of the engine during running under a multitude of operating and atmospheric conditions.
- The aerodynamic qualities of the inlet and exhaust passages within an engine's cylinder-head as measured on a specifically designed "flow bench".

PART 1. MEASUREMENT INSTRUMENTS FOR LIQUID FUEL CONSUMPTION

The great curse of liquid fuel handling and consumption measurement is vapor and air, whether entrained in the liquid flow or as a bubble creating an airlock somewhere in the system; the more complex the tubing systems and the higher the liquid temperatures, the worse are the problems. All systems must be designed and installed to minimize vapor formation and allow release of any air or vapor that does get into them.

For all liquid fuel consumption measurement it is critical to control fuel temperature within the metered system so far as is possible; therefore, most modern cells have well-integrated temperature control and measurement systems installed as close to the engine as is practical. The condition of the fuel returned from the engine can cause significant problems as it may return at pulsing pressure, considerably warmer than the control temperature and containing vapor bubbles. This means that the volume and density of the fluid can be variable within the measured system, and that this variability has to be reduced so far as is possible by the overall metering system and the level of measurement uncertainty taken into account (also see Chapter 8, "Engine Fuel Temperature Control" section).

Whatever type of instrument is used in the measurement of liquid fuels they must be capable of handling the range of liquids used without suffering chemical attack to any of its internals and seals. It is also vital that the meters and the local pipe systems in which they are installed should be capable of complete draining to avoid cross-contamination when changing fuels; however, many large operations tend to avoid system poisoning by the use of cells dedicated to only one fuel type.

There are two generic types of fuel gauge intended for cumulative measurement of liquid fuels on the market:

- *Volumetric gauges,* which usually measure the number of engine revolutions, at a constant power output, taken to consume a known volume of fuel either from containment of known volume or as measurement of flow through a measuring device.
- *Gravimetric gauges,* which usually measure the number of engine revolutions, at a constant power output, taken to consume a known mass of fuel.

Figure 14.1 shows a volumetric gauge in which an optical system gives a precise time signal at the start and end of the emptying of one of a choice of

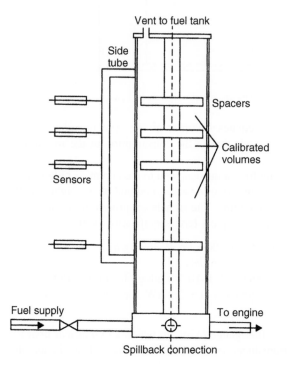

FIGURE   14.1   Volumetric fuel consumption gauge.

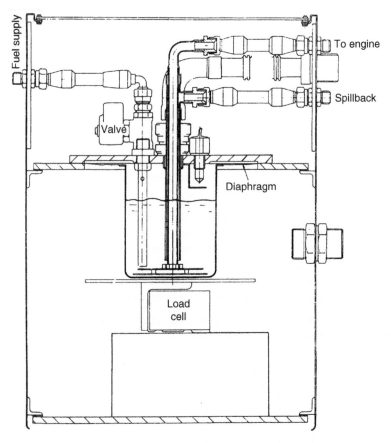

FIGURE 14.2   Gravimetric, direct weighing, fuel gauge.

calibrated volumes. This signal actuates a counter, giving a precise value for the number of engine revolutions made during the consumption of the measured volume of fuel.

Figure 14.2 shows a gravimetric gauge designed to meter a mass rather than a volume of fuel, consisting essentially of a vessel mounted on a weighing cell from which fuel is drawn by the engine. The signals are processed in the same way as for the volumetric gauge, a typical base specification being:

- Measuring ranges: 0–150 kg/h, 0–200 kg/h, 0–360 kg/h
- Fuels: petrol and diesel fuels and biofuels
- Computer interface: serial RS232C, 1–10 V analog plus digital I/O
- Electrical supply required: typically 0.5–2.5 kW for measuring and fuel conditioning.

A further type of gravimetric fuel gauge exists, in which a cylindrical float is suspended from a force transducer in a cylindrical vessel. The change in flotation force is then directly proportional to the change of mass of fuel in the vessel.

These gravimetric gauges, also known as fuel balances, have the advantage over the volumetric type that the metered mass may be chosen at will, and a common measuring period may be chosen, independent of fuel consumption rate. The specific fuel consumption is derived directly from only three measured quantities: the mass of fuel consumed, the number of engine revolutions during the consumption of this mass, and the mean torque. The disadvantage of a fuel balance is that it cannot be used continuously in the same way as a flowmeter, but the inherently greater accuracy of cumulative fuel consumption measurement is due to the averaging of three direct measurements while rate measurements involve four or, in the case of volumetric meters, five measured quantities.

All gauges of this type have to deal with the problem of fuel spillback from fuel injection systems, and Figure 14.3 shows a circuit incorporating a gravimetric fuel gauge that deals with this matter. When a fuel consumption measurement is to be made, a solenoid valve diverts the spillback flow, normally returned to the header tank during an engine test, into the bottom of the fuel gauge. It is not satisfactory to return the spillback to a fuel filter downstream of the fuel gauge, since the air and vapor always present in the spilled fuel lead to variations in the fuel volume between fuel gauge and engine, and thus to incorrect values of fuel consumption.

Mass Flow or Consumption Rate Meters

There are many different designs of rate meter on the market and the choice of the most suitable unit for a given application is not easy. The following factors need to be taken into account:

- Absolute level of accuracy
- Turn-down ratio
- Sensitivity to temperature and fuel viscosity
- Pressure difference required to operate
- Wear resistance and tolerance of dirt and bubbles
- Analog or impulse-counting readout
- Suitability for stationary/in-vehicle use.

Positive-Displacement Flowmeters

Positive-displacement meters provide high accuracy, ±0.1% of actual flow rate is claimed, and good repeatability. Their accuracy is not affected by pulsating flow and they do not require a power supply nor lengths of straight pipe upstream and downstream in their installation. Their main disadvantage is the appreciable pressure drop, which may approach 1 bar, required to drive the metering unit.

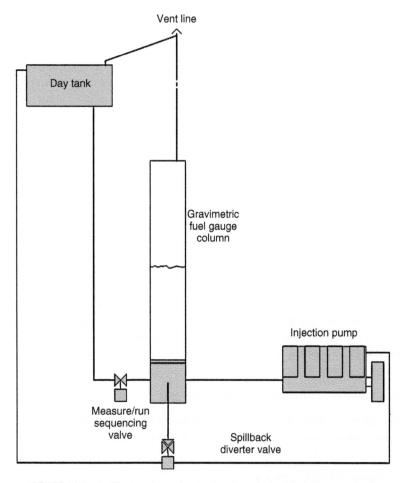

FIGURE 14.3 Spillback and associated valves in a gravimetric fuel gauge circuit.

Several designs of positive-displacement (volumetric) fuel gauges make use of a four-piston metering unit with the cylinders arranged radially around a single-throw crankshaft. Crankshaft rotation is transmitted magnetically to a pulse output flow transmitter. Cumulative flow quantity and instantaneous flow rate are indicated and these meters are suitable for in-vehicle use. A turn-down ratio as high as 100:1 is claimed in some designs.

Coriolis Effect Flowmeters

The market-leading mass flowmeters now tend to make use of the Coriolis effect, in which the fuel is passed through a pair of vibrating U-tubes (Figure 14.4).

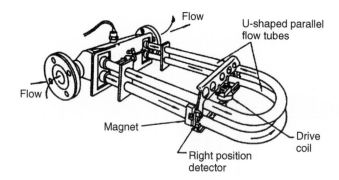

FIGURE 14.4 Coriolis effect flowmeter.

When the design consists of two parallel tube loops, flow is divided into two streams by a splitter near the meter's inlet and is recombined at the exit. The tubes are vibrated by an electromagnetic driver that forces the tubes towards and away from one another at a fixed frequency. Two sensors, positioned on opposite sides of the loop, detect the position, velocity, or acceleration of the tubes, creating a sinusoidal voltage output. When there is no flow in through the tubes the vibration caused by the driver results in identical displacements at the two sensing points. When flow is present, the Coriolis forces act to produce a secondary twisting vibration, resulting in small phase differences in the relative motions that are detected at the sensing points. The deflection of the tubes caused by the Coriolis force only exists when both axial fluid flow and tube vibration are present, and the meters are designed so that identical phased vibration detected by the detectors equals zero flow and the difference in phase, induced by flow, is calibrated to read as the volume of liquid flowing through the unit.

Coriolis effect devices, when built into an electronically controlled, fuel mass flow metering package, have a very high turn-down factor, and can have (depending on seal materials, etc.) multi-fuel density capacity, allowing a single unit to cover a wide range of engines, typically, in the automotive range, from a small three-cylinder gasoline engine up to 600 kW diesel units. The electronic controls also allow flow to be updated at rates measured in kilohertz. Calibration of meters using the Coriolis effect is not straightforward and requires a special rig, or the unit being returned to the manufacturer's service base.

Spillback

If it is ever required to measure spillback as a flow separate from engine consumption, it may be necessary to use two sensing units, one in the flow line and one in the spill line; this increases cost and complexity.

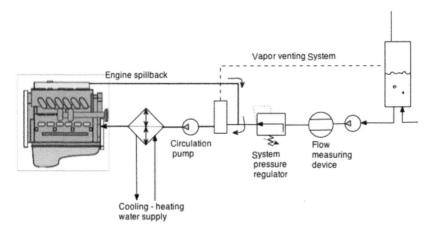

FIGURE 14.5   A simplified schematic diagram of an engine fuel system in which the spillback is recirculated and flow into the loop recorded as engine consumption.

In most cases engine tests are required to record engine fuel consumption and therefore the fuel meter is on the supply side of a recirculating loop circuit containing a volume of fuel from which the engine draws and to which spillback is returned, as shown in Figure 14.5. Change in the volume of fuel in the recirculating loop, from whatever cause, will be recorded as consumption, which is why stable, continuously circulating fuel has to be ensured by a pump of a greater flow rate than consumption. The recirculating fuel in the loop has to be temperature conditioned and fitted with some means of detecting excessive aeration so that a venting sequence can be initiated.

Fuel Consumption Measurements: Gaseous Fuels

Metering of gases is a more difficult matter than liquid metering, partially because the pressure difference available to operate a meter is limited. The density of a gas is sensitive to both pressure and temperature, both of which must be known when, as is usually the case, the flow is measured by volume. The traditional domestic gas meter contains four chambers separated by bellows and controlled by slide valves. Successive increments in volume metered are quite large, so that instantaneous or short-term measurements are not possible.

Other meter types, capable of indicating smaller increments of flow, make use of rotors having sliding vanes or meshing rotors. In the case of natural gas supplied from the mains flow, measurement is a fairly simple matter. Pressure and temperature at the meter must be measured, and accurate data on gas properties (density, calorific value, etc.) will be available from the supplier.

Measurement of liquefied petroleum gas consumption is less straightforward, since this is stored as a liquid under pressure and is vaporized, reduced in pressure, and heated before reaching the engine cylinders. The gas meter must be installed in the line between the "converter" and the engine, and it is essential to measure the gas temperature at this point to achieve accurate results.

Effect of Fuel Condition on Engine Performance

The effect of the condition (pressure, temperature, humidity, and purity) of the combustion air that enters the engine is discussed later in this chapter. In the case of a spark-ignition engine the maximum power output is more or less directly proportional to the mass of oxygen contained in the combustion air, since the air/fuel ratio at full throttle is closely controlled.

In the case of the diesel engine the position is different and less clear-cut. The maximum power output is generally determined by the maximum mass rate of fuel oil flow delivered by the fuel injection pump operating at the maximum fuel stop position. However, the setting of the fuel stop determines the maximum volumetric fuel flow rate and this is one reason why fuel temperature should always be closely controlled in engine testing.

Fuel has a high coefficient of cubical expansion, lying within the range 0.001–0.002 per °C (compared with water, for which the figure is 0.00021 per °C). This implies that the mass of fuel delivered by a pump is likely to diminish by between 0.1% and 0.2% for each degree rise in temperature, not a negligible effect.

A further fuel property that is affected by temperature is its viscosity, significant because in general the higher the viscosity of the fuel, the greater the volume that will be delivered by the pump at a given rack setting. This effect is likely to be specific to a particular pump design; as an example, one engine manufacturer regards a "standard" fuel as having a viscosity of 3 cSt at 40 °C and applies a correction of 2.5% for each centistoke departure from this base viscosity, the delivered volume increasing with increasing viscosity.

As a further factor, the specific gravity (density) of the fuel will also affect the mass delivered by a constant-volume pump, though here the effect is obscured by the fact that, in general, the calorific value of a hydrocarbon fuel falls with increasing density, thus tending to cancel the change out.

Measurement of Lubricating Oil Consumption

Oil consumption measurement is of increasing importance since it is considered to be an important component in the production of engine exhaust emissions and in particulates, but it is one of the most difficult measurements associated with engine testing within the test cell.

Oil consumption rate is very slow relative to the quantity in circulation. It is also quite variable between engine designs and highly dependent on the engine's operating regime. GM have stated that for all 1996–2004 passenger cars and gasoline-powered light-duty trucks under 8500 lb their accepted rate of oil consumption is 0.946 liter (1 qt) in 3200 km (2000 mi) [1] and while consumption by 2010 passenger cars will generally be lower than this, the running of a test to simulate vehicle mileage of 2000 miles would generally be considered as being too expensive for wide use.

Oil consumption of modern (2010) US-produced heavy-duty diesel engines is believed to be about 1:1000 of fuel consumption. This means that test durations would have to be of long duration before meaningful results, from measurements based on changes in oil volume, are obtained.

There are several oil consumption meter systems on the international market based on a procedure that is an automated drain and weigh method. By connections into the engine's sump drain hole or dipstick hole, the sump is arranged to drain to a separate receiver, the contents of which are monitored, either by weighing or by depth measurement. An alternative method is to draw off oil down to a datum sump level and weigh it before and after a test period.

Difficulties with all such systems include:

- The quantity of oil adhering to the internal surfaces of the engine is very sensitive to temperature, as is the volume in transit to the receiver. This can lead to large apparent variations in consumption rate.
- Volumetric changes due to temperature soak of the engine and fluids.
- Apparent consumption is influenced by fuel dilution and by any loss of oil vapor.
- The rate of oil consumption tends to be very sensitive to conditions of load, speed, and temperature. There is also a tendency to medium-term variations in apparent rate, due to such factors as "ponding up" of oil in return drains and accumulation of air or vapor in the circuit.

For a critical analysis of the dry sump method, see Ref. [2], which also describes a statistical test procedure aimed at minimizing these random errors.

Near real-time continuous measurement of oil consumption is now possible by the detection of products of its combustion in the exhaust gas stream. One method is based on lubricant labeling using radiotracer compounds, where oil consumption measurement is achieved by trapping and monitoring tracer residues in the exhaust line. The use of radiotraceable isotopes is a technique that is increasingly used in tribology research (see Chapter 20) but is restricted in its use to specialist laboratories.

An alternative near real-time oil consumption measurement is achieved through the monitoring of the SO_2 concentration in the exhaust stream. The Da Vinci Lubricant Oil Consumption (DALOC) system [3] uses an ultraviolet fluorescence detector and pretreats the exhaust sample in order to remove NO, which would create unwanted fluorescence, and convert any unburnt sulfur species into SO_2. Clearly the fuel used must be of a known, constant, and very low sulfur content so that variation in readings are truly indicative of lubricating oil consumption. As with exhaust emission benches, the DALOC system can be repeatedly calibrated using an SO_2 calibration gas (Figure 14.6).

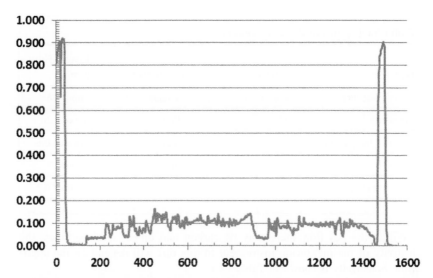

FIGURE 14.6 Concentration of SO_2 in an engine exhaust flow measured as a proxy for oil consumption during the running of a Tier 4 EPA "non-road" transient cycle. The vertical scale units are parts per million by volume, and the peaks at the start and end are due to calibration gas being injected into the system as a quality check. *(Courtesy of Da Vinci Emissions Services Ltd.)*

Measurement of Crankcase Blow-By

In all reciprocating internal combustion engines there is a flow of gas into and out of the clearances between piston top land, ring grooves, and cylinder bore. In an automotive gasoline engine these can amount to 3% of the combustion chamber volume. Since in a spark-ignition engine the gas consists of unburned mixture that emerges during the expansion stroke too late to be burned, this can be a major source of hydrocarbon (HC) emissions and represents a loss of power. Some of this gas will leak past the rings and piston skirt as blow-by into the crankcase. It is then vented back into the induction manifold and to this extent reduces the HC emissions and fuel loss, but has an adverse effect on the lubricant. A second possible source of blow-by gas is from the oil drain a turbocharger system feeds into the crankcase void; to isolate this source and to measure the volumes produced requires a separate metering device.

Blow-by flow is highly variable over the full range of an engine's performance and life; therefore, accurate blow-by meters will need to be able to deal with a wide range of flows and with pulsation. Instruments based on the orifice measurement principle, coupled with linearizing signal conditioning (the flow rate is proportional to the square root of the differential pressure), are good at measuring the blow-by gas in both directions of flow that can occur when there is heavy pulsation between pressure and partial vacuum in the crankcase.

TABLE 14.1 Types of Blow-By Meters

Type	Typical Accuracy	Dirt Sensitivity	Lowest Flow (l/min)
Positive impeller	2% FS	Medium	Approx. 6
Hot wire	2%	High	Approx. 28
Commercial gas meter	1% FS	High	Claimed 0.5
Vortex	1%	High	Approx. 7
Flow through an orifice	1%	Low	Claimed 0.2

Within the gas flow there may be carbon and other "dirt" particles; the sensitivity of the measuring instrument to this dirt will depend on the application. Sensitivity is shown in Table 14.1.

Crankcase blow-by is a significant indicator of engine condition and should preferably be monitored during any extended test sequence. An increase in blow-by can be a symptom of various problems such as incipient ring sticking, bore polishing, deficient cylinder bore lubrication, or developing turbocharger problems.

PART 2. MEASUREMENT OF AIR CONSUMPTION AND GAS FLOWS

The accurate measurement of the air consumption of an internal combustion engine is a matter of some complexity but is of great importance. The theory also has relevance to gas flow measurement in emissions testing; therefore, the subject is covered in some detail in the following text.

Properties of Air

Air is a mixture of gases with the following approximate composition:

Major constituents	By mass	By volume
Oxygen, O_2	23%	21%
Nitrogen, N_2	77%	79%

plus trace gases and water vapor of variable amount, usual range 0.2–2.0% of volume of dry air.

The amount of water vapor present depends on temperature and prevailing atmospheric conditions. It can have an important influence on engine performance, notably on exhaust emissions, but for all but the most precise work its influence on air flow measurement may be neglected.

The relation between the pressure, specific volume, and density of air is described by the gas equation:

$$p_a = 10^5 = \rho R T_a \tag{14.1}$$

where R, the gas constant, has the value for air, $R = 287$ J/kg·K.

A typical value for air density in temperate conditions at sea level would be $\rho = 1.2$ kg/m^3.

Air Consumption, Condition, and Engine Performance

The internal combustion engine is essentially an "air engine" in that air is the working fluid; the function of the fuel is merely to supply heat. There is seldom any particular technical difficulty in the introduction of sufficient fuel into the working cylinder but the attainable power output is strictly limited by the charge of air that can be aspirated.

It follows that the achievement of the highest possible volumetric efficiency is an important goal in the development of high-performance engines, and the design of inlet and exhaust systems, valves and cylinder passages represents a major part of the development program for engines of this kind.

The standard methods of taking into account the effects of charge air condition as laid down in European and American Standards are complex and difficult to apply, and are mostly used to correct the power output of engines undergoing acceptance or type tests. A simplified treatment, adequate for most routine test purposes, is given below.

The condition of the air entering the engine is a function of the following parameters:

1. Pressure
2. Temperature
3. Moisture content
4. Impurities.

For the first three factors standard conditions, according to European and American practice, are:

Atmospheric pressure 1 bar (750 mmHg)
Temperature 25 °C (298 K)
Relative humidity 30%.

Variations in Engine Performance Due to Atmospheric Pressure

Since the volumetric efficiency of an engine tends to be largely independent of the air supply pressure, the mass of air consumed tends to vary directly with the density, which is itself proportional to the absolute pressure, other conditions remaining unchanged. Since the standard atmosphere = 1 bar, we may write:

$$\rho_t = \rho_n P \tag{14.2}$$

where ρ_t = density under test conditions (kg/m^3), ρ_n = density under standard conditions (kg/m^3), and P = atmospheric pressure under test conditions (bar).

It follows that a change of 1% or 7.5 mmHg corresponds to a change in the mass of air entering the engine of 1%. For most days of the year the (sea level) atmospheric pressure will lie within the limits 750 mmHg ± 3%, say between 775 mmHg (103 kPa) and 730 mmHg (97 kPa), with a corresponding percentage variation in charge air mass of 6%.

It is common practice to design the test cell ventilation system to maintain a small negative pressure in the cell, to prevent fumes entering the control room, but the level of depression is unlikely to exceed about 50 mm water gauge (50 Pa), equivalent to a change in barometer reading of less than 0.37 mmHg. Clearly the effect on combustion air flow, if the engine is drawing air from within the cell, is negligible.

The barometer also falls by about 86 mmHg (11.5 kPa) for an increase in altitude of 1000 m (the rate decreasing with altitude) (see Table 14.2). This indicates that the mass of combustion air falls by about 1% for each 90 m (300 ft) increase in altitude, a very significant effect.

Variations in charge air pressure have an important knock-on effect: the pressure in the cylinder at the start of compression will in general vary with the air supply pressure and the pressure at the end of compression will change in the same proportion. This can have a significant effect on the combustion process.

TABLE 14.2 Variation in Atmospheric Pressure with Height Above Sea Level

Altitude (meters)	Fall in Pressure (bar)
0	0
500	0.059 (5900 Pa)
1000	0.115
1500	0.168
2000	0.218
3000	0.312
4000	0.397 (39.7 kPa)

Variations in Engine Performance Due to Air Temperature

Variations in the temperature of the air supply have an effect of the same order of magnitude as variations in barometric pressure within the range to be expected in test cell operation. Air density varies inversely with its absolute temperature:

$$\rho_t = \rho_n \cdot \frac{298}{T_t + 273}$$

The temperature at the start of compression determines that at the end. In the case of a naturally aspirated diesel engine with a compression ratio of 16:1 and an air supply at 25 °C, the charge temperature at the start of compression would typically be about 50 °C. At the end of compression the temperature would be in the region of 530 °C, increasing to about 560 °C for an air supply temperature 10 °C higher. The level of this temperature can have a significant effect on such factors as the NO_x content of the exhaust, which is very sensitive to peak combustion temperature. The same effect applies, with generally higher temperatures, to turbocharged engines.

Variations in Engine Performance Due to Relative Humidity

Compared with the effects of pressure and temperature the influence of the relative humidity of the air supply on the air charge is relatively small, except at high air temperatures. The moisture content of the combustion air does, however, exert a number of influences on performance. Some of the thermodynamic properties of moist air have been discussed under the heading of psychometry (Chapter 5), but certain other aspects of the subject must now be considered.

The important point to note is that unit volume of moist air contains less oxygen in a form available for combustion than the same volume of dry air under the same conditions of temperature and pressure. Moist air is a mixture of air and steam: while the latter contains oxygen it is in chemical combination with hydrogen and is thus not available for combustion. The European and American Standards specify a relative humidity of 30% at 25 °C. This corresponds to a vapor pressure of 0.01 bar, thus implying a dry air pressure of 0.99 bar and a corresponding reduction in oxygen content of 1% when compared with dry air at the same pressure. Figure 14.7 shows the variation in dry air volume expressed as a percentage adjustment to the standard condition with temperature for relative humidities ranging from 0% (dry) to 100% (saturated).

The effect of humidity becomes much more pronounced at higher temperatures; thus, at a temperature of 40 °C the charge mass of saturated air is 7.4% less than for dry air. The effect of temperature on moisture content (humidity) should be noted particularly, since the usual method of indicating specified moisture content for a given test method, by specifying a value of relative humidity, can be misleading. Figure 14.7 indicates that at a temperature of 0 °C the difference between 0% (dry) and 100% (saturated) relative humidity represents a change of only 0.6% in charge

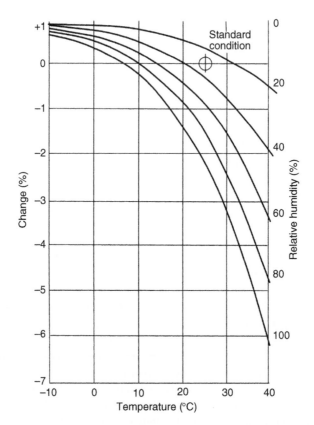

FIGURE 14.7   Variation of air charge mass with relative humidity at different temperatures.

mass; at −10 °C it is less than 0.3%. It follows that it is barely worthwhile adjusting the humidity of combustion air at temperatures below (perhaps) 10 °C.

The moisture content of the combustion air has a significant effect on the formation of NO_x in the exhaust of diesel engines. The SAE procedure for measurement of diesel exhaust emissions gives a rather complicated expression for correcting for this. This indicates that, should the NO_x measurement be made with completely dry air, a correction of the order of +15% should be made to give the corresponding value for a test with moisture content of 60% relative humidity.

These various adjustments or "corrections" to the charge mass are of course cumulative in their effect and the following example illustrates the magnitudes involved.

Consider:

A. A hot, humid summer day with the chance of thunder.
B. A cold, dry winter day of settled weather.

TABLE 14.3

Condition	A	B
Pressure	0.987 bar, 740 mmHg	1.027 bar, 770 mmHg
Temperature	35 °C	10 °C
Relative humidity	80%	40%
Pressure	$\dfrac{1}{0.987} = 1.0135$	$\dfrac{1}{1.027} = 0.9740$
Temperature	$\dfrac{308}{298} = 1.0336$	$\dfrac{283}{298} = 0.9497$
Relative humidity from Figure 14.4	+3.48% = 1.0348	0%
Total adjustment	1.0840, +8.4%	0.9233, −7.7%

The conditions are summarized in Table 14.3. The "adjustment" indicates the factor by which the observed power should be multiplied to indicate the power to be expected under "standard" conditions. We see that under hot, humid, stormy conditions the power may be reduced by as much as 15% compared with the power under cold anticyclonic conditions, a not at all negligible adjustment.

It should be pointed out that the power adjustment calculated above is based on the assumption that charge air mass directly determines the power output, but this would only be the case if the air/fuel ratio were rigidly controlled (as in spark-ignition engines with precise stoichiometric control). In most cases a reduction, for example, in charge air mass due to an increase in altitude will result in a reduction in the air/fuel ratio, perhaps with an increase in exhaust smoke, but not necessarily in a reduction in power. The correction factors laid down in the various Standards take into account the differing responses of the various types of engine to changes in charge oxygen content.

It should also perhaps be pointed out that the effects of water injection are quite different from the effects of humidity already present in the air. Humid air is a mixture of air and steam; the latent heat required to produce the steam has already been supplied. When, as is usually the case, water is injected into air, leaving the turbocharger at a comparatively high temperature, the cooling effect associated with the evaporation of the water achieves an increase in charge density that much outweighs the decrease associated with the resulting steam.

Variations in Engine Performance Due to Impurities

Finally, it should be mentioned that pollution of the combustion air can result in a reduction in the oxygen available for combustion. A likely source of such pollution may be the ingestion of exhaust fumes from other engines or from neighboring industrial processes such as paint plant, and care should also be taken in the siting of air intakes and exhaust discharges to ensure that, even under unfavorable conditions, there cannot be unintended[1] exhaust recirculation. Exhaust gas has much the same density as air and a free oxygen content that can be low or even zero, so that, approximately, 1% of exhaust gas in the combustion air will reduce the available oxygen in almost the same proportion.

The Airbox Method of Measuring Air Consumption

The simpler methods of measuring air consumption involve drawing the air through some form of measuring orifice and measuring the pressure drop across the orifice. It is good practice to limit this drop to not more than about 125 mm H_2O (1200 Pa). For pressures less than this, air may be treated as an incompressible fluid, with much simplification of the air flow calculation.

The velocity U developed by a gas expanding freely under the influence of a pressure difference Δp, if this difference is limited as above, is given by:

$$\frac{\rho U^2}{2} = \Delta p, \quad U = \sqrt{\frac{2\Delta p}{\rho}} \tag{14.3}$$

Typically, air flow is measured by means of a sharp-edged orifice mounted in the side of an airbox, coupled to the engine inlet and of sufficient capacity to damp out the inevitable pulsations in the flow into the engine, which are at their most severe in the case of a single-cylinder four-stroke engine (Figure 14.8). In the case of turbocharged engines the inlet air flow is comparatively smooth and a well-shaped nozzle without an airbox will give satisfactory results.

The air flow through a sharp-edged orifice takes the form sketched in Figure 14.9. The coefficient of discharge of the orifice is the ratio of the transverse area of the flow at plane a (the vena contracta) to the plan area of the orifice. Tabulated values of C_d are available in BS 1042, but for many purposes a value of $C_d = 0.60$ may be assumed.

We may easily derive the volumetric flow rate of air through a sharp-edged orifice as follows:

Flow rate = Coefficient of discharge × Cross-sectional area of orifice

× Velocity of flow

$$Q = C_d \frac{\pi d^2}{4} \sqrt{\frac{2\Delta p}{\rho}}$$

$$\tag{14.4}$$

---

1. Intended, and therefore carefully metered, recirculation of exhaust gas (EGR) is covered in Chapter 16.

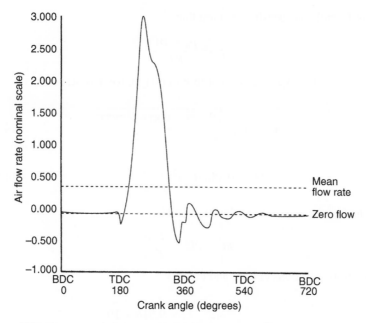

FIGURE 14.8   Induction air flow, single-cylinder, four-stroke diesel engine.

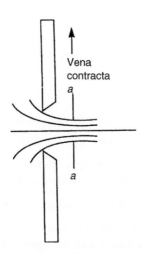

FIGURE 14.9   Flow through a sharp-edged orifice.

Noting from the air density equation that

$$\rho = \frac{p_a \times 10^5}{RT_a}$$

and assuming that Δp is equivalent to h mm H_2O, we may write:

$$Q = C_d \frac{\pi d^2}{4} \sqrt{\frac{2 \times 9.81h \times 287T}{p_a \times 10^5}} \qquad (14.5a)$$

$$Q = 0.1864 C_d d^2 \sqrt{\frac{hT_a}{p_a}} \, m^3/s \qquad (14.5b)$$

To calculate the mass rate of flow, note that:

$$\dot{m} = \rho Q = \frac{p_a Q}{RT_a}$$

giving, from equations (14.5a) and (14.5b):

$$\dot{m} = C_d \frac{\pi d^2}{4} \sqrt{\frac{2 \times 9.81h \times 10^5}{287T_a}} \qquad (14.6a)$$

$$\dot{m} = 64.94 C_d d^2 \sqrt{\frac{hp_a}{T_a}} \qquad (14.6b)$$

Equations (14.5b) and (14.6b) give the fundamental relationship for measuring air flow by an orifice, nozzle, or venturi.

Sample Calculation

If:

$C_d = 0.6$, $D = 0.050$ m, $H = 100$ mm H_2O, $T_a = 293$ K (20°C),
$P_a = 1.00$ bar,

then:

$$Q = 0.04786 \, m^3/s \, (1.69 \, ft^3/s)$$
$$\dot{m} = 0.05691 \, kg/s \, (0.1255 \, lb/s)$$

To assist in the selection of orifice sizes, Table 14.4 gives approximate flow rates for orifices under the following standard conditions:

$H = 100$ mm H_2O, $T_a = 293$ K (20°C), $p_a = 1.00$ bar.

A disadvantage of flow measurement devices of this type is that the pressure difference across the device varies with the square of the flow rate. It follows that a turn-down in flow rate of 10:1 corresponds to a reduction in pressure

TABLE 14.4 Approximate Air Flow Rates for
Orifice Sizes 10–150 mm

Orifice Diameter (mm)	Q (m³/s)	$\dot{m}$ (kg/s)
10	0.002	0.002
20	0.008	0.009
50	0.048	0.057
100	0.19	0.23
150	0.43	0.51

difference of 100:1, implying insufficient precision at low flow rates. It is good practice, when a wide range of flow rates is to be measured, to select a range of orifice sizes, each covering a turn-down of not more than 2.5:1 in flow rate.

Air Consumption of Engines: Approximate Calculations

The air consumption of an engine may be calculated from:

$$V = \eta_v \frac{V_s}{K} \frac{n}{60} \ \text{m}^3/\text{s} \tag{14.7}$$

where $K = 1$ for a two-stroke and 2 for a four-stroke engine.

For initial sizing of the measuring orifice η_v, the ratio of the volume of air aspirated per stroke to the volume of the cylinder may be assumed to be about 0.8 for a naturally aspirated engine and up to about 2.5 for supercharged engines.

Sample Calculation

Consider a single-cylinder four-stroke engine, swept volume 0.8 liter, running at a maximum of 3000 rev/min, naturally aspirated:

$$V = 0.8 \times \frac{0.0008}{2} \times \frac{3000}{60} = 0.16 \ \text{m}^3/\text{s}$$

Suitable orifice size, from Table 14.4, is 30 mm.

Connection of Airbox to Engine Inlet

It is essential that the configuration of the connection between the airbox and the engine inlet should model as closely as possible the configuration of the air intake arrangements in service. This is because pressure pulsations in the inlet can have a powerful influence on engine performance, in terms of both volumetric efficiency and pumping losses.

The resonant frequency of a pipe of length L, open at one end and closed at the other, equals $a/4L$, where "a" is the speed of sound, roughly 330 m/s. Thus, an inlet connection 1 meter long would have a resonant frequency of about 80 Hz. This corresponds to the frequency of intake valve opening in a four-cylinder four-stroke engine running at 2400 rev/min. Clearly such an intake connection could disturb engine performance at this speed. In general the intake connection should be as short as possible.

The Viscous Flow Air Meter

The viscous flow air meter, invented by Alcock and Ricardo in 1936, was for many years the most widely used alternative to the airbox and orifice method of measuring air flow. In this device the measuring orifice is replaced by an element consisting of a large number of small passages, generally of triangular form. The flow through these passages is substantially laminar, with the consequence that the pressure difference across the element is approximately directly proportional to the velocity of flow, rather than to its square, as is the case with a measuring orifice.

This has two advantages. First, average flow is proportional to average pressure difference, implying that a measurement of average pressure permits a direct calculation of flow rate, without the necessity for smoothing arrangements. Second, the acceptable turn-down ratio is much greater.

The flowmeter must be calibrated against a standard device, such as a measuring orifice.

Lucas–Dawe Air Mass Flowmeter

This device depends for its operation on the corona discharge from an electrode coincident with the axis of the duct through which the air is flowing. Air flow deflects the passage of the ion current to two annular electrodes and gives rise to an imbalance in the current flow that is proportional to air flow rate.

An advantage of the Lucas–Dawe flowmeter is its rapid response to changes in flow rate, of the order of 1 ms, making it well suited to transient flow measurements, but it is sensitive to air temperature and humidity, and requires calibration against a standard.

Positive-Displacement Flowmeters

As rotation occurs, successive pockets of air are transferred from the suction to the delivery side of the flowmeter, and the flow rate is proportional to rotor speed. Some of these flowmeters operate on the principle of the Roots blower. Advantages of the positive-displacement flowmeter are accuracy, simplicity, and good turn-down ratio. Disadvantages are cost, bulk, relatively large pressure drop, and sensitivity to contamination in the flow.

Hot Wire or Hot Film Anemometer Devices

The principle of operation of these popular devices is based on the cooling effect caused by gas flow on a heated wire or film surface. The heat loss is directly proportional to the air mass flow rate, providing the flow through the device is laminar. The advantages of these designs are their reliability and good tolerance to contaminated air flows. Their disadvantages include the possibility of condensation of moisture on the temperature detector and coating or material build-up on the sensor. Calibration at site is virtually impossible so they are supplied with a maker's certification. Probably the best known device of this type in European test cells is the Sensyflow™ range made by ABB.

Flow Benches

These devices are designed for measuring the dynamic resistance to air flow imposed by an engine component, usually the intake or exhaust ports in a cylinder-head. Essentially a measured volume of air is drawn through the unit under test (UUT) and the pressure drop created by the UUT is recorded by a calibrated manometer.

Flow benches are much favored by motor sport engine tuners and can be built by any competent workshop; however, the variability in quality and design of these units makes any objective correlation between different units very difficult. Commercially available units range significantly in complexity and price, but industrial standard flow benches such as those made by companies such as SuperFlow® and for major R&D laboratories by AVL (AVL Tippelmann) are the best known and are designed for use on a wide number of engine components in addition to cylinder-heads. The top-of-the-range designs can be used the test the flow through not only engine gas ports, but other components including catalytic converters, and are capable of integrating signals from multiple local velocity probes and from devices measuring air swirl and tumble, thus giving a detailed analysis of air flow through the UUT.

The steady state of pressure and known temperature conditions used in flow bench measurement bear no relation to the dynamics of air, or exhaust gas flow in a running engine; however, changes in engine breathing through reworking of the intake ducting, valve layout, and head design can be judged and recorded by the use of such devices.

NOTATION USED IN AIR CONSUMPTION CALCULATIONS

Atmospheric pressure	p_a (bar)
under test conditions	p (bar)
Atmospheric temperature	T_a (K)
Test temperature	T_t (°C)
Density of air	ρ (kg/m³)
under standard conditions	ρ_n (kg/m³)
under test conditions	ρ_t (kg/m³)

(Continued)

Pressure difference across orifice	Δp (Pa), h (mm H_2O)
Velocity of air at contraction	U (m/s)
Coefficient of discharge of orifice	C_d
Diameter of orifice	d (m)
Volumetric rate of flow of air	Q (m^3/s)
Mass rate of flow of air	$\dot{m}$ (kg/s)
Constant, 1 for two-stroke and 2 for four-stroke engines	K
Engine speed	n (rev/min)
Number of cylinders	N_c
Swept volume, total	V_s (m^3)
Air consumption rate	V (m^3/s)
Volumetric efficiency of engine	η_v
Volume of airbox	V_b (m^3)
Gas constant	R (J/kg·K)

REFERENCES

[1] Fluid Flow Measurement Standards, BS 1042/ISO 5167.

[2] P.-W. Johren, B.A. Newman, Evaluating the Oil Consumption Behaviour of Reciprocating Engines in Transient Operation (1988). SAE Paper 880098.

[3] K. Froelund et al., An accelerated testing approach for lubricant oil consumption reduction on an EMD 710 diesel engine, 2010 ASME ICE Conference (2010), September.

FURTHER READING

BS 5514 Parts 1 to 6. Reciprocating Internal Combustion Engines: Performance.

BS 7405. Guide to Selection and Application of Flowmeters for the Measurement of Fluid Flow in Closed Conduits.

Engine Oil Consumption Information, GM Bulletin No. 01-06-01-011A.

J.B. Heywood, Internal Combustion Engine Fundamentals, McGraw-Hill, Maidenhead, 1988.

L.J. Kastner, The airbox method of measuring air consumption, Proc. I. Mech. E. 157 (1947).

E. Loy Upp, P.J. LaNasa, Fluid Flow Measurement: A Practical Guide to Accurate Flow Measurement, Gulf Professional Publishing, 2002. ISBN-13: 978–0884157588.

C.R. Stone, Airflow Measurement in Internal Combustion Engines (1989). SAE Technical Paper Series 890242.

Chapter 15

The Combustion Process and Combustion Analysis

Engine Testing. DOI: 10.1016/B978-0-08-096949-7.00015-7

INTRODUCTION

The processes taking place in the cylinder and combustion chamber are obviously central to the performance of the internal combustion engine; in fact, the rest of the engine may be regarded merely as a device for managing these processes and for extracting useful work from them. With the upsurge in development of hybrid vehicles and the corresponding research into internal combustion engines that are optimized for low emission performance within alternative hybrid configurations, combustion analysis has been brought back into the forefront of engine testing. For many test engineers the remit of combustion analysis has widened considerably to cover not only Otto cycle SI and CI reciprocating engines that are being "downsized" and highly boosted, but also linear piston engines and those operating homogeneous charge compression ignition (HCCI) cycles and new technologies such as laser-initiated ignition are now on the horizon. While electronic controls have made such radical engine designs realistic propositions in our attempts to develop ever more fuel-efficient vehicles, significant new problems have appeared such as low-speed pre-ignition (LSPI or "super-knock"), discussed later in this chapter.

The term "combustion analysis" (CA) has become somewhat interchangeable with that of "engine indicating" (EI); the latter and older term covers both combustion analysis and a number of other engine phenomena that occur at the same time scale as combustion, such as injector needle movement. In this book, the term "engine indicating" is used to cover the general task of high-speed engine data analysis and the term "combustion analysis" is used in its specific sense.

SINGLE-CYLINDER RESEARCH AND "OPTICAL" ENGINES

Developments of downsized, supercharged engine variants and improvements in laser imaging technology are giving new work to single-cylinder research engine designers and users. Single-cylinder research engines have been made for many years and have enabled cylinder-head and valve design to be refined before implementation into multicylinder engines. The newer optical versions allow visualization and measurement of combustion phenomena in the engine while running under load and (usually) up to 5000 rpm. Various components such as the cylinder walls and piston crown of the fully adaptable engines are constructed from a "glass" and, using optical diagnostic techniques, a number of important phenomena may be observed and photographed by high-speed cine cameras, including:

- Combustion, flame growth, and flame propagation, used not only in testing cylinder-head design, but also comparison of flame propagation with different fuel compositions.
- Flow velocity and particle size can be measured by a non-intrusive phase Doppler anemometer (PDA).

- Concentrations in fuel sprays can be measured using laser-induced fluorescence (LIF).
- In synchronization with in-cylinder pressure sensing, flame signals before and during irregular combustion phases, such as those during LSPI, can be detected and visualized.

Because single-cylinder research engines are designed to be quickly adapted to test combustion spaces of differing geometries, they are to be found in the test cells of most major technical universities worldwide, in the form of an integrated engine and DC dynamometer unit. Due to the expense of new research engines and the mechanical durability of the older models, there are many cases of new instrumentation being integrated with older engine units.

Currently (2011) one of the best known and widely used models, the Ricardo "Hydra", is made in the following major subtypes to address specific areas of automotive engine research:

- Port injection gasoline
- Direct injection gasoline
- Indirect injection (IDI) diesel
- Direct injection (DI) diesel
- Twin-mechanical variable lift (TMVL)
- Controlled auto-ignition (CAI, aka HCCI)
- Camless hydraulic valve actuation (HVA).

Before discussing the problems raised by testing new engine designs, the fundamentals of the internal combustion process should briefly be considered.

FUNDAMENTAL INFLUENCES ON COMBUSTION

At a fundamental level the performance of an internal combustion engine is largely determined by the events taking place in the combustion chamber and engine cylinder. These is turn are influenced by a large number of factors:

- Configuration of the combustion chamber and cylinder-head
- Flow pattern (swirl and turbulence) of charge entering the cylinder, in turn determined by design of induction tract and inlet valve size, shape, location, lift, and timing
- Ignition and injection timing, spark plug position and characteristics, injector and injection pump design, location of injector
- Compression ratio
- Air/fuel ratio
- Fuel properties
- Mixture preparation
- Exhaust gas recirculation (EGR)
- Degree of cooling of chamber walls, piston, and bore
- The presence or absence of particulates from previous cycles.

For those readers not familiar with the details of the combustion process of the spark-ignition (SI) or compression-ignition (CI) engine, they are summarized below.

Combustion in the Conventional Gasoline Engine

1. A mixture of gas and vapor is formed, either in the induction tract, in which the fuel is introduced either by a carburettor or injector, or in the cylinder in the case of direct injection.
2. Combustion is initiated at one or more locations by an electric spark triggered by the engine management system.
3. After a delay, the control of which still presents problems, a flame is propagated through the combustible mixture at a rate determined, among other factors such as fuel chemistry, by the air motion in the cylinder.
4. Heat is released progressively with a consequent increase in temperature and pressure. During this process the bulk properties of the fluid change as it is transformed from a mixture of air and fuel to a volume of combustion products. Undesirable effects, such as pre-ignition, excessive rates of pressure rise, late burning and detonation or "knock", may be present.
5. Heat is transferred by radiation and convection to the surroundings.
6. Mechanical work is performed by expansion of the products of combustion.

The development of the combustion chamber, inlet and exhaust passages, and fuel supply system of a new engine involve a vast amount of experimental work, some on flow rigs that model the geometry of the engine, most of it on the test bed.

Combustion in the Conventional Diesel Engine

1. Air is drawn into the cylinder, without throttling, but frequently with pressure charging. Compression ratios range from about 14:1 to 22:1, depending on the degree of supercharge, resulting in compression pressures in the range 40–60 bar and temperatures from 700 to 900 °C.
2. Fuel is injected at pressures that have increased in recent years, leading to common rail diesels, now fitted with piezoelectric injectors, having fuel pressures over 1800 bar. Charge air temperature is well above the auto-ignition point of the fuel. Fuel droplets vaporize, forming a combustible mixture that ignites after a delay that is a function of charge air pressure and temperature, droplet size and fuel ignition quality (cetane number).

   Note: This simplified description is of a single injection event; however, most automotive diesels are now calibrated for multiple injection events per combustion cycle (see "Fuel Rail Pressure and 'Shot Volume' Measurement" later in this chapter).
3. Fuel subsequently injected is ignited immediately, and the progress of combustion and pressure rise is to some extent controlled by the rate of

injection. Air motion in the combustion chamber is organized to bring unburned air continually into the path of the fuel jet.

4. Heat is transferred by radiation and convection to the surroundings.

5. Mechanical work is performed by expansion of the products of combustion.

Choice of combustion "profile" involves a number of compromises. A high maximum pressure has a favorable effect on fuel consumption but increases NO_x emissions, while a reduction in maximum pressure, caused by retarding the combustion process, results in increased particulate emissions. *Unlike the spark-ignition engine, the diesel must run with substantial excess air to limit the production of smoke and soot.* A pre-combustion chamber engine can run with a lambda ratio of about 1.2 at maximum power, while a direct injection (DI) engine requires a minimum excess air factor of about 50%, roughly the same as the maximum at which a spark-ignition engine will run. It is this characteristic that has led to the widespread use of pressure charging, which achieves a reasonably specific output by increasing the mass of the air charge.

Large industrial or marine engines invariably use direct injection; in the case of vehicle engines the indirect injection or pre-chamber engine is being abandoned in favor of direct injection because of the better fuel consumption and cold starting performance of the DI engine. Compact, very high pressure, piezoelectric injectors and electronic control of the injection process have made these developments possible.

Effects on Combustion Process of Air/Fuel Ratio

Several definitions are important:

- *Mixture strength.* A loose term usually identified with air/fuel ratio and described as "weak" (excess air) or "rich" (excess fuel).
- *Air/fuel ratio.* Mass of air in charge to mass of fuel.
- *Stoichiometric air/fuel ratio* (sometimes known as "correct" air/fuel ratio). The ratio at which there is exactly enough oxygen present for complete combustion of the fuel. Most gasolines lie within the range 14:1 to 15:1 and a ratio of 14.51:1 may be used as a rule of thumb. Alcohols, which contain oxygen in their make-up, have a much lower value in the range 7:1 to 9:1.
- *Excess air factor or "lambda" (λ) ratio.* The ratio of actual to stoichiometric air/fuel ratio. The range is from about 0.6 (rich) to 1.5 (weak). Lambda ratio has a great influence on power, fuel consumption, and emissions.
- *The equivalence ratio ϕ* is the reciprocal of the lambda ratio and is preferred by some authors.

A standard test is to vary the air/fuel ratio over the whole range at which the engine is able to run, keeping throttle opening and speed constant. The results are often presented in the form of a "hook curve" (Figure 15.1), which shows

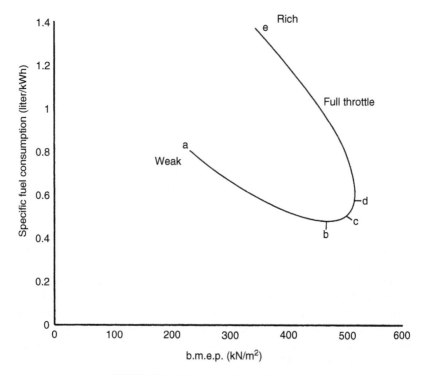

FIGURE 15.1 Hook curve for gasoline engine.

the relation between specific fuel consumption and b.m.e.p. over the full range of mixture strengths.

If such a test is carried out on an optical engine, the following changes are observed:

- At mixture strength corresponding to maximum power and over a range of weaker mixtures, combustion takes place smoothly and rapidly with a blue flame that is extinguished fairly early in the expansion stroke.
- With further weakening, combustion becomes uneven and persists throughout the expansion stroke. "Popping back" into the induction manifold may occur.
- As we continue towards richer mixtures, the combustion takes on a yellow color, arising from incandescent carbon particles, and may persist until exhaust valve opening. This may lead to explosions in the exhaust system.

The following features of Figure 15.1 call for comment:

- Point **a** corresponds to the weakest mixture at which the engine will run. Power is much reduced and specific fuel consumption can be as much as twice that corresponding to best efficiency.

- Point **b** corresponds to the best performance of the engine (maximum thermal efficiency). The power output is about 95% of that corresponding to maximum power.
- Point **c** corresponds to the stoichiometric ratio.
- Point **d** gives maximum power, but the specific consumption is about 10% greater than at the point of best efficiency. It will be evident that a prime requirement for the engine management system must be to operate at point **b** except when maximum power is demanded.
- Point **e** is the maximum mixture strength at which the engine will run.

It is possible to produce similar curves for the whole range of throttle positions and speeds, and hence to derive a complete map of optimum air and fuel flow rates as one of the bases for development of the engine management system, whether it be a traditional carburettor or a computer-controlled injection system.

If at the same time that the hook curve is produced the air flow rate is measured, the same information may be presented as curves of power output and specific consumption against air/fuel ratio or λ (see Figure 15.2), which corresponds to Figure 15.1.

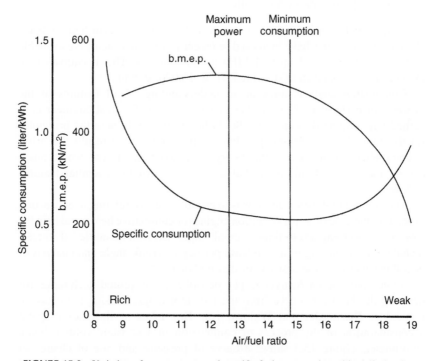

FIGURE 15.2   Variation of power output and specific fuel consumption with air/fuel ratio.

Engine Indicating (EI) Measurements

Many special techniques have been developed for the study of the combustion process. The oldest is direct observation of the process of flame propagation, using high-speed photography through a quartz window. More recent developments include the use of flame ionization detectors (FIDs) to monitor the passage of the flame and the proportion of the fuel burned; laser Doppler anemometry is also used.

The standard tool for the study of the combustion process is the cylinder pressure indicator, the different attributes of which are described later in this chapter. Along with cylinder pressure, a variety of quantities must be measured, using appropriate transducers, either on a time base or synchronized with crankshaft position. They may include:

- Fuel line pressure and needle lift
- Cylinder pressure
- Ionization signals
- Crank angle
- Time
- Ignition system events
- Inlet and exhaust pressures
- Various (quasi-static) temperatures.

Each signal calls for individual treatment and appropriate recording methods. Data acquisition rates have increased in recent years and systems are available with a sampling rate of up to 1 MHz on 16 channels. This is equivalent to cylinder pressures taken at $0.1°$ intervals at 16,000 rev/mm.

Combustion analysis aims at an understanding of all features of the process, in particular of the profile of heat release. The test engineer concerned with engine development should be familiar with both the principles involved and the considerable problems of interpretation of results that arise. A comprehensive account of the theory of combustion analysis would much exceed the scope of this book, but a description of the essential features follows.

The aim of combustion analysis is to produce a curve relating mass fraction burned (in the case of a spark-ignition engine) or cumulative heat release (in the case of a diesel engine) to time or crank angle. Derived quantities of interest include rate of burning or heat release per degree crank angle and analysis of heat flows based on the first law of thermodynamics.

Stone and Green-Armytage [1] describe a simplified technique for deriving the burn rate curve from the indicator diagram using the classical method of Rassweiler and Withrow [2]. This takes as the starting point a consideration of the process of combustion in a constant-volume bomb calorimeter. Figure 15.3 shows a curve of pressure and rate of change of pressure against time. It is then assumed that mass fraction burned and

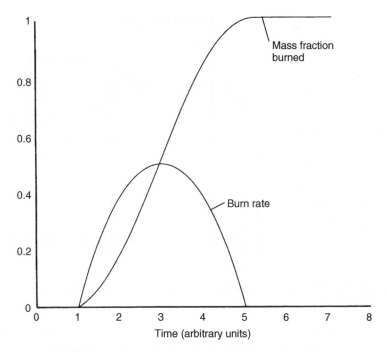

FIGURE 15.3 Combustion in a constant-volume bomb calorimeter.

cumulative heat release, at any stage of the process, are directly proportional to the pressure rise.

Combustion in the engine is assumed to follow a similar course, the difference being that the process does not take place at constant volume. There are three different effects to be taken into account:

- Pressure changes due to combustion
- Pressure changes due to changes in volume
- Pressure changes arising from heat transfer to or from the containing surfaces.

Figure 15.4 shows a pressure–crank angle diagram and indicates the procedure. The starting point is the "motored" or "no burn" curve, shown dotted. This curve used to be obtained by interrupting the ignition or injection for a single cycle and recording the corresponding pressure diagram; certain engine indicators are able to do this. With the more common availability of motoring dynamometers, it is possible to obtain the curve by motoring the unfired engine. An alternative, less accurate, method is to fit a polytropic compression line to the compression curve before the start of combustion and to extrapolate this on the assumption that the polytropic index n_c in the expression $pv^{n_c} = $ constant remains unchanged throughout the remainder of the compression stroke.

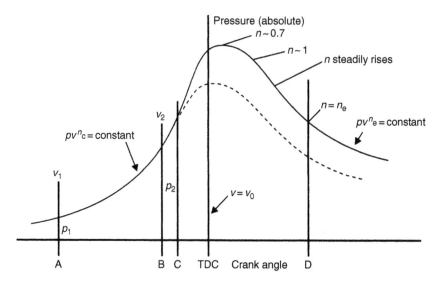

FIGURE 15.4 Pressure–crank angle diagram showing derivation of mass fraction burned.

If A and B are two points on the compression line at which volumes and (absolute) pressures are respectively v_1, v_2 and p_1, p_2, then the index of compression is given by:

$$n_c = \frac{\log(p_1/p_2)}{\log(v_2/v_1)} \tag{15.1}$$

The value of the index n_c during compression is likely to be in the region of 1.3, to be compared with the value of 1.4 for the adiabatic compression of air. The lower value is the result of heat losses to the cylinder walls and, in the case of spark-ignition engines, to the heat absorbed by the vaporization of the fuel. The start of combustion is reasonably well defined by the point C at which the two curves start to diverge.

It is now necessary to determine the point D in Figure 15.4 at which combustion may be deemed to be complete. Various methods may be used, but perhaps the most practical one is a variation on the method described above for determining the start of combustion. Choose two points on the expansion line sufficiently late in the stroke for it to be reasonable to assume that combustion is complete but before exhaust valve opening: 90° and 135° after top dead center (TDC) may be a reasonable choice. The polytropic index of expansion (in the absence of combustion) n_e is calculated from equation (15.1). This is also likely to be around 1.3, except in the special case when burning continues right up to exhaust valve opening, as in a spark-ignition engine burning a weak mixture.

The next step is to calculate the polytropic index of compression/expansion for successive intervals, typically one degree of crank angle, from the start of combustion. Again, equation (15.1) is used, inserting the pressures and volumes at the beginning and end of each interval. The value of n varies widely as combustion continues, but towards the end of the process it converges on the value n_e determined above.

The point D at which the two indices become equal is generally ill-defined, as there is likely to be considerable scatter in the values derived for successive intervals, but fortunately the shape of the burn rate curve is not very sensitive to the position chosen for point D.

The curve of heat release or cumulative mass fraction burned is derived as follows. Figure 15.5a shows diagrammatically an element of the indicator diagram before TDC but after the start of combustion. During this interval the pressure rises from p_1 to p_2 while the pressure rise corresponding to the (no combustion) index n_c is from p_1 to p_0. It is then assumed that the difference between p_2 and p_0 represents the pressure rise due to combustion. It is given by:

$$\Delta p = p_2 - p_0 = p_2 - p_1 \left(\frac{v_1}{v_2}\right)^{n_c} \tag{15.2}$$

It is now necessary to refer this pressure rise to some constant volume, usually taken as v_0, the volume of the combustion chamber at TDC. It is assumed that the pressure rise is inversely proportional to the volume:

$$\Delta p_c = \Delta p \frac{(v_1 + v_2)}{2v_0} \tag{15.3}$$

(a)   (b)

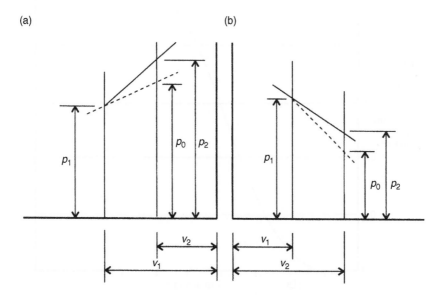

FIGURE 15.5   Derivation of mass fraction burned: (a) compression; (b) expansion.

Essentially the same procedure is followed during the expansion process (Figure 15.5b), except that here there is a pressure fall corresponding to the (no combustion) index of expansion n_e. Δp and Δp_e are calculated from equations (15.2) and (15.3) as before.

Finally, Δp_c is summed for the whole combustion period and the resulting curve (Figure 15.6) is taken to represent the relation between mass fraction burned or cumulative heat release and crank angle. The "tail" of the curve, following the end of combustion, point D, will be horizontal if the correct value of n_e has been assumed. If n_e has been chosen too low it will slope downwards and if too high upwards.

As with much combustion analysis work, there are a number of assumptions implicit in this method, but these do not seriously affect its value as a development tool. The technique is particularly valuable in the case of the DI diesel engine, in which it is found that the formation of exhaust emissions is very sensitive to the course of the combustion process, which may be controlled to some extent by changing injection characteristics and air motion in the cylinder. Figure 15.7 shows the time scale of a typical diesel combustion process.

Figure 15.8 shows an indicator diagram taken from a small single-cylinder diesel engine and the corresponding heat release curve derived by means of a computer program modeled on the method of calculation described above.

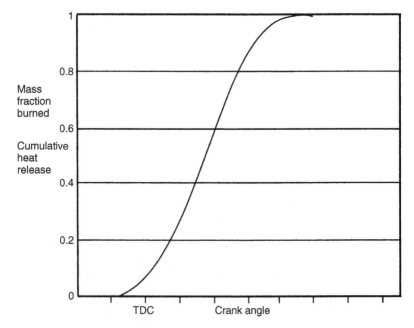

FIGURE 15.6 Mass fraction burned plotted against crank angle.

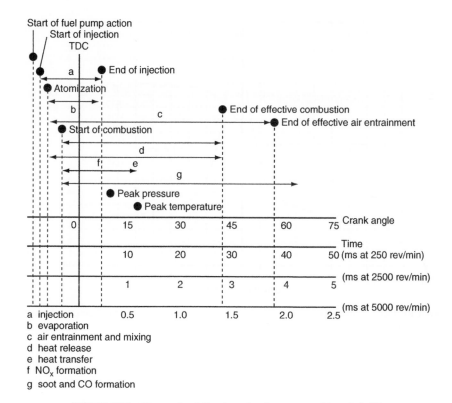

Start of fuel pump action
Start of injection
TDC

a ● End of injection

● Atomization

b   c   ● End of effective combustion
                                → ● End of effective air entrainment

● Start of combustion

d

f   e

g

● Peak pressure
    ● Peak temperature

| 0 | 15 | 30 | 45 | 60 | 75 Crank angle |

Time
| 10 | 20 | 30 | 40 | 50 (ms at 250 rev/min) |

(ms at 2500 rev/min)
| 1 | 2 | 3 | 4 | 5 |

(ms at 5000 rev/min)
a injection   0.5   1.0   1.5   2.0   2.5
b evaporation
c air entrainment and mixing
d heat release
e heat transfer
f NO_x formation
g soot and CO formation

FIGURE 15.7 Time scale of diesel combustion process. *(From Ref. [3].)*

The start of combustion, at 9° before TDC, is clearly defined and lags the start of injection by 6°. Combustion is complete by about 60° after TDC and the derived value of the (no combustion) index of expansion in this case was 1.36 (smaller engines generally display a higher index, owing to the higher heat losses in proportion to volume). Figure 15.9 shows a similar trace taken from a larger automotive diesel by a combustion analysis instrument.

TOTAL AND INSTANTANEOUS ENERGY RELEASE

Many of the required results used by the engine developer are derived from the fundamental calculations relating to energy release in the cylinder. This can be expressed as total energy per cycle (expressed as a pressure value × mean effective pressure, MEP) or in the calculation of instantaneous energy release with respect to crank angle (either as a rate or integral curve).

These calculations are well known and supported by established theory. Experimenters may adjust or adapt the calculations according to engine type or application. Any assumptions made in these calculations, to simplify them and

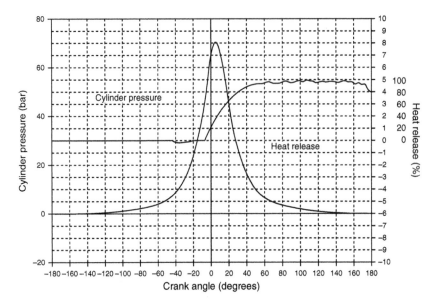

FIGURE 15.8 Cylinder pressure diagram and heat release curve for a small single-cylinder diesel engine.

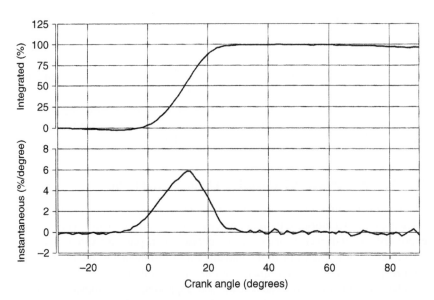

FIGURE 15.9 Diagram showing instantaneous and integrated energy release vs. crank angle. (*Source: AVL.*)

reduce calculation time, can be adjusted according to preference and laboratory procedure. However, this can lead to the analysis becoming somewhat subjective; it is therefore important to know which theory and calculation method has been adopted when comparing results.

CYCLIC ENERGY RELEASE—MEAN EFFECTIVE PRESSURE (INDICATED, GROSS, AND PUMPING)

Cylinder pressure data can be used to calculate the work transfer from the gas to the piston. This is generally expressed as the indicated mean effective pressure (IMEP) and is a measure of the work output for the swept volume of the engine, The result is a fundamental parameter for determining engine efficiency as it is independent of speed, number of cylinders, and displacement of the engine.

The IMEP calculation is basically the enclosed area of the high-pressure part of the $P–V$ diagram and can be derived via integration:

$$IMEP = \int P \, dV$$

This effectively gives energy released or gross MEP (gross work over the compression and expansion cycle, GMEP). Integration of the low-pressure (or gas exchange) part of the cycle gives the work lost during this part of the process (pumping losses, also known as PMEP). Subtraction of this value from the gross IMEP gives the net IMEP (NMEP) or actual work per cycle.

Therefore, it can be stated that:

$$NMEP = GMEP - PMEP$$

The most important factor to consider when measuring IMEP is the TDC position (see "Exact Determination of True Top Dead Center Position" section in this chapter).

Note that, in order to optimize the calculation, it is not necessary to acquire data at high resolution. Measurement resolution of a maximum of one degree crank angle is sufficient; higher than this does not improve accuracy and is a waste of system resources.

Also important for accurate IMEP calculation are the transducer properties, mounting location, and stability of the transducer sensitivity during the engine cycle.

INSTANTANEOUS ENERGY RELEASE

The cylinder pressure data, in conjunction with cylinder volume, can be used to extrapolate the instantaneous energy release from the cylinder with respect to crank angle. There are many well-documented theories and proposals for this but all are fundamentally similar and stem from the original work carried out by Rassweiler and Withrow.

The calculations can range from the quite simple with many assumptions, to very sophisticated simulation models with many variables to be defined and boundary conditions to be set.

The fundamental principle to the calculation of energy release relies on the definition of the compression and expansion process via a polytropic exponent (ratio of specific heats under constant pressure and volume conditions for a given working fluid). If this can be stated with accuracy then a motored pressure curve can be extrapolated for the fired cycle. The fired curve can then be compared with the motored curve and the difference in pressure between them, at each angular position, and with respect to cylinder volume at that position, allows calculation of energy release.

A typical simplified algorithm to calculate energy release is:

$$Q_i = \frac{K}{K-1}[\kappa \times p_i \times (V_{i+n} - V_{i-n}) + V_i \times (p_{i+n} - p_{i-n})] \qquad (15.4)$$

This calculation can be executed quickly within a computerized data acquisition system but it does not account for wall or blow-by losses in the cylinder. Also, it assumes that the polytropic index is constant throughout the process. These compromises can affect accuracy in absolute terms but in relative measurement applications this simplified calculation is well proven and established in daily use.

Under conditions where engine development times are short, algorithms have to be simplified such that they can be executed quickly. Most important for the engine developer are the results that can be extracted from the curves, which give important information about the progress and quality of combustion.

COMPUTERIZED ENGINE INDICATING TECHNOLOGY AND METHODOLOGY

The modern engine indicating or combustion analyzer creates data that is larger by at least an order of magnitude when compared with the log point data measured and stored by a typical test cell automation system. Therefore, appropriate methodology for handling, reducing, and analyzing this data is an essential function of EI equipment.

The combustion analysis device is the reciprocal of the test bed with respect to raw and calculated data. The ratio of formula to channels is far higher for a combustion analyzer, and an efficient way of creating and standardizing thermodynamic calculations is another important part of the overall efficiency of the process.

Key to the whole process is the measurement of pressure within the combustion chamber during the complete four-stroke or two-stoke process. This process, in exact detail, can be highly variable between cylinders and even within cycles inside the same cylinder, hence the need for high levels of data analysis to enable mean performance levels to be determined.

The tools used to measure cylinder pressure are pressure transducers that work on the piezoelectric principle wherein the change of pressure creates a change in circuit charge.

The safe working maximum pressure of cylinder pressure transducers is usually between 200 and 300 bar, with an (occasional) overload capacity of up to 350 bar. For conventional gasoline engines the expected maximum pressure measured would be about 140 bar and for diesel engines 180 bar. The 200 bar rating therefore leaves sufficient working "headroom" for sensor protection against damage from (unintended) exposure to engine knock. For the study of engine knock, and now super-knock, special transducers are available.

The cylinder pressure transducer circuit is quite unlike that of any other sensor or transducer commonly seen in an engine test environment and it is therefore important that users fully understand the special handling requirements of this technique.

BASIC CIRCUIT AND OPERATION OF PRESSURE MEASUREMENT CHAIN

Piezoelectric combustion pressure sensors are always used with appropriate signal conditioning in combustion measurement applications. The signal conditioning required for a charge signal is a charge amplifier and this consists of a high-gain inverting voltage amplifier with a metal-oxide semiconductor field-effect transistor (MOSFET) or junction gate field-effect transistor (JFET) at its input to achieve high insulation resistance.

The purpose of the charge amplifier is to convert the high-impedance charge input into a usable output voltage. The basic circuit diagram is shown in Figure 15.10.

A charge amplifier consists of a high gain amplifier and a negative feedback capacitor (C_G). When a charge is delivered from a piezoelectric pressure transducer (PT), there is a slight voltage increase at the input of the amplifier (A). This increase appears at the output substantially amplified and inverted. The negatively biased negative feedback capacitor (C_F) correspondingly taps charge from the input and keeps the voltage rise small at the amplifier input.

At the output of the amplifier (A) the voltage (V_O) sets itself so that it picks up enough charge through the capacitor to allow the remaining input voltage to result in exactly V_O when amplified by A. Because the gain factor of A is very large (up to about 100,000), the input voltage remains virtually zero. The charge output from the pressure transducer is not used to increase the voltage at the input capacitances, it is drawn off by the feedback capacitor.

With sufficiently high open-loop gain, the cable and sensor capacitance can be neglected; therefore, changes in the input capacitance due to different cables with different cable capacitance (CC) have virtually no effect on the

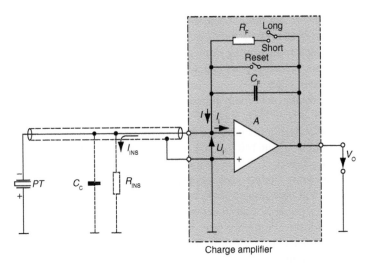

FIGURE 15.10   Basic charge amplifier circuit. *(AVL List.)*

measurement result. This leaves the output voltage dependent only on the input charge and the range capacitance. That is:

$$V_O = Q/C_F$$

An output is produced only when a change in state is experienced, thus the piezoelectric transducer and charge amplifier cannot perform true static measurements. This is not an issue in normal combustion measurements, as they require a highly dynamic measurement chain to adequately capture all aspects of the phenomenon. Special considerations are required for calibration of the measurement chain and in data analysis.

While the modern charge amplifier measurement chain (Figure 15.11) is a robust and well-proven measurement, and the technique is almost universally adopted for combustion pressure measurement, there are important issues and measurement characteristics to consider:

- *Drift.* This is defined as an undesirable change in output signal over time that is not a function of the measured variable. In the piezoelectric measurement system there is always a cyclic temperature drift that is a function of the actual transducer design (see section on "Cylinder Pressure Transducers" later in this chapter). Electrical drift in a charge amplifier can be caused by low insulation resistance at the input.
- *Time constant.* The time constant of a charge amplifier is determined by the product of the range capacitor and the time constant resistor. It is an important factor for assessing the capability of a piezoelectric measurement system for measurement of very slow engine phenomena (cranking speed) without any significant errors due to the discharge of the capacitor. Many

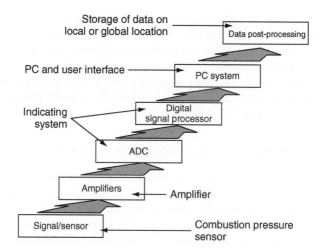

FIGURE 15.11 Typical workflow process through combustion analysis.

charge amplifiers have selectable time constants that are altered by changing the time constant resistor.

Drift and time constant simultaneously affect a charge amplifier's output. One or the other will be dominant. There are a number of methods available in modern amplifier technology to counteract electrical drift and modern charge amplifier technology contains an electronic drift compensation circuit.

If it is possible to maintain extremely high insulation values at the pressure transducer, amplifier input, cabling and associated connections, leakage currents can be minimized and high-quality measurements can be made. In order to do this, though, the equipment must be kept clean and free of dirt and grease to laboratory standards. In practice, a normal engine test cell environment does not provide this level of cleanliness and this method of preventing drift is not a practical proposition. Operation of the amplifier in short-time constant mode means that the drift due to input offset voltage can be limited to a certain value and thus drifting into saturation can be prevented.

- *Filtering.* Electrical filtering is generally provided in the charge amplifier to eliminate certain frequencies from the raw measured data. Typically, high-pass filters are used to remove unwanted lower frequency components; for combustion measurement, this may be necessary in a knock measurement situation where only the high-frequency components are of interest.

The high-pass filter will allow optimization of the input range on the measurement system to ensure the best possible analog-to-digital conversion of the signal. Low-pass filters can be implemented to remove unwanted high-frequency, interference signal content from the measurement signal, such as structure-borne noise signals from the engines that are transmitted to the transducer.

It is important to remember when using electrical filters that a certain phase shift will always occur and that this can cause errors that must be considered. For example, any phase shift has a negative effect on the accuracy of the IMEP determination when a low-pass filter is used. The higher the engine speed, the higher the lowest permitted filter frequency. As a rule, in order to avoid unacceptable phase shift, the main frequency of the cylinder pressure signal should not be more than 1% of the filter frequency.

GROUND LOOPS IN INDICATION EQUIPMENT

These are discussed in Chapter 5 and can be a significant problem for piezo-electric measurement chains (due to the low-level signal) and can be extremely difficult to detect. It is therefore prudent to take appropriate precautions to prevent them and hence avoid their effects. The simplest way is to ensure all connected components are at a common ground through interconnection with an appropriate cable; this is more difficult to implement if equipment is not permanently installed at the test cell.

Ground isolated transducers and charge amplifiers can be bought that will eliminate the problem, but it is important to note that only one or the other of these solutions should be used. If used together the charge circuit will not be complete and hence will not function!

Modern amplifier technology uses developments in digital electronics and software to correct, or allow correction of, signal changes occurring during the test period. The amplifier settings are software driven by way of the instrumentation PC, which means that settings can be adjusted during measurements without entering the test cell. Local intelligence at the amplifier provides automatic gain factor calculation from sensor sensitivity, pressure range, and output voltage, thus providing optimum adaptation of the input signal with nearly infinitely variable gain factors. This optimizing information can be stored on board the amplifier complete with sensor parameters.

Modern amplifiers are physically small, which allows mounting very close to the actual sensors, and this means short cables between sensors and amplifier and, therefore, the best protection from electromagnetic interference. The direct connection of piezo sensors without interconnection reduces signal drift and possible problems with grease and humidity on intermediate connectors.

"EXACT" DETERMINATION OF TRUE TOP DEAD CENTER POSITION

The test engineer in possession of modern equipment continues to be faced with what appears to be a rather prosaic problem: the determination of the engine's top dead center point. This is a more serious difficulty than may be at first apparent. An electronic engine indicator records cylinder pressure in terms of crank angle, and measurements at each tenth of one degree of rotation are commonly available. However, to compute indicated power it is necessary to transform the cylinder pressure–crank angle data to a basis of cylinder pressure–piston stroke. This demands a very accurate determination of crank angle at the top dead center position of the piston.

If the indicator records top dead center 1° of crankshaft rotation ahead of the true position the computed IMEP will be up to 5% greater than its true value. If the indicator records TDC 1° late computed IMEP will be up to 5% less than the true value.

Top dead center can now be accurately determined, for a particular running condition, by using a special capacitance sensor such as the AVL 428 unit, which is mounted in a spark plug hole of one cylinder while running the engine on the remaining cylinders. The TDC position thus determined is then electronically recorded by using the angle of rotation between TDC and a reference pulse, of a rotational encoder attached to the crankshaft. Another technique makes use of recording the cylinder pressure peak from a cylinder pressure transducer, but without such special transducers and the required software to match the shaft angle, precise determination of geometrical TDC is not easy. The usual and time-honored method is to rotate the engine to positions equally spaced on either side of TDC, using a dial gauge to set the piston height, and to bisect the distance between these points. The rotation position of TDC determined this way ignores dynamic effects and even then has to be fixed on the flywheel in such a way as to be recognized by a pickup.

This traditional method leads to several sources of error:

- Difficulty of carrying out this operation with sufficient accuracy.
- Difficulty of ensuring that the signal from the pickup, when running at speed, coincides with the geometrical coincidence of pin and pickup.
- Torsional deflections of the crankshaft, which are always appreciable and are likely to result in discrepancies in the position of TDC when the engine is running, particularly at cylinders remote from the flywheel.

In high-performance engines the TDC position does not remain entirely constant during the test run due to physical changes in shape of the components; in these cases a TDC position specific to a particular steady state may have to be calculated.

COMBUSTION ANALYSIS ROLES IN IC ENGINES FOR HYBRID VEHICLES

In serial hybrids, also known as extended range electric vehicles (EREVs), the internal combustion engine has the role of powering an alternator rather than propelling the vehicle, and can therefore be designed and calibrated quite differently from convention propulsion engines. A number of four-stroke gasoline engines, some in three cylinder configurations running with modified combustion cycles, are being developed for hybrid vehicles where low NO_x emission and fuel consumption are more important than power density. The

Atkins Cycle engine, used in a number of vehicles, has an operating cycle during which the intake valve is held open to allow a reverse flow of intake air into the intake manifold, which reduces the compression ratio but leaves the expansion ratio unchanged. It is probable that the optimization, for minimum fuel consumption and exhaust emissions within hybrid configurations, of these modified Otto cycle engines will tend to dominate CA work for years to come.

The most radical design currently being developed is the *linear* or *free-piston engine*, wherein the linear motion of the two opposing pistons shuttles back and forth inside a chamber and the pistons, equipped with rows of rare-earth magnets, pass within coils to create an electrical current. The theoretical advantages of the free-piston concept include:

- A simple design with few moving parts, giving a compact engine with low maintenance costs and reduced frictional losses.
- The variable compression ratio, achieved without change in the mechanical configuration, allows operation optimization for all operating conditions and multi-fuel operation.
- The design is well suited for HCCI operation.
- A fast power stroke expansion enhances fuel/air mixing and reduces the time available for heat transfer losses and the formation of temperature-dependent emissions such as nitrogen oxides (NO_x).

The main challenge for the free-piston engine is engine control, particularly the influence of cycle-to-cycle variations in the combustion process and engine performance during transient operation.

Homogeneous charge compression ignition (HCCI), also known as *controlled auto-ignition* (CAI), engines are based on another technology being pushed by hybrid vehicle development and with combustion control problems yet to solve. The advantages of HCCI engines are worth some effort in realizing:

- Fuel is burned quicker, reducing heat loss, and at lower temperature producing lower NO_x compared with conventional SI engines.
- Lean combustion means an increase (15% claimed) in fuel efficiency compared with conventional SI engines.
- The induction system is throttleless, so the engine has lower frictional pumping losses.

However, the higher cylinder pressure means HCCI engine have to have a stronger construction than conventional SI engines and the combustion control problems are significant.

"Knock" Sensing

One of the most important boundary conditions that an engine control system must be programmed to avoid is "knock". This destructive, spontaneous ignition of unburnt gas in the cylinder can commonly be induced by (over) advancing the ignition timing and is detected in the engine test cell by:

- A characteristic pressure pulse signature, the normal method used by CA equipment using engines fitted with a cylinder pressure transducer.
- An accelerometer tuned to look for the vibration caused by knock; this is the common method used by an ECU since the transducer is appropriate for use in production vehicles.
- Microphone and operator detection is still a valid method but probably only appropriate to specialist motor sport test facilities, where low engine numbers and rapidly changing ECU maps do not allow for standard models to be formed.

Low-Speed Pre-Ignition (LSPI) or "Super-Knock"

Super-knock seems to have recently appeared to prove to engine development and test engineers involved with engine downsizing the truth in the words of Sir Harry Ricardo, who wrote in 1936:

An engine may be regarded as a creature of infinite and dogged cussedness, but entirely lacking in a sense of humour.

LSPI events are infrequent, random occurrences that happen at low speed and high torque and lead to a very heavy knock, which can cause catastrophic engine damage in only a few cycles. When using normally rated combustion pressure measurement, the biggest problem is to capture the event without, or before, destroying either the sensor or the engine. Some optical work has suggested that there are localized signs of glowing in the charge just before the event, but the cause or causes are not yet fully understood. Theories currently include:

- That downsize engines are operating at thermal and chemical limits such that there is little tolerance for avoidance of abnormal combustion events.
- The role and interaction of different fuels and lubricants.
- Deposit particles being drawn into the combustion chamber and, upon compression, creating hot spots that cause pre-ignition.

LSPI suppression and avoidance strategies being researched are almost as numerous as the theories concerning its cause, and range from cooled EGR systems to fuel and lubrication chemistry, all requiring combustion analysis in test cells.

INTEGRATION OF COMBUSTION ANALYSIS EQUIPMENT WITHIN THE TEST CELL

In many test cells in the world, combustion analysis equipment may be a temporary visitor, brought in when required and operated in parallel to the test bed automation system. Temporary installations may give problems and inconsistencies due to signal interference and poor event alignment in separate data acquisition systems. It is now common to find dedicated engine calibration cells with combustion analysis and other high-speed data recording equipment as a permanent part of an optimized installation.

There may be fewer than 10 specialist suppliers of complete IE hardware and software suites worldwide and probably only half of these are also specialist CA transducer manufacturers. Any search of the internet will result in discovery of many universities and private test houses that have assembled their own analysis systems from commercially available components and even more who have used standard software tools such as the MATLAB suite to develop their own diagnostics packages. Currently there is no ASAM standard for the storage of raw CA data so it tends to be stored within structured binary files designed by the makers of the analysis equipment.

As with many other aspects of engine testing, the correlation of results produced on different systems, some of which are not well integrated, ranges from difficult to self-deluding.

Whatever the level of sophistication, the physical equipment involved in engine calibration and CA work can be considered as a high-speed data acquisition, storage, manipulation, and display system. It will consist of three subsystems:

- Engine-rigged transducers, some of which may require special engine preparation such as machining of heads for pressure transducers and mounting of an optical speed encoder.
- Interconnection loom, which, in the case of mobile systems, will have to pass through the firewall between cell and control room.
- Data acquisition and analysis equipment mounted in the control room, either fully integrated with the test control system or at least sharing a common time and crank position flag.

HIGH-END ENGINE CALIBRATION CELLS

Let us consider the operational problem faced by the cell operators of an OEM test cell carrying out the optimization of an engine control map. There are three distinct subsystems producing and displaying data while the engine is running:

- The *test cell automation system* that is controlling the engine performance and monitoring and protecting, by way of alarm logic, both engine and test facility operation.
- The *engine ECU* that is controlling the fueling and ignition using engine-derived parameters to calculate essential variables.

- The *combustion analysis system* that is using special (nonvehicular) transducers to directly measure cylinder pressure against crankshaft rotation and is able to derive and display such values as:
  - mean effective pressures, IMEP, PMEP, etc.
  - location of peak pressure and pressure rise rates
  - mass fraction burn parameters
  - polytropic index
  - injector needle movement and injected fuel mass
  - heat release, etc.

As is the case with exhaust emissions equipment, the set-up and operation of CA equipment and analysis of the data requires training and experience. The test equipment industry has produced integrated suites of instrumentation and software that, to a degree, can reduce the skill levels required to design the test sequences and to efficiently locate the boundary operating conditions of the test engine, but appropriate training remains essential.

Cells that carry out engine calibration work such as that described above have to be designed from the beginning as an integrated system if they are to work efficiently. All the data having different and various native abscissae such as time base, crank angle base, and engine cycle base, communicated via different software interfaces (COM, ASAP, CAN bus, etc.), have to be correctly aligned, processed, displayed, and stored, which is not an easy task. In fully automated beds it is common to have upwards of six 17″ monitors on the operator's desk.

Along with the pressure transducers required for cylinder pressure, additional transducers are required to measure events and consequences of the combustion that occurs on the same time scale, such as injector needle operation and fuel rail pressure (see Figure 15.12). Pressure waves, reflecting and rebounding within the common fuel rail, can produce unexpected effects at individual injectors and need to be monitored on the common time base.

OCCASIONAL USE OF COMBUSTION ANALYSIS IN TEST CELLS

The operator of an engine test required to carry out occasional map modification work or performance testing will use combustion analysis equipment that is often, although not always, mounted on a wheeled trolley. Test sequences designed to support CA or engine indicating work are often run under manual control to locate the problem areas of engine performance that require analysis.

The running of the required cable looms and large multiway plugs, from cell to control desk, can give fire safety problems when having to be passed through the cell to control room firewall and resealed; consideration should be given to the permanent installation of such looms in the case of multiples of cells that each have occasional use of the same CA equipment.

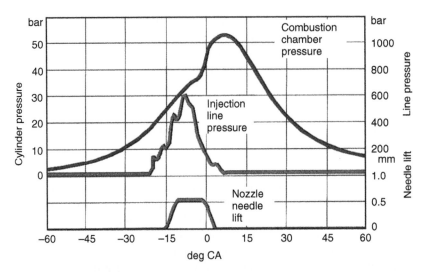

FIGURE 15.12 Typical engine indicating display parameters.

Typically, engine indicating plant will have a "pseudo-oscilloscope" display, quite separate from the cell control system display; therefore, desk space for two operators and an absolute minimum of two flat display screens should be catered for in the control room layout.

ENGINE INDICATING PRESSURE TRANSDUCERS (EIPTs)

Cylinder pressure measurements play the key role in any engine indicating and combustion analysis work, but since modern engine research requires more than peak cylinder pressure to be measured against crank angle, pressure transducers are also made for the following roles in engine indicating work:

- Cylinder pressure for R&D work, optimized for precision of measurement.
- Cylinder pressure for the in-service monitoring of non-automotive engines, optimized for durability.
- Intake and exhaust manifold pressure for R&D work, optimized to handle the temperature and pressure ranges of each work environment.
- Fuel line pressure for R&D work for pressure pulse measurement up to 4000 bar.

Cylinder Pressure Transducers

Most transducers are of the piezoelectric type and require their signals to be processed by matched charge amplifiers. However, there are now dynamic pressure transducers available that work by the deflection of an optic fiber, in the range of temperatures and pressures that make them suitable for CA work. Although these devices are not yet used in mainstream CA work they can be made down to diameters of around 2 mm so they may find a place in "difficult to locate" cylinder-head installations.

Up to the early 1990s commercially available pressure transducers were more temperature sensitive, both in terms of cyclic effects and absolute maximum working temperature, than units available today. The temperature effects are due to exposure to the combustion process and most would fail if required to operate above 200 °C, measured at the diaphragm; consequently, most EIPTs were water cooled.

Water-cooled EIPTs are still required for some situations, including turbocharged engines, and it is vital that the cooling system meets the following conditions:

- The transducer cooling system, while in use, must be integrated with the control system shutdown circuit to ensure that the cooling is running during the complete testing cycle, including engine start and cool down. Transient dips in the coolant supply pressure will risk permanent loss of the transducers.
- The water should be distilled (deionized) and filtered. The transducers have very small passages that will become ineffective if blocked by scale.
- The deionized water should be supplied at a constant pressure that is as low as possible to ensure flow; this is to avoid changes in internal transducer pressure that could corrupt output signal.

Uncooled EIPTs are continually being developed; currently models will operate to 400 °C, which covers the majority of naturally aspirated automotive work outside specialist and motor sport development work.

An important characteristic of any EIPT is its *cyclic temperature drift*, which is the range of measurement error due to the heating of the transducer's diaphragm over the cylinder's working cycle. The lower the drift value, the higher the unit accuracy; the figures quoted are typically better than ± 0.3 to ± 0.6 bar depending on the design and size. Since there is no standard procedure to define how cyclic drift has to be measured, it makes a direct comparison between the values measured by transducers from different manufacturers almost impossible, which is why almost all test facilities use all their CA equipment in the complete transducer–cable–amplifier–analysis chain from just one manufacturer. There is some similarity with use of exhaust emission equipment, where direct result comparisons across manufacturers can be difficult in detail.

The choice of pressure transducer from the wide range now commercially available [4,5] needs to be made with great care, as the correct mounting in the engine and its integration within the calibrated system are vital ingredients in obtaining optimum results.

Mounting of the Cylinder Pressure Transducer

The precision of pressure measurement is critically dependent on the location and mounting of the probe in the cylinder-head. The fundamental choice facing the test engineer is whether the cylinder-head can be machined with a hole terminating at a suitable location; if not, then the only choice is to use either a transducer designed to replace the glow plug in a diesel or a special combined transducer–spark plug in an SI engine (see Figure 15.13).

If the head can be machined and there is space and material in the engine head structure to support the transducer, there are three types that might be used: probe types that are usually the thinnest, plug types, and the most common, threaded types. Note that some pressure transducer manufacturers supply the cutting tools designed to machine the recesses required to house the specific model.

Threaded EIPTs are made in various sizes, usually ranging from 5 mm up to 18 mm. They are normally inserted into the combustion chamber so that their tips are flush with the parent combustion chamber material.

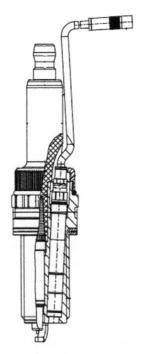

FIGURE 15.13 Part section through a typical combined spark plug and cylinder pressure transducer.

FUEL RAIL PRESSURE AND "SHOT VOLUME" MEASUREMENT

Common rail (CR) systems have enabled significant reduction of exhaust emissions and, of particular importance to the small diesel car market, lowered engine noise or the "diesel rattle". CR systems and electronic management have allowed the flexible division of fuel injection into several separate events during any one combustion cycle. The pre, main and post injections, variable in timing and duration, produce engine performance that can better match the requirements of the driver while remaining within permitted emission limits over the full range of vehicle use.

A passenger car or truck diesel engine that met Euro 3 emissions levels would typically require two or three fuel injections per combustion cycle, but the same engines having to meet Euro 4 and Euro 5 limits might require four or more injection events.

Recent control strategies might require two early or pilot injections, to reduce combustion noise, followed by the main injection, which is then followed by one or two post injections to help control of particulate emissions. A further late injection may follow late in the cycle to enable the regeneration of exhaust gas after treatment devices.

Each injection pulse in this complex, multi-injection strategy creates a pressure wave within the common rail pipe, waves that can combine to create transient pressures, above (additive) or lower (subtractive) than that regulated by the CR system. This means that, for example, a main injection event with a duration of 500 μs that would be calibrated to inject 12 mm^3 of fuel could inject 1 or 2 mm^3 more or less, depending on the coincidence of pressure waves within the rail at the time of injection. In other words, when more than one injection per combustion cycle takes place, the size of the second injection is influenced by the first injection or possibly by injection events of other cylinders. Some of the injection volumes are as small as 1 mm^3 so the pressure wave effect has to be measured during the combustion analysis and engine calibration process. The fuel consumption devices discussed in Chapter 14 only measure the volume of fuel consumed over many injection cycles but there are instruments such as the AVL Shot To Shot™ PLU 131 Flow Sensor that are able to measure the volume of individual injection events (shots) down to volumes of 0.3 mm^3. Such devices are normally used in conjunction with single-cylinder research engines or spray chambers.

SPEED/CRANK ANGLE SENSORS

For all tests involving engine indicating work the engine will have to be rigged, at the crankshaft free end, with an optical shaft encoder. The installation and alignment of the encoder components and the drive shaft have to be done with care and using the correct components to suit the application. There are a number of different designs, the most robust being totally enclosed units, but these require suitable space and access to the free crankshaft end. Where access is very restricted, an open disk design may be required.

The devices usually consist of a static pickup that is able to read positional data from an engine shaft-mounted disk that is etched with, typically, 3600 equally spaced lines producing that number of square wave pulses per revolution in the device output. There may be a number of missing lines that enable a discrete angular position to be established as an event trigger. The disks are normally made of a glass material except in cases where very high shock forces are present and steel disks with lower pulse counts may be required. Some crank angle encoders have a second output that determines direction of rotation.

CALCULATION FOR COMBUSTION ANALYSIS TEST RESULTS

Calculation of critical results from the large store of raw combustion data is an essential part of the analysis and forms the basis of intelligent data reduction. The actual results, which are derived from the raw measured data and calculated curves, can be grouped logically into direct and indirect results.

Direct results are derived directly from the raw cylinder pressure curve. Most of them can also be calculated before the end of the complete engine cycle (the author avoids the use here of the term "in real time"). Generally, any signal error will create a result in a calculation error of similar magnitude. Typical direct result calculations are:

- Maximum pressure and position
- Maximum pressure rise and position
- Knock detection
- Misfiring
- Combustion noise analysis
- Injection/ignition timing
- Cyclic variation of above values.

These results will be acquired from data with acquisition resolution appropriate to the task. Resolution of pressure and pressure rise will typically be at one degree crank angle; certain values will need higher resolutions due to higher frequency components of interest (combustion knock or noise) or the need for accurate determination of angular position (injection timing).

Indirect results are more complex as they are derived by calculation from the raw pressure curve and associated data. They are reliant on additional information like cylinder volume, plus other parameters such as the polytropic exponent. These results are much more sensitive to correct set-up of the equipment and the error in a signal acquired is multiplied considerably when passed into a calculation. Hence the result error can be greater by an order of magnitude. Typical indirect results are:

- Heat release calculation (dQ, integral)
- Indicated mean effective pressure
- Pumping losses, PMEP
- Combustion temperature
- Burn rate calculation
- Friction losses
- Energy conversion
- Gas exchange analysis
- Residual gas calculation
- Cyclic variation of above values.

Using modern computer systems, these results, although derived through complex calculations, do not place a heavy demand on modern acquisition hardware, nor do they require high resolutions with respect to crank angle (typically one degree is sufficient) or analog-to-digital conversion (12- to 14-bit). The user demand has always been for more and faster calculating power to execute the algorithms and return results and statistics as quickly as possible. Each year physically smaller and faster data processing and display equipment is brought to the market, but the need for training in its cost-effective use does not diminish.

For the combustion analysis system, important features are:

- Real-time or online calculation and display of all required results for screen display or transfer to the test bed through the software interface or as analog voltage values.
- Features in the user interface for easy parameterization of result calculations, or for definition of new, customer-specific result calculation methods without the requirement for detailed programming knowledge.
- Flexibility in the software interface for integration of the combustion analyzer to the test bed, allowing remote control of measurement tasks, fast result data transfer for control mode operation, and tagging of data files such that test bed and combustion data can be correlated and aligned in post-processing.
- Intelligent data model that allows efficient packaging and alignment of all data types relevant in a combustion measurement (e.g. crank angle data, cyclic data, time-based data, engine parameters, etc.).

Most importantly, engine indicating and combustion analyzers are always very sensitive to correct parameterization and the fundamentals, that is:

- Correct determination of TDC (particularly for IMEP)
- Correct definition of the polytropic exponent (particularly for heat release)
- Correct and appropriate method of zero level correction.

NOTATION

Pressure, beginning of interval	p_1 (bar)
Pressure, end of interval	p_2 (bar)
Pressure, end of interval, no burn	p_0 (bar)
Volume, beginning of interval	v_1 (m^3)
Volume, end of interval	v_2 (m^3)
Volume at top dead center	v_0 (m^3)
Pressure change due to combustion	Δp (bar)
Pressure change, normalized to TDC	Δp_c (bar)
Polytropic index, general	n
Polytropic index, compression	n_c
Polytropic index, expansion (no burn)	n_e

REFERENCES

[1] C.R. Stone, D.I. Green-Armytage, Comparison of methods for the calculation of mass fraction burnt from pressure–time diagrams, Proc. I. Mech. E. 201 (D1) (1987).

[2] G.M. Rassweiler, L. Withrow, Motion pictures of engine flame propagation. Model for S.I. engines, SAE J. 42 (1938) 185–204 (translation).

[3] Z. Bazari, A D.I. Diesel Combustion and Emission Predictive Capability for Use in the Cycle Simulation (1992). SAE Paper No. 920462.

[4] AVL Product Catalog, Pressure Sensors for Combustion Analysis. Available at: www.avl.com.

[5] Kistler document, Engine Combustion Analysis—Engine Pressure Measurement for Research and Development. Available at: www.kistler.com.

FURTHER READING

B. Hsu, Practical Diesel Engine Combustion Analysis, SAE International, 2002. ISBN- 13: 978-0768009149.

Measurement and Observation Analysis of Combustion in Engines (Seminar), Wiley-Blackell, 1994. ISBN-13: 978-0852989302.

D.R. Rogers, Engine Combustion: Pressure Measurement and Analysis, SAE International, 2010.

and test engineers in the twenty-first century is the reduction of vehicle fuel consumption without degrading the user's perception of performance.

The EU has agreed that average CO_2 emissions from new passenger cars should not exceed 120 g/km by 2012; this corresponds to fuel consumption of 4.5 liters per 100 km for diesel cars and 5 liters per 100 km for petrol cars. The "average" is manufacturer specific; therefore, if a manufacturer continues to produce high-powered passenger cars with commensurately high CO_2 emissions they have to also produce (or perhaps "badge") a low emitting vehicle to bring down their average.

Given the finite hydrocarbon resources in the world and the increasing cost of their refined products, such a development policy finds much favor with the customer, while the same may not be said of complex exhaust post-processing systems that have little sales appeal and add cost and vulnerability to the vehicle.

Legislative pressure is not only bearing down on engine developers but also on their customers, directly through differential tax rates on vehicles based on their emissions and indirectly through taxes on fuel.

The subject of engine and vehicle emissions is extremely wide reaching and continues to generate a vast number of technical papers each year, a summary of which is well outside the scope of this book. The various harmful results of atmospheric pollution on the environment in general and human health in particular are now widely recorded and all readers of this edition will be aware of them. The human health problems caused by polyaromatic hydrocarbons (PAHs) have given rise to standards being set on the fuel quality in Europe (Directive 2009/30/EC) and various national regulatory bodies. The US Environmental Protection Agency (EPA) list of the PAH group includes 16 named compounds, the best known being benzene, formaldehyde, and 1,3-butadiene, many of which are known or suspected carcinogens.

More recently there has also been increased concern about the dangers of very small carbon particulates that have been proved to harm the human respiratory and cardiovascular systems and are linked to asthma. The smaller particles are the most damaging and current targets focus on particles less than 10 μm in diameter (PM 10), particularly the "nanoparticles" below 50 nm in size that are able to penetrate to the surface of the mammalian lung. Development in test instrumentation has been targeted at the problem of measuring the mass and counting the number of particulates in any given exhaust gas flow; these are discussed later in this chapter.

Automotive development aimed at the reduction of hydrocarbon fuel consumption has been developing new generations of hybrid and electric vehicles. With hybrids we have seen development of "downsized" power units and a resurgence of interest and use of Atkins cycle (naturally aspirated) and Miller cycle (supercharged) four-stroke engines, which offer higher efficiency and lower NO_x output but lower power density (hybrids are briefly discussed in Chapter 13).

EMISSION LEGISLATION, CERTIFICATION, AND TEST PROCESSES

Since so much of the operational strategy of automotive test facilities is determined by current and planned environmental protection legislation, it is necessary for all test engineers to be aware of the legislative framework under which their home and export markets work. The details within the first of the "pillars" shown in Figure 16.1 are defined by the democratic processes in Europe, Japan, and the USA, processes that are by definition uncoordinated and that have created situations where implementation dates of similar legislation across the world are in some doubt.

Although emission instrumentation has undergone significant development and levels of pollutants produced by vehicles and engines are much reduced, the basic form of much legislative testing has remained very similar to the original EPA methodology of the 1970s.

In general all engine emission legislation consists of the four component parts shown diagrammatically in Figure 16.1:

- Test limits, which define the maximum allowed emission of the regulated components in the engine exhaust. For light-duty vehicles the limit is expressed in mass per driving distance (g/km), for heavy-duty and off-road (including marine) engines the limits are expressed in mass per unit of work (g/kWh). Limits for particulates are expressed both in mass (mg/km) and number (number/km, often written as #/km).
- Test cycles, which describe the operation of the tested vehicle or engine. For light-duty vehicles it simulates the actual driving on the road, in that it defines a vehicle velocity profile over the test time. For heavy-duty and off-road engines where only the engine is tested on a dynamometer, the test cycle defines a speed and torque profile over the test time.

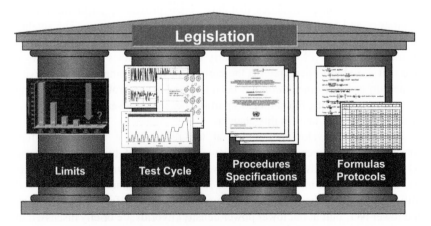

FIGURE 16.1 The four pillars of emission legislation.

- Test procedures, which define in detail how the test is executed and which exhaust sample collection and measurement methods are used. They state which test systems have to be used and define the test conditions. Probably the best known example of a legislative procedure is the use of sample bags in which an accumulation of diluted sample of exhaust gases resulting from a drive cycle is stored for analysis of content and concentration. Figure 16.2 summarizes diagrammatically the scope of US Federal emission test codes.

- Formulas, which are used in calculating the results, and the protocols covering the manner in which they are reported.

There is now legislation covering every type of machine powered by an IC engine, from grass strimmers and chainsaws to the largest off-road tractor units and cruise ships.

The basic methodology used has been, and still is, based on running the engine through a strictly defined test sequence either in a test cell or by driving the test vehicle through a prescribed "drive cycle" on a chassis dynamometer.

During the test a proportion of the exhaust gases produced during specific stages of the test is analyzed, according to the procedures, and the amount and proportions of the key pollutants are determined according to the formulas and protocols.

The drive cycles used have been criticized for having little resemblance to real-life driving and therefore being incapable of producing realistic results. There has been some development in drive cycles but those that have been developed with higher rates of acceleration and speed based on empirical road

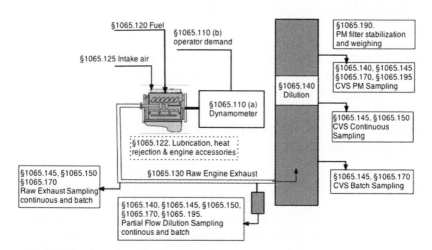

FIGURE 16.2 An example of the scope of legislation as a representation of the Code of Federal Regulations (CFR) Part 1065: Engine-Testing Procedures. This diagram illustrates the equipment specified by the legislation for an engine test cell incorporating a CVS system and gives references to sections in CFR 1065. *(Taken from US GPO e-CFR Data; current as of March 7, 2011.)*

data such as the European ARTEMIS driving cycles (urban, rural road, and motorway) have not yet been legislatively adopted. The more dynamic and demanding the drive cycle, the more difficult it is to get good test repeatability and correlation across the world; therefore, drive cycles tend to remain the same but the limits of pollutants get tighter.

For the emission testing of heavy-duty vehicles there is a major initiative to gain agreement on two "World Harmonized Cycles" (World Harmonized Stationary Cycle (WHSC) and World Harmonized Transient Cycle (WHTC)). At the time of writing (early 2011) no agreement has been reached and it seems likely that they will be implemented first by the EU, followed by Japan. In the USA there is, in some politically powerful groups, strong antipathy towards environmental protection and towards the agencies charged with implementing environmental polices such as the EPA; therefore, progress of emission legislation is as difficult to predict as is the ebb and flow of party politics in western democracies.

In legislative policy documentation there is often reference to vehicle classifications that are based on levels of emissions. The following abbreviations, some of which will appear later in this chapter, are the most commonly used:

Low emission vehicle	LEV
Ultra low emission vehicle	ULEV
Super ultra low emission vehicle	SULEV
Partial zero-emissions vehicle	PZEV
Advanced technology PZEV	AT-PZEV
Transitional low emission vehicle	TLEV
Zero emission vehicle	ZEV

Note: To qualify as a PZEV, a vehicle must meet the SULEV standard and, in addition, have zero evaporative emissions plus an extended (15-year/150,000-mile) warranty on its emission-control components.

AUXILIARY ENGINE-DRIVEN UNITS, FITTED OR OMITTED FOR THE EMISSIONS TEST

As mentioned in Chapter 11, engines in the test cell cannot always be rigged with all of the auxiliary units with which they are fitted in their final working configuration, yet these devices will absorb power and therefore have some effect on the engine's emissions during a test cycle. Some of these devices are running a vehicular system and are not part of the base engine; these would include units such as air-pumps for trailer braking and air-conditioning units.

Legislation covers this eventuality. World Harmonized Cycles legislation contains the following instruction concerning the fitting, or removal, of auxiliary, power-absorbing, mechanisms that are fitted to the HD engine when submitted to emission testing:

If it is inappropriate to install the auxiliaries needed for engine operation on the test bench, the power absorbed by them shall be determined and subtracted from the measured engine power over the whole engine speed range of the WHTC and over the test speeds of the WHSC.

Where the auxiliaries that are needed only for operation of the vehicle cannot be removed, the power absorbed by them may be determined and added to the measured engine power over the whole engine speed range of the WHTC and over the test speeds of the WHSC. If this value is greater than 3 per cent of the maximum power at the test speed it shall be verified by the type approval or certification authority.

This instruction is typical of most worldwide emission legislation covering heavy-duty and off-road vehicle engines.

Note on "cycle beating": Most powertrain development work is aimed at producing systems that are compliant with both the detail and the intent of emissions legislation. However, there is an alternative development strategy: that of developing a powertrain that meets only the details of the test cycles and whose performance outside and beyond the (benign) legislative cycles is significantly worse than the law intends. Such cycle beating practices are incorporated in the design and function of several, currently available, high-powered diesel and gasoline cars. This volume does not cover details of cycle beating mechanisms but draws the reader's attention to legislation drawn up in Europe designed to outlaw "defeat devices". The draft treaty outlaws "any element of design which senses temperature, vehicle speed, engine rotational speed, transmission gear, manifold vacuum or any other parameter for the purpose of activating, modulating, delaying or deactivating the operation of any part of the emission control system, that reduces the effectiveness of the emission control system under conditions which may reasonably be expected to be encountered in normal vehicle operation and use."

LEGISLATION CLASSIFICATIONS

Much of the test subject classification in automotive emission legislation is by vehicle size rather than engine size and type. The main classifications are:

- Light duty—gasoline
- Light duty—diesel
- Heavy duty.

Each of these categories has legislative pollutant limits, test methodologies, and instrumentation designed specifically for it.

In Europe, as in the rest of the world, the pollutants limits for all three automotive categories have developed, and Tables 16.1 and 16.2 are included to illustrate their progressive reduction over a period of 20 years.

TABLE 16.1 European Emission Limits for Gasoline Passenger Cars (Car M1)

Stage	Date	CO (g/km)	HC (g/km)	HC + NO$_x$ (g/km)	NO$_x$ (g/km)	PM (g/km)	PN (#/km)
Euro 1	07-1992	2.72	—	0.97	—	—	—
Euro 2	01-1996	2.20	—	0.5	—	—	—
Euro 3	01-2000	2.30	0.20	—	0.15	—	—
Euro 4	01-2005	1.0	0.10	—	0.08	—	—
Euro 5	09-2009	1.0	0.10 (0.068)	—	0.06	0.005* (0.0045)	—
Euro 6	09-2014	1.0	0.10 (0.068)	—	0.06	0.005* (0.0045)	6×10^{12} (applies to DI engines only)

* Value in parentheses obtained using PMP measurement procedure.

TABLE 16.2 EU Emission Standards for Diesel Passenger Cars (Cat M1)

Stage	Date	CO (g/km)	HC (g/km)	HC + NO$_x$ (g/km)	NO$_x$ (g/km)	PM (g/km)	PN (#/km)
Euro 1	07-1992	2.72	—	0.97	—	0.14	—
Euro 2, IDI	01-1996	1.0	—	0.70	—	0.08	—
Euro 2, DI	01-1996	1.0	—	0.90	—	0.10	—
Euro 3	01-2000	0.64	—	0.56	0.50	0.05	—
Euro 4	01-2005	0.50	—	0.30	0.25	0.025	—
Euro 5a	09-2009	0.50	—	0.23	0.18	0.005* (0.0045)	—
Euro 5b	09-2011	0.50	—	0.23	0.18	0.005* (0.0045)	6×10^{11}
Euro 6	09-2014	0.05	—	0.17	0.08	0.005* (0.0045)	6×10^{11}

* Value in parentheses obtained using PMP measurement procedure.

To gain type approval, since Euro 1, vehicles have had to be tested for emissions, not only to gain type approval as new vehicles, but also after a number of miles in service. The "mileage accumulation dynamometers" developed for preparing vehicles for such testing are described in Chapter 17. Current EU rules state that the original equipment manufacturer (OEM) is responsible for assuring that vehicles meet emission standards for 80,000 kilometers; in the US, the rules use 160,000 km.

AFTER-MARKET EMISSION LEGISLATION

In addition to the prime legislation covering the type approval and certification of new engines and vehicle systems, in most countries of the developed world there is emission legislation covering the condition of cars in the population as they age. Such emission tests are usually required annually as part of a road-worthiness check and in order for the vehicle to be licensed for use on public roads; they may range from a single visual smoke check, through a check at fast idle for levels of CO and HCs, to a test under light load on a rolling road that checks CO, CO_2, and HCs.

All the emission tests are carried out using tail-pipe probes. The application of the legislation is made quite complicated because of the range of ages and types of vehicles that may be tested, from rare veteran cars, through owner-built kit cars (in the UK) to the latest supercar.

FUEL STANDARDS

At the core of the chemistry of hydrocarbon fuel combustion and the resulting emissions is the chemical make-up of the fuels themselves; these vary considerably country to country and season to season but to meet the low emission limits a minimum standard of fuel is required. In Europe, by 2000 a minimum diesel cetane number of 51 was required. The maximum diesel sulfur content was 350 ppm in 2000 and 50 ppm in 2005. The maximum gasoline sulfur content was 150 ppm in 2000 and 50 ppm in 2005. The so-called sulfur-free diesel and gasoline fuels, having ≤10 ppm of sulfur, were available from 2005, and became mandatory from 2009 onwards. The European Standard for Gasoline is EN 228 (2004) and for Diesel is EN 590. Heavy fuel oils are covered later in this chapter.

BASIC CHEMISTRY OF INTERNAL COMBUSTION ENGINE EMISSIONS

The two chemical processes that are shown in Figure 16.3 are the complete and incomplete combustion of the hydrocarbon fuel in air (treated here as an oxygen/nitrogen mixture). The production of carbon dioxide in this process is of concern since it is classed as a "greenhouse" gas but can only be reduced by

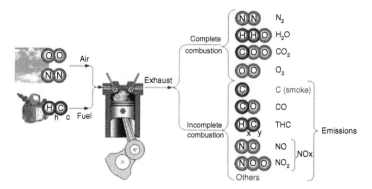

FIGURE 16.3 Gaseous components of combustion processes.

an increase in the overall efficiency of the engine and vehicle, thereby producing lower fuel consumption per work unit.

The emission gases and particulates covered by legislation are produced by incomplete combustion and are:

- Carbon monoxide (CO), a highly toxic, odorless gas.
- Carbon (C) experienced in the form of smoke and soot particles.
- Hydrocarbons (HCs) or total hydrocarbons (THCs) formed by unburnt fractions of the liquid fuel and oil.
- Nitric oxide (NO) and nitrogen dioxide (NO_2), together considered as NO_x.

The physics and chemistry covering the development of particles between being formed in the combustion chamber and circulated in the atmosphere is complex and has to be emulated within the systems designed to measure exhaust emissions. The main phases of particle formation are shown in Figure 16.4; it is the products of these processes that form the gases and particles that may be absorbed by life forms "on the street".

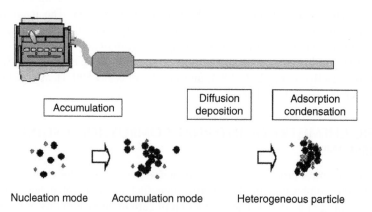

FIGURE 16.4 Phases of particle formation.

EMISSIONS FROM SPARK-IGNITION ENGINES

Perhaps the most characteristic feature of exhaust emissions, particularly in the case of the spark-ignition engine, is that almost every step that can be devised to reduce the amount of any one pollutant has the effect of increasing some other pollutant.

Figure 16.5 shows the effect of changing air/fuel ratio on the emission of the main pollutants: CO, NO_x, and unburned hydrocarbons. It will be apparent that one strategy is to confine combustion to a narrow window, around the stoichiometric ratio, and a major thrust of vehicle engine development since the mid 1990s has concentrated on the so-called "stoichiometric" engine, used in conjunction with the three-way exhaust catalyzer, which converts these pollutants to CO_2 and nitrogen.

We have already seen in Chapter 15 that a spark-ignition engine develops maximum power with a rich mixture, $\lambda \sim 0.9$, and best economy with a weak mixture, $\lambda \sim 1.1$, and before exhaust pollution became a matter of concern it was the aim of the engine designer to run as close as possible to the latter condition except when maximum power was demanded. In both conditions the emissions of CO, NO_x, and unburned hydrocarbons are high. However, the catalyzer requires precise control of the mixture strength to within about $\pm 5\%$ of stoichiometric and this requirement has been met by the introduction of fuel injection systems with elaborate arrangements to control fuel/air

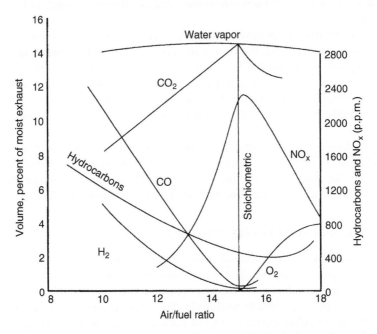

FIGURE 16.5 Relation between exhaust emissions and air/fuel ratio for gasoline engines.

mixture (lambda closed-loop control). In addition, detailed development of the inlet passages in the immediate neighborhood of the inlet valve and the adoption of four-valve heads has been aimed at improving the homogeneity of the mixture entering the cylinder and developing small-scale turbulence: this improves the regularity of combustion, itself an important factor in reducing emissions.

Spark plug design and location, duration, and energy of spark, as well as the use of multiple spark plugs, all affect combustion and hence emissions, while spark timing has a powerful influence on fuel consumption as well as emissions. Here again conflicting influences come into play. Delaying ignition after that corresponding to best efficiency reduces the production of NO_x and unburned hydrocarbons, due to the continuation of combustion into the exhaust period, but increases fuel consumption.

Exhaust gas recirculation (EGR), involving the deflection of a proportion of the exhaust into the inlet manifold, reduces the NO_x level, mainly as a result of reduction in the maximum combustion temperature. The level of NO_x production is very sensitive to this temperature (this incidentally has discouraged development of the so-called adiabatic engine, in which fuel economy is improved by reducing in-cylinder heat losses). It also acts against the pursuit of higher efficiency by increasing compression ratios.

EGR control calibration requires that the volume and content of the recirculated portion of the exhaust flow is measured as a distinct channel for the gas analyzer system.

An alternative line of development is concerned with the lean-burn engine, which operates at λ values of 1.4 or more, where NO_x values and fuel consumption are acceptable but HC emissions are then high and the power output per unit swept volume is reduced; also, light load and idling running conditions tend to be irregular. The lean-burn engine depends for its performance on the development of a stratified charge, usually with in-cylinder fuel injection.

Development of exhaust after-treatment systems (covered later in this chapter) is a continuing process, at present particularly concerned with reducing the time taken for the catalyzer to reach operating temperature, important because emissions are particularly severe during cold starts.

EMISSIONS FROM DIESEL ENGINES

The composition of the emissions from a diesel engine, like that of the spark-ignition engine, depends on the engine design, its operating condition, and the composition of the fuel used, the latter property being particularly important since it can be very variable worldwide. The sulfur content of diesel fuel is responsible for sulfur dioxide (SO_2) in the exhaust. In the developing world sulfur content may be higher than the ≤ 10 ppm required in Europe; an obstacle to reduction is the substantial cost of the necessary refinery modifications.

The diesel engine presents rather different problems from the spark-ignition engine because it always operates with considerable excess air, so that CO emissions are not a significant problem, and the close control of air/fuel ratio, so significant in the control of gasoline engine emissions, is not required. On the other hand, particulates are much more of a problem, and NO_x production is substantial.

The indirect injection (pre-chamber) engine performs well in terms of NO_x emissions; however, the fuel consumption penalty associated with indirect injection has resulted in a general move to direct injection (DI). The development of the "common rail" fuel systems is associated with a sharp increase in injection pressures, now commonly between 1500 and 2000 bar, and the "shaping" of the fuel injection or the use of a multipulse injection strategy.

NO_x emissions are very sensitive to maximum cylinder temperature and to the excess air factor. This has prompted the use of EGR to become universally required for engines meeting the equivalent of Euro 4 emission standards.

Diesel Particulate Emissions

Individual diesel exhaust particles may be highly complex coagulations of compounds, as shown in Figure 16.6. Around 90% of particles emitted by a modern automotive diesel may be below 1 μm in size, which challenges the tools used to measure their presence.

It was understood from the early days of emission studies that some of the chemical reactions resulting from in-cylinder combustion continue as the gases, and particularly "soot" particles, pass through the exhaust system and into the atmosphere; therefore, the pollution products have to be subjected to realistic post-combustion mixing with air before being analyzed.

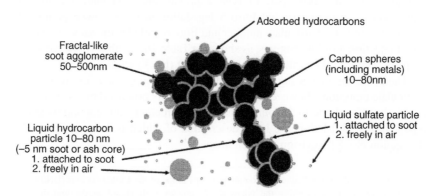

FIGURE 16.6 Schematic representation of diesel particles. *(Kulijk–Foster.)*

TABLE 16.3 Euro 4 to 6 Standards for Particulates Covering Light-Duty Vehicles

Measurement	Engine	Euro 4	Euro 5b	Euro 6
Particulate mass	GDI	No limit	4.5 mg/km	4.5 mg/km
Particulate mass	Diesel	25 mg/km	4.5 mg/km	4.5 mg/km
Particulate number	GDI	No limit	No limit	6×10^{11}/km
Particulate number	Diesel	No limit	6×10^{11}	6×10^{11}/km
Year of implementation		2005	2011	2014

PRINCIPLES OF PARTICULATE EMISSIONS MEASUREMENT

Particulates, when they appear to the human observer, are called "smoke". Smoke colors are indicative of the dominant source of particulate:

- Black = "soot" or more accurately carbon, which typically makes up some 95% of diesel smoke either in elemental, the majority, or organic form.
- Blue = hydrocarbons, typically due to lubricating oil burning due to an engine fault.
- White = water vapor, typically from condensation in a cold engine or coolant leaking into the combustion chambers. White smoke is not detected by conventional tail-pipe smoke meters.
- Brown = NO_2, maybe detected in exhaust of heavy fuel engines.

Legislative Limits and Measurement of Particulate Mass (PM) and Number (PN)

Particulate mass, measured by deposits on filter paper (described below) and visible smoke, has been part of type approval and after-market emission legislation for some years, but Euro 5 legislation also lays down permitted maximum limits for particulate number and Euro 6 includes the same limits for direct injection gasoline engines, as shown in Table 16.3. Table 16.4 covers the proposed (mid 2011) particulate figures for HD (truck) engines.

There are an increasing number of methods being used for measuring particulate emissions; however, the numerical results produced by most cannot readily be related one to the other. The first two methods are not approved by type approval legislation, the latest description of which is found in UN/ECE Regulation No. 83 Revision 3 Amendment 2:[1]

---

1. Regulation No. 83 of the Economic Commission for Europe of the United Nations (UN/ECE). Uniform provisions concerning the approval of vehicles with regard to the emission of pollutants according to engine fuel requirements, Revision 3.

TABLE 16.4 Proposed Limits for Euro 6 Covering HD Commercial Vehicles

Measurement	Euro 4	Euro 5	Euro 6
Particulate mass (g/kWh)	0.02/0.03	0.02/0.03	0.01
Particulate number (#/kWh)			? 66% reduction of Euro 5 norms expected
NO_x (g/kWh)	3.5	2	0.4

1. Opacity meters measure the opacity of the undiluted exhaust by the degree of obscuration of a light beam. These devices are able to detect particulate levels in gas flow at lower levels than the human eye can detect; they are widely used in QA testing and in service on ships. The value output is normally the percentage of light blocked by the test flow with zero being clean purge air and 100% being very thick black smoke. A secondary output giving an absorption factor "k" m^{-1} may be given, which allows some degree of comparison between devices since it removes the effect of different distances between light source and sensor.

2. Smoke meters that perform the measurement of the particulate content of an undiluted sample of exhaust gas by drawing it through a filter paper of specified properties; the filtered soot causes blackening on the filter paper, which is detected by a photoelectric measuring head and evaluated in the microprocessor to calculate the result in the form of "filter smoke number" (FSN) specific to the instrument maker or mg/m^3.

3. Particulate samplers that measure the actual mass of particulates trapped by a filter paper during the passage of a specified volume of diluted exhaust gas. The relevant standard is typically contained in BS ISO 16183:2002. There are a number of instruments on the market, all of which require climatically controlled laboratory test filter weighing facilities.

4. *Particle number.* Due to developments in laser and microphone technology, particulate counting, based on either a laser-based condensation particle counter (CPC; Horiba) or the photoacoustic principle (PP; AVL), is practical; very sensitive, reliable, and cost-effective instruments are now being produced.

   The CPC is an optical particle counter; however, the particles have first to be enlarged by a condensing vapor to form easily detected droplets.

   The measurement principle behind the PP is based on the periodic irradiation of an absorbing sample (soot particle), in the near infrared region of the spectral range. The resulting periodic heating and cooling of the absorber induces a periodic expansion and contraction of the carrier gas; this produces pressure waves to develop around the particles that are detected as sound waves. The reflection of the pressure waves, from the ends of a resonant measuring cell whose length is tuned to the irradiation period, causes a standing wave to form. The more particles that are inside the measurement tube, the greater the amplitude will be of the standing wave; this is picked up by a supersensitive microphone and is directly proportional to the soot concentration. The measurement of soot concentrations below 5 $\mu g/m^3$ with a time resolution of below 1 second is claimed.

Particle mass. Instruments giving a real-time measurement of soot concentration in the exhaust tend to use either the diffusion charging (DC) principle or photoacoustic principle and are capable of displaying results in mg/m^3. Because of the wide use of diesel particulate filters (DPFs) the sensitivity of these instruments has to be extremely high when working downstream of the DPF; the best units have resolution limits of the order of 5 $\mu g/m^3$ and can be adapted to measure up- and downstream of the DPF in order to measure filter efficiency.

PRINCIPLES OF MEASUREMENT AND ANALYSIS OF GASEOUS EMISSIONS

The description of instrumentation that follows lists only the main types of instrument used for measuring the various gaseous components of exhaust emissions. Developments continue in this field, notably in the evolution of instruments having the shortest possible response time. Legislation prescription of approved instrumentation and test methods always lags behind their development, therefore producing a dichotomy between test equipment used in research and that used for type approval testing.

Nondispersive Infrared Analyzer (NDIR)

These are called "nondispersive" because all the polychromatic light from the source passes through the gas sample before going through a filter in front of the sensor, whereas "dispersive" instruments, found in analytical laboratories, filter the source light to a narrow frequency band before the sample.

The CO_2 molecule has a very marked and unique absorbence band of infrared (IR) light that shows a dominant peak at the 4.26 μm wavelength that the instrument sensor is tuned to detect and measure. By selecting filters sensitive to other wavelengths of IR it is possible to detect other compounds such as CO and other hydrocarbons at around 3.4 μm (see FTIR below). Note that the measurement of CO_2 using an NDIR analyzer is cross-sensitive to the presence of water vapor in the sample gas.

Fourier Transform Infrared Analyzer (FTIR)

This operates on the same principle as the NDIR, but performs a Fourier analysis of the complete infrared absorption spectrum of the gas sample. This permits the measurement of the content of a large number of different components. The method is particularly useful for dealing with emissions from engines burning alcohol-based fuels, since methanol and formaldehyde may be detected.

Chemiluminescence Detector (CLD)

Chemiluminescence is the phenomenon by which some chemical reactions produce light. The reaction of interest to exhaust emissions is:

$$NO + O_3 \rightarrow NO_2 + O_2 \rightarrow NO_2 + O_2 + photon$$

The nitrogen compounds in exhaust gas are a mixture of NO and NO_2, described as NO_x.

In the detector the NO_2 is first catalytically converted to NO and the sample is reacted with ozone, which is generated by an electrical discharge through oxygen at low pressure in a heated vacuum chamber. The light is measured by a photomultiplier and indicates the NO_x concentration in the sample.

A great deal of development work continues to be carried out to improve chemical reaction times, which are highly temperature dependent, and so to shorten instrument response times.

Flame Ionization Detector (FID)

The FID has a very wide dynamic range and high sensitivity to all substances that contain carbon. The operation of this instrument, shown schematically in Figure 16.7, depends on the production of free electrons and positive ions that takes place during the combustion of hydrocarbons. If the combustion is arranged to take place in an electric field, the current flow between anode and cathode is closely proportional to the number of carbon atoms taking part in the reaction. In the detector the sample is mixed with hydrogen and helium and burned in a chamber that is heated to prevent condensation of the water vapor formed. A typical, sample to measurement, response time is 1–2 seconds.

Fast FID, Cutter FID, and GC-FID

These are a miniaturized development of the FID instrument that is capable of a response time measured in milliseconds and may thus be used for in-cylinder and exhaust port measurements. Cutter-FID and GC-FID analyzers are used to measure methane (CH_4). In the fast FID nonmethane hydrocarbons are cata-lyzed into CH_4 or CO_2 and measured.

Paramagnetic Detection (PMD) Analyzer

PMD analyzers are used to measure oxygen in the testing of gasoline engines. They work due to the fact that oxygen has a strong paramagnetic susceptibility. Inside the measuring cell, the oxygen molecules are drawn into a strong

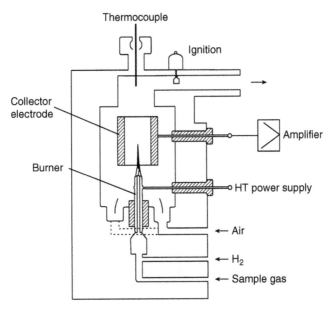

FIGURE 16.7   Flame ionization detector.

inhomogeneous magnetic field where they tend to collect in the area of strongest flux and physically displace a balanced detector whose deflection is proportional to oxygen concentration. Since NO_x and CO_2 show some paramagnetic characteristics, the analyzer has to be capable of compensating for this interference.

Mass Spectrometer

These devices, not yet specified in any emission legislation, are developing rapidly and can distinguish most of the components of automotive engine exhaust gases; currently they remain an R&D tool, but may represent the future technology of general emission measurement.

Response Times

All the above-mentioned instruments are required to operate under one of two quite different conditions, depending on whether the demand is for an accurate measurement of a sample collected over a fairly long time interval, as in the case of the various statutory test procedures, or for measurement in real time, as is required in development work.

These two sets of requirements are conflicting: Analyzers for steady-state work must be accurate, sensitive and stable, and thus tend to have slow response times and to be well damped. Analyzers for transient work do not require such a high standard of accuracy but must respond very quickly, preferably within a few milliseconds. Reduction of response time is a prime objective of instrument development in this field.

A further problem in transient emission testing is concerned with the time and distance lags associated with the positioning of the exhaust gas sampling points. As any search will reveal, this matter is discussed in many SAE papers concerning the reconstruction of the true signal from the instrument signals, taking into account sampling delays and instrument response characteristics.

EXHAUST "AFTER-TREATMENT" CATALYSTS AND FILTERS, OPERATION AND TESTING

Three-Way Catalytic Converters

Spark-ignition engines in modern powertrains are fitted with three-way catalytic converters that carry out three simultaneous chemical reactions:

- The reduction of NO_x to nitrogen and oxygen
- The oxidation of carbon monoxide to carbon dioxide
- The oxidation of unburnt hydrocarbons to carbon dioxide and water.

These three reactions generally occur most efficiently when the catalytic converter receives exhaust from an engine running between λ 14.6 and 14.8 for gasoline and require very tight close loop control in order to maintain optimum efficiency over an acceptably long service life. In some designs variable air injection takes place between the first (NO_x reduction) and second (HC and CO oxidation) stages of the converter.

The reactions depend on the composition of the exhaust gas and critically on the temperature of the internal surfaces of the converter. The "light-off" temperature at which the reactions start is around 300 °C and the highest conversion rate requires temperatures in the 400–700 °C range, but prolonged exposure to temperatures over 800 °C will lead to thermal degradation. The problem of overheating has led to designs in which the three-way catalytic converters are built as two separate units: the first, which is optimized for the highest temperature operation and fast light-off, is installed close to the engine; the second, physically larger, unit is installed further downstream under the vehicle floor in full air-cooling flow and designed to have as low a light-off temperature as possible.

The period between a cold start and light-off temperature being reached is when the majority of polluting emissions are recorded; many passive and active strategies are being developed to reduce both the light-off temperature and the engine warm-up time.

Catalytic converters all have the same three major components installed within the containment "can", which are:

- A core or substrate. This is commonly made up of a ceramic honeycomb but folded stainless steel foil is also used.

- A washcoat. This is a coating on the walls of the core designed to create the maximum possible surface area on which the chemical reactions may take place. It is usually formed by an alumina and silica mixture.
- The catalyst. The catalytic material deposited on the washcoat will be one or more precious metals. Platinum and rhodium are the reducing catalysts, used in the first stage of three-way converters, while platinum and palladium act as oxidizing catalysts used in the subsequent stages.

A three-way catalyst may be deactivated by meltdown of the substrate, thermal aging, or by chemical poisoning. Testing of the thermal and chemical stability of catalytic units may be carried out either in an engine test cell as part of a vehicle exhaust system, or in a special thermal test bench, or in a laboratory furnace at high temperature in a varied controlled gas atmosphere.

The use of an engine test cell for the catalyst aging and testing process can be a long and therefore expensive procedure, but it is the closest to real life in service. It is commonly carried out by Tier 1 suppliers, in custom-built test cells using client's engines.

A common form of damage suffered by a catalyst, particularly the first NO_x reduction stage, results from the exposure to a strongly oxidizing gas flow at high temperatures; conditions that can be encountered during vehicle braking, with the fuel cut, after running at high power; and conditions that can be repeatedly simulated in an engine test cell.

The high-temperature gas flow test bench may be the best way to simulate meltdown. Testing of catalyst poisoning can use an engine cell or a flow bench. There are a large number of compounds known to cause problems but the sources are sometimes difficult to locate and include agents such as the gasoline additive methylcyclopentadienyl manganese tricarbonyl (MMT); other common sources are lubrication oil additives, silicone[2] (which usually comes from coolant leakage), and lead compounds.

An important feature of the catalytic converter is the ability to store oxygen during the inevitable and small excursions of the lambda ratio into a "lean" phase; this allows continuous full operation during a short "rich" phase.

The correct and continuous operation of the lambda sensor is crucial to the whole closed-loop emissions control system and, like catalytic converters, they are vulnerable to thermal degradation and poisoning. The one sensor, fitted before the first catalyst, is being replaced in many systems by two-sensor systems (pre- and post-cat) and, in SULEV vehicles, by three or more sensors that use a complex cascade control.

---

2. The use of silicone tubing in engine rigging and test cell pipework is often banned by users of test facilities running catalytic aging or test work.

Diesel Oxidizing Catalysts (DOCs)

Like other catalysts the DOC consists of a metal canister containing a honey-comb-like substrate through which the exhaust gases flow. It is installed after the exhaust manifold and turbocharger. It uses oxygen in the (lean) diesel engine exhaust gas stream to convert carbon monoxide to carbon dioxide and hydrocarbons to water and carbon dioxide (see Figure 16.8). A DOC will suffer from thermal degradation if subjected to a prolonged period of running above 650 °C but, since diesel engines have intrinsically "cool" exhaust gases, deterioration is not likely to take place under normal operating conditions. Much of the in-cell and vehicle testing in the last few years has been related to "retro-fit" DOC kits and it has been found that removal of around 80% of the soluble organic fraction produces reductions in total PM emissions of some 30–50%. The back-pressure imposed by such a device may be around 2.5 kPa

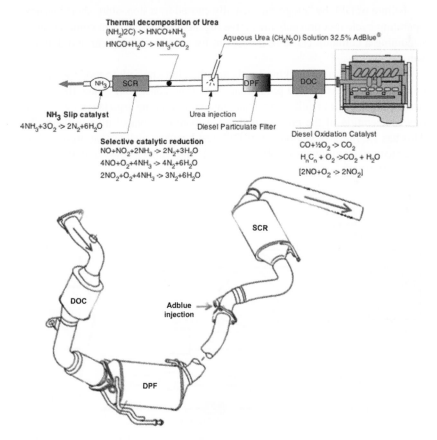

FIGURE 16.8 (Top) Diagram showing different modules and reactions, in their normal relative positions, that may be found in the CI (diesel) engine exhaust systems of vehicles of both HD or LV classifications. (Bottom) A typical passenger car layout, pre-silencer (muffler) unit.

(10″ of water) on a large truck engine, but has the virtue of being constant rather than the increasing back-pressure of a DPF as it loads up (see below).

Diesel Particulate Filter (DPF)

A DPF, by the nature of its role, increases exhaust back-pressure as the particles are captured within its matrix. The in-vehicle method of cleaning the filter is called "regeneration" and may be carried out passively or actively. Passive regeneration requires the unit's internals to be at a high enough temperature during regular use to "burn off" the deposits, usually a minimum of 350 °C.

Active regeneration is triggered when the exhaust back-pressure across the DPF is sensed to have reached a first critical point, whereupon the ECU initiates a fuel injection routine designed to increase the exhaust temperature in the DPF to over 600 °C in order to oxidize the particulate deposits.

Failure of DPF by "clogging", usually caused by a stop–start, short journey, operating cycle that prevents the active regeneration cycle from completing, is becoming a common cause of complaint by some passenger car and delivery van drivers.

Simulating partial blocking levels and the clogging failure mode within an engine test cell has proved problematic and time consuming, since the rate of particulate deposit is quite slow from an engine running legislative cycles and any acceleration of the process requires a modified control map in the ECU. This has led to the development of DPF test benches such as that produced by Cambustion (http://www.cambustion.com) and shown diagrammatically in Figure 16.9.

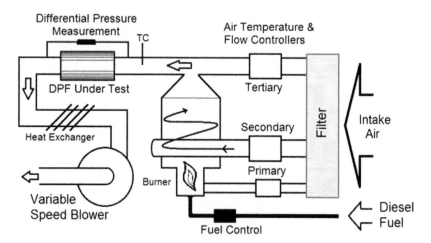

FIGURE 16.9 Diagrammatic representation of a diesel particulate filter test bench. Such test equipment improves test control and reduces test times compared with engine/dynamometer testing of the DPF. *(Courtesy of Cambustion© Ltd.)*

The main operational features of this fully automated unit are that the diesel is burned under controlled, stable conditions and is unaffected by the DPF's back-pressure. The rate of soot deposition can be controlled by adjusting the air flow rates, which allows for repeatable DPF loading conditions. Systems such as this are designed to measure filtration efficiency and to regenerate the DPF under operator-controlled conditions, making the test methodology cheaper and faster than using an engine in a test cell.

Selective Catalytic Reduction (SCR)

SCR catalysts are similar in basic construction to the three-way type. Currently the most common and effective SCR is the ammonia-SCR system, which reacts ammonia (NH_3) with the NO_x to form nitrogen (N_2) and water. There are three reaction pathways, which are shown under the "SCR" graphic in Figure 16.8.

The alternative hydrocarbon-SCR (lean NO_x reduction) systems use hydrocarbons as the reductant. The hydrocarbon may be that occurring in the exhaust gas ("native") or it may be added to the exhaust gas. This has the advantage that no additional reductant fluid source need be carried, but these systems do not currently have the performance of ammonia-SCR systems.

Ammonia-SCR systems use AUS32, which is a nontoxic 32.5% solution of urea (CH_4N_2O) in demineralized water sold in Europe under the trade name of "AdBlue®".

AdBlue is a trademark held by the German Association of the Automobile Industry (VDA) and is produced to ISO 22241 specifications, which is important as impurities or contamination will damage the SCR. In North America the fluid is currently called diesel exhaust fluid (DEF). In all cases the solution is injected in metered doses upstream of the SCR catalyst.

A few words of caution to test houses using urea injection:

- AdBlue, like any AUS32 fluid, is very susceptible to contamination from foreign matter and absorption of ions from containment material; it is also slightly corrosive for some metals, so must be treated strictly according to instructions and kept in approved containers. Contamination can lead to SCR deactivation.
- There is a developing and apparently legal trade in diluting and repackaging AdBlue then selling it as "automotive urea solution". Test results are completely invalidated if a solution of unknown concentration is used, so the lesson is to buy only from authorized dealers supplying fluid in original AdBlue containers.

AdBlue is held in a tank/pump unit and dosed upstream of the SCR system at a rate equivalent to 3–5% of diesel consumption. Vehicle SCR systems, both HD and LV, that are designed to meet Euro 6 and (US EPA 2010 USA) LEV-Bin 5 limits are currently in use throughout Europe, Japan, Australasia, and the USA.

All European truck manufacturers currently offer SCR equipped models, and retro-fitted kits are likely to become widely marketed, requiring an extensive amount of optimization and testing of multiple vehicle systems to be carried out.

Research into the dosing of the SCR system and calibration of the ECU requires accurate measurement of the liquid volume being injected. Devices such as the AVL "Shot To Shot"™ PLU 131 injector flow measuring device are available for this purpose.

Ammonia Slip Catalyst (ASC)

Due to the highly transient nature of automotive engine use in service, it is possible for an SCR system to release unreacted ammonia from the tail pipe; this is known as "ammonia slip". This eventuality has now to be recognized by the vehicle's OBD system as a fault; the legislative limit in the EU is 10 ppm. To deal with this problem ASC units have been developed and are installed, close coupled, to the downstream side of the SCR; the chemical reaction converts ammonia to nitrogen and water.

TESTING OF POST-COMBUSTION EXHAUST SYSTEMS AND DEVICES

The majority of engine test cells used in the testing of catalysts and exhaust systems are specifically designed for the purpose, having large and easily adaptable floor space able to contain whole vehicle exhaust systems and deal with high heat input into the cell air.

The various modules forming the exhaust system have to be installed in the appropriate positions relative to each other and the engine. The greatest threat to the thermal durability of an SCR is periodic burning off of the DPF by fuel injection and raising its temperature above 600 °C, which causes irreversible damage called sintering.

The high-temperature catalytic units need to use the actual vehicle connection to the engine and while some modification of the pipes joining other components can take place, great care has to be exercised with the pipe section joining of the AdBlue injection unit and the SCR due to the critical mixing taking place.

The position of the exhaust components downstream of the highest temperature units is carefully arranged in vehicles to expose them to an optimum cooling air flow when the vehicle is moving. In the test cell it may be necessary either to shield exhaust sections or to use spot cooling in order to obtain optimum performance and avoid overheating. It is important that in rigging the exhaust components the forces applied by their mounting and containment do not exceed that for which they are designed; where practical, the in-vehicle fixing points must be used with the loads imposed as in the vehicle. In this type of test work the engine tends to be considered as a gas generator and the test sequences prescribing engine speed and dynamometer load are designed specifically to provide exhaust at the flows and temperatures required to simulate the key life regime of the catalytic converters and filters. Catalytic aging cells invariably run continuously and are therefore designed for unmanned or limited supervision running, but with stringent fire safety precautions because of the high surface temperatures involved.

INTEGRATION AND MANAGEMENT OF EXHAUST EMISSION INSTRUMENTATION

The great majority of engine test work is concerned with the taking of measurements that are in principle quite simple, even though great skill may be needed to ensure the measurements give the right answer. Where emissions measurements are concerned, we are forced to move into a totally different and very sophisticated field of instrumentation engineering. The apparatus makes use of subtle and difficult techniques borrowed from the field of physics and the chemical laboratory.

It is very desirable, for all but the simplest garage-type emissions apparatus, that every test organization has one or more technicians who are specially trained to take technical and logistical responsibility (ownership) for the maintenance and calibration of the whole emissions measurement chain.

There are significant subsystems that need to be housed in engine or vehicle test facilities, each of which requires some special health and safety considerations due to the handling of hot and toxic gases within a building space. These will be considered later in this chapter.

The full flow of exhaust gases have to be safely ducted and samples for analysis have to be removed in the correct amounts at critical points in any vehicular after-treatment system.

It is important that water, which is a product of the combustion process, is not allowed to condense within the sampling system. Some of the compounds produced by the partial combustion of fuels are soluble in water, so it is possible to lose a significant amount of these compounds in the condensation on the equipment surfaces.

To prevent condensation in critical parts of the exhaust sampling system they will be electrically heated; for example, the sampling system for THCs will consist of heated probe, sampling line filter, and sampling pump, all of which are required to prevent the temperature of the equipment being lower than the dew point of the exhaust gas.

CALIBRATION, SPAN GASES, STORING GAS DISTRIBUTION SYSTEM

In order to function correctly exhaust gas analyzers have to be calibrated regularly, using gases of known composition. The principle of calibration is similar to that of any transducer in that the zero point is set, in this case by purging with pure nitrogen (sometimes referred to as "zero air"), then the 100% value is set with a gas of that composition, known as a "span gas". Good practice dictates that a gas of intermediate composition is then used as a check of system linearity. The whole routine may be highly automated by a calibration routine built into the analyzer control unit. After calibration the gas supply system is purged of the gases, particularly of NO_x, to prevent "crosstalk" and degrading in the lines.

Many R&D analyzers are capable of part-spanning in order to change their measuring sensitivity and can be calibrated by using known span gases, precisely diluted with zero air using a device known as a "gas divider". Alternatively, they may be calibrated directly using a bottled span gas of the correct composition. If using the gas divider method, it is important that the same gas (N_2 or artificial air) is used in the dilution as that which was used as the zero gas.

In addition to calibration or span gases, the facility will require operational gases such as hydrogen, plus an oxygen source as fuel for any flame ionization device (FID) and, in some cases, test gases for such routines as a critical flow check (see CVS).

Table 16.5 shows a possible list of gases, including a set of span gases, required in a modern emissions test facility. All these gases have to be stored at high pressure, usually between 150 and 200 bar, and distributed, at a regulated pressure of about 4 bar, to the analysis devices within the facility. The pipework system used in the distribution has to be chemically inert and medically clean to prevent compromising the composition of the calibration gases. For a basic single-fuel emission cell, the minimum number of 40- or 50-liter gas storage bottles[3] containing individual gases is probably 12, and a typical diesel/gasoline certification cell may have upwards of 20 gas lines feeding its analyzer system.

It is not uncommon in mixed fuel facilities with two cells to have 20 calibration bottles and four operating gases connected at any one time.

Each gas bottle has to be delivered, connected, disconnected, and removed in its cycle of use on site; therefore, the positioning of the gas store has some important logistical considerations.

Clearly the distance between the gas bottles, the pressure regulation station, and analyzers directly affects the cost of the gas distribution system. The practical requirements of routing upwards of 12 quarter-inch or 6 mm OD stainless steel pipes on a support frame system within a new facility is an important design and cost consideration. The length of gas lines may be sufficient to create an unacceptable pressure drop, which may require transmission at an intermediate pressure and regulation local to the analyzers.

---

3. A 40-liter bottle operating at 150 bar has a capacity of around 6000 liters of gas at the analyzer pressure and a 50-liter bottle (normal size for synthetic air and nitrogen) around 7500 liters.

TABLE 16.5 A List of Typical Gases That May Be Used in the Operation of Basic Engine Emission Certification Cell

Gases	Used As	Concentration	For Analyzer Type
Synthetic air (2)	Burner air	21% O_2, 79%N_2	FID
H_2/He mixture	Burner gas	40% H_2, 60% He	FID
C_3H_8 in synth. air	Test gas	99.95%	External bottle with CFO system
O_2	Operating gas	99.98% purity	CLD
CO in N_2	Span gas	50 ppm	CO_{low}
C_3H_8 in synth. air	Span gas	100 ppm	FID
C_3H_8 in synth. air	Span gas	10,000 ppm	FID
NO in N_2	Span gas	100 ppm	CLD
NO in N_2	Span gas	5000 ppm	CLD
CO in N_2	Span gas	500 ppm	CO_{low}
CO in N_2	Span gas	10%	CO_{high}
CO_2 in N_2	Span gas	5%	CO_2
CO_2 in N_2	Span gas	20%	CO_2
O_2 in N_2	Span gas	20%	CO_2
Synthetic air (1)	Zero gas	Oil free	FID
N_2	Zero gas	CO free	NO, CO_{low}, CO_{high}, CO_2, O_2

The storage of the gas bottles should be in a well-ventilated secure store where the bottles are not subjected to direct solar heating and thus remain under 50 °C. For operational reasons in very cold climates the bottle store needs to be kept at temperatures above freezing that allows bottle handling and operation without regulator malfunction.

The store commonly takes the form of one or more roofed "lean-to" structures on the outside wall of the facility, in temperate climates having steel mesh walls and always positioned for easy access of delivery trucks. Storage of gas bottles above ground within buildings creates operational and safety problems that are best avoided.

Pressure regulators may be either wall mounted with a short length of high-pressure (up to 200 bar) flexible hose connecting the cylinder or, more rarely, directly cylinder mounted. The regulators handling NO_x gases need to be constructed of stainless steel; other gases can be handled with brass and steel regulators.

The gas lines are usually made of stainless steel specifically manufactured for the purpose; sometimes, and always in the case of those transporting NO_x, the lines are electropolished internally to minimize degradation and reaction between gas and pipe. Pipes have to be joined by orbital welding[4] or fittings made expressly for the purpose. The pipes should terminate near to the emission bench position in a manifold of isolating valves and self-sealing connectors for the short flexible connector lines to the analyzer made of inert Teflon or PTFE, ideally not more than 2 meters long.

Local regulations and a planning permission requirement may apply to the construction of a bottle store.

GAS ANALYZERS AND "EMISSION BENCHES"

Although often used interchangeably with "emission bench", the term "analyzer" should refer to an individual piece of instrumentation dedicated to a single analytical task installed within an emission bench that contains several such units. As illustrated in Figure 16.10, most analyzers fit within a standard 19-inch rack and are typically 4 HU high; most are capable of being set for measuring different ranges of their specific chemical. Table 16.6 shows what analyzers might be installed within a high-quality bench built to measure raw gas from engines tested under ISO 8178-1:2006 cycles (plant illustrated in Figure 16.11).

CONSTANT-VOLUME SAMPLING (CVS) SYSTEMS

In both diesel and gasoline testing we need to dilute the exhaust with ambient air and prevent condensation of water in the collection system. It is necessary to measure or control the total volume of exhaust plus dilution air and collect a continuously proportioned volume of the sample for analysis. The use of a full-flow CVS system is mandatory in some legislation, particularly that produced by the EPA. It may be assumed that this will change over time with developing technology and "mini-dilution systems" will become widely allowed.

CVS systems consist of the following major component parts:

- A tunnel inlet air filter in the case of diesel testing or a filter/mixing tee in the case of gasoline testing, which mixes the exhaust gas and dilution air in

---

4. Orbital welding uses a gas–tungsten arc-welding (GTAW) process that prevents oxidation and minimizes surface distortion of pipe interior.

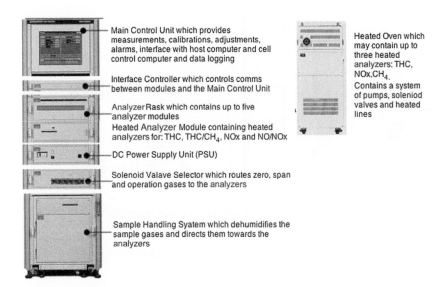

Main Control Unit which provides measurements, calibrations, adjustments, alarms, interface with host computer and cell control computer and data logging

Interface Controller which controls comms between modules and the Main Control Unit

Analyzer Rask which contains up to five analyzer modules

Heated Analyzer Module containing heated analyzers for: THC, THC/CH$_4$, NOx and NO/NOx

DC Power Supply Unit (PSU)

Solenoid Valave Selector which routes zero, span and operation gases to the analyzers

Sample Handling Syatem which dehumidifies the sample gases and directs them towards the analyzers

Heated Oven which may contain up to three heated analyzers: THC, NOx,CH$_4$.

Contains a system of pumps, soleniod valves and heated lines

FIGURE 16.10 The typical contents and layout of the two major instrumentation modules of a modern emission bench. The large 19-inch rack unit is normally housed in the control room, while the oven will be in the test cell. Diagram based on a Horiba 7000 unit and used with the permission of Horiba.

a ratio usually about 4:1. There is also a sample point to draw off some of the (ambient) dilution air for later analysis from a sample bag.

- The dilution tunnel, made of polished stainless steel and of sufficient size to encourage thorough mixing and to reduce the sample temperature to about 125 °F (51.7 °C). It is important to prevent condensation of water in the tunnel so in the case of systems designed for use in climatic cells the air will be heated before mixing with gas or the tunnel is outside the cell.

TABLE 16.6 Typical Analyzer Measuring Minimum and Maximum Ranges

Analyzer		Low Range	High Range
NDIR	CO_{low}	0–100 ppm	0–3000 ppm
	CO_{high}	0–1%	0–12%
	CO_2	0–2%	0–20%
	NO	0–100 ppm	0–5000 ppm
PMD	O_2	0–10%	0–25%
FID heated	THC	0–100 ppm	0–10,000 ppm
CLD	NO_x	0–100 ppm	0–5000 ppm

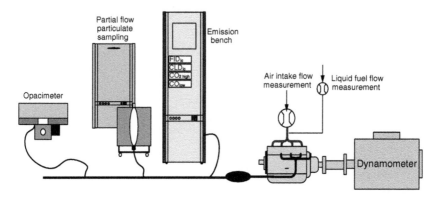

FIGURE 16.11 Typical emission plant used in ISO 8178-1:2006 engine emission cycles (diesel).

- The bag sampling unit; a proportion of the diluted exhaust is extracted for storage in sample bags.
- The gas sample storage bag array.
- The critical flow venturi, which controls and measures the flow of gas that the turboblower draws through the system.
- The blower that induces the air flow through the system, usually mounted well away from the work spaces due to the noise produced.
- The analyzer and control system by which the mass of HC, CO, NO_x, CO_2, and CH_4 is calculated from gas concentration in the bags, the gas density, and total volume, taking into account the composition of the dilution air component.

In the case of diesel testing a part of the flow is taken off to the dilution sampler containing filter papers for determining the mass of particulates over the test cycle.

A typical layout of a CVS system based on a 2 × 2 chassis dynamometer is shown in Figure 16.12. The arrangement shown is for a cell certifying vehicles, either diesel or gasoline, to Euro 3 or Euro 4 regulation. The system shown includes critical flow venturi (marked CFV), through which the gas/air is sucked by a fixed speed "blower" fitted with an outlet silencer. Some systems control flow with a variable speed blower and a different type of "subsonic" venturi.

US Federal regulations require that the flow and dilution rates can be tested by incorporation of a critical flow orifice (CFO) check. This requires a subsidiary stainless steel circuit that precisely injects 99.95% propane as a test gas upstream of the mixing point. The CVS system dilutes this gas according to the flow rate setting, allowing the performance to be checked at a stabilized temperature.

The full-flow particulate tunnel for heavy vehicle engines is a very bulky device. The modules listed above may be dispersed within a test facility

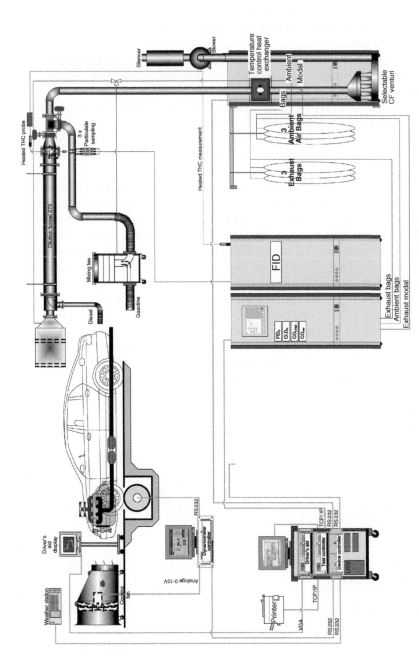

FIGURE 16.12 The major components of a Euro IV CVS system and their interconnections, installed within a chassis dynamometer facility.

because of the constraints of the building. The tunnel has to be in the cell near the engine and there are some legislative requirements concerning the distance from the tail pipe and the dilution point, ranging from 6.1 meters for light-duty to about 10 meters for heavy-duty systems.

Similarly, there are limits in some legislation concerning the transit time for HC samples from the sample probe to the analyzer of 4 seconds, which will determine the position within the building spaces of the FID unit.

The analyzers may be in the control room, the venturi and sample bags within another room adjoining the control room. Turboblowers are commonly placed on the roof since they often produce high noise levels. Alternatively, for light vehicle engines the analyzers, CVS control, and bag storage may be packaged in one, typically four-bay, instrument cabinet.

The so-called "mini-dilution tunnel" has been developed to reduce the problems inherent in CVS systems within existing test facilities; however, users must check if its use is authorized by the legislation to which they are working. It is necessary in this case to measure precisely the proportion of the total exhaust flow entering the tunnel. This is achieved by the use of a specially designed sampling probe system that ensures that the ratio between the sampling flow rate and the exhaust gas flow rate is kept constant.

In all CVS systems exhaust gases may be sent to the analyzers for measurement either from the accumulated volume in the sample bags or directly from the vehicle, so-called "modal" sampling.

In an advanced test facility such as those involved with SULEV development, there will be three to five or more tapping points created in the vehicle system, from which gas may be drawn for analysis, such as:

1. Exhaust gas recirculation sample (EGR)
2. Before the vehicle catalytic converter (pre-CAT)
3. After the vehicle catalytic converter (post-CAT)
4. Tail-pipe sample (modal)
5. Diluted sample (sample bags).

From 2 and 3 the efficiency of the catalyst(s) may be calculated.

In addition to the tapping points listed above, there may be several more taking samples from the CVS tunnel for particulate measurement.

HEALTH AND SAFETY IMPLICATIONS AND INTERLOCKS FOR CVS SYSTEMS

The following three fault conditions are drawn from experience and should be included in any operational risk analysis of a new facility:

1. Where the exhaust dilution air is drawn from the test cell and in the fault condition of the blower working and the cell ventilation system intake being out of action, or switched off, a negative cell pressure can be created that

will prevent the opening of cell doors. The cell ventilation system should therefore be interlocked with the blower so as to prevent this occurring.

2. In the fault condition of the blower being inoperative or the airflow too low due to too small a flow venturi being used, the hot exhaust gas can back flow into the filter and cause a fire. A temperature transducer should be fitted immediately downstream of the dilution tunnel inlet filter and connected to an alarm channel set at about 45 °C.

3. Gas analyzers that have to discharge their calibration gases at atmospheric pressure should have an extraction cowl over them to duct the gases into the ventilation extract; such an extract should be interlocked with the analyzer, preferably by a flow sensor.

TEMPERATURE SOAK AREAS FOR LEGISLATIVE TESTING

A feature of many legislative, chassis dynamometer-based, test procedures is that the vehicle has to soak at a uniform temperature before commencement. The two temperatures stipulated are between 20 and 30 °C for "ambient" tests and −7 °C for cold-start tests, and the soak periods are usually 12 hours long. This requirement needs the efficient use of a climatically controlled building space with temperature recording and where vehicles are not subjected to direct solar radiation. There are several strategies that may be adopted by vehicle test facilities having to deal with multiple vehicles in order to minimize the footprint of the soak area.

For cold-start testing, vehicles, additional to the one rigged for the first test, may be parked inside the chassis cell; they may then be moved onto the dynamometer by using special wheeled vehicle skids. ISO refrigerated shipping containers that can be parked close to the cell access doors may be used to soak vehicles providing that the transit time, from container to cell, is as short as possible and without engine rotation. In ambient soak areas it may be possible to use vertical stacking frames to store cars two or three high and test them in order of access.

THE EUROPEAN EXHAUST EMISSIONS TEST PROCEDURE (PASSENGER CARS)

EEC Directive 90/C81/01 stipulates the use of the ECE + EUDC test cycle, which is also known as the MVEG-A cycle, for emission certification and type approval of light-duty vehicles in Europe.

The entire cycle includes four ECE segments, ECE-15, repeated without interruption, followed by one EUDC segment. The pre-test cold soak is at 20–30 °C for a minimum of 6 hours. The engine starts the first cycle at 0 seconds (no idle) and the emission sampling begins at the same time. This modified cold-start procedure is also referred to as the New European

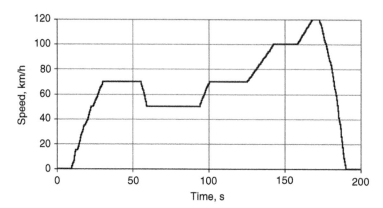

FIGURE 16.13 EUDC cycle.

Driving Cycle or NEDC. The temperature for Type IV approval, cold soak, and cold start for gasoline engines has been lowered to $-7\,°C$ since 2002 and may be referred to as the modified NEDC. Emissions are collected in bags during the urban drive cycle (UDC) and the extra urban cycle (EUC) according to the "constant-volume sampling" technique, analyzed, and expressed in g/km for each of the pollutants.

The extra urban driving cycle (EUDC) segment has been added after the fourth ECE cycle to account for more aggressive, high-speed driving modes (Figure 16.13). The maximum speed of the EUDC cycle is 120 km/h. An alternative EUDC cycle for low-powered vehicles has also been defined with a maximum speed limited to 90 km/h.

There have been criticisms of these test procedures on account of their lack of severity, in particular of the modest acceleration rates, and it has been claimed that they underestimate emissions by 15–25% compared with more realistic driving at the same speed.

THE US FEDERAL LIGHT-DUTY EXHAUST EMISSION TEST PROCEDURE (FTP-75)

FTP-75 is a more complex procedure than the European test, and is claimed to more realistically represent actual road conditions. The cycle is illustrated in Figure 16.14 and, in contrast to the European, embodies a very large number of speed changes. The cycle has three separate phases:

1. A cold-start (505-second) phase known as bag 1
2. A hot-transient (870-second) phase known as bag 2
3. A hot-start (505-second) phase known as bag 3.

During a 10-minute cool-down between the second and third phases, the engine is switched off. The 505-second driving sequences of the first and third phases

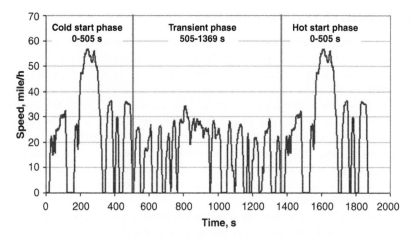

FIGURE 16.14 The FTP-75 test cycle.

are identical. The total test time for the FTP-75 is 2457 seconds (40.95 minutes), the top speed is 56.7 mph, the average speed is 21.4 mph, and the total distance covered is 11 miles.

The SC03 Supplemental Federal Test Procedure (SFTP), see Figure 16.15, has been introduced to represent the engine load and emissions associated with the use of air-conditioning units in vehicles certified over the FTP-75 test cycle. The cycle represents a 3.6-mile (5.8-km) route with an average speed of 21.6 mph (34.8 km/h), maximum speed 54.8 mph (88.2 km/h), and a duration of 596 seconds.

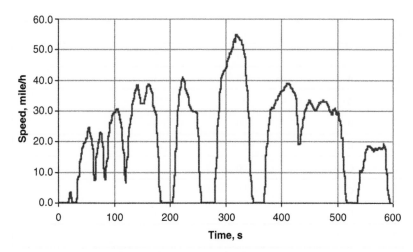

FIGURE 16.15 The Supplemental Federal Test Procedure.

HEAVY-DUTY TEST PROCEDURES

In the USA heavy-duty truck engines built from 2007 onwards have to comply with emission limits that have been tightened to such a degree that they will not be achieved without exhaust treatment containing SCR and particulate filtration.

The European Stationary Cycle (ESC; also known as OICA/ACEA cycle) engine test is for certification of HD truck diesel engines; it is carried out in a dynamometer test cell and the engine is run through a sequence of steady-state modes (Table 16.7). The speeds have to be held to within ± 50 rpm and the specified torque held to within $\pm 2\%$ of the maximum torque at the test speed. Emissions are measured during each mode and averaged over the cycle using a set of weighting factors.

In the ESC test sequence the stage speeds have to be calculated for each engine type. The high-speed n_{hi} is that at 70% of the declared maximum net power, which is the highest engine speed where this power value occurs on the power curve. The low speed n_{lo} is that at 50% of the declared maximum net power, which is the lowest engine speed where this power value occurs on the

TABLE 16.7 European 13-Stage ESC Heavy-Duty Engine/Dynamometer Test Sequence

Mode	Engine Speed	Load (%)	Weight Factor (%)	Duration (min)
1	Low idle	0	15	4
2	A	100	8	2
3	B	50	10	2
4	B	75	10	2
5	A	50	5	2
6	A	75	5	2
7	A	25	5	2
8	B	100	9	2
9	B	25	10	2
10	C	100	8	2
11	C	25	5	2
12	C	75	5	2
13	C	50	5	2

power curve. Then the engine speeds A, B, and C to be used during the test can be calculated from the following formulas:

$$A = n_{lo} + 0.25(n_{hi} - n_{lo})$$
$$B = n_{lo} + 0.50(n_{hi} - n_{lo})$$
$$C = n_{lo} + 0.75(n_{hi} - n_{lo})$$

The US EPA has introduced *not-to-exceed (NTE)* emission limits and testing requirements to make sure that heavy-duty engine emissions are controlled over "real-life" conditions.

Testing does not involve a specific driving cycle of any specific length; rather it involves driving of any type that could occur within the bounds of the NTE control area, including operation under steady-state or transient conditions and under varying ambient conditions. Emissions are averaged over a minimum time of 30 seconds and then compared to the applicable NTE emission limits. This legislation has required extensive negotiation with manufacturers in order to arrive at appropriate use of the NTE control area by particular truck types.

EXHAUST EMISSIONS OF MARINE PROPULSION DIESEL ENGINES

Pollution rules concerning marine diesel engines are contained in the "International Convention on the Prevention of Pollution from Ships", known as MARPOL 73/78, to which regular versions or "Annexes" are negotiated and issued. The latest, MARPOL Annex VI, came into force in July 2010 and sets limits on NO_x and SO_x emissions from ship exhausts. It also prohibits deliberate emissions of ozone-depleting substances.

The International Maritime Organization (IMO) emission standards are usually referred to as Tier I, II, and III standards. NO_x emission limits are set for diesel engines depending on the engine maximum operating speed (n, rpm), as shown in the table of Figure 16.16 and presented graphically below. Tier I and Tier II limits are global, while the Tier III standards apply only in NO_x emission control areas. Two sets of emission and fuel quality requirements are defined by Annex VI: global requirements and more stringent requirements applicable to ships in Emission Control Areas (ECA), which include most of the North America coastal waters and, in Europe, the North and Baltic Seas.

ISO 8178 AND TESTING EMISSIONS OF NON-ROAD VEHICLE ENGINES

Marine engine emissions are tested on various ISO 8178 cycles (see Table 16.8). The E2, E3 cycles are used for various types of propulsion engines, D2 for constant speed auxiliary engines, and C1 for variable speed and load auxiliary engines.

Tier	Date	NOx Limit, g/kWh		
		n < 130	$130 \leq n < 2000$	$n \geq 2000$
Tier I	2000	17.0	$45 \cdot n^{-0.2}$	9.8
Tier II	2011	14.4	$44 \cdot n^{-0.23}$	7.7
Tier III	2016†	3.4	$9 \cdot n^{-0.2}$	1.96

† In Nox Emission Control Areas (Tier II standards apply outside ECAs).

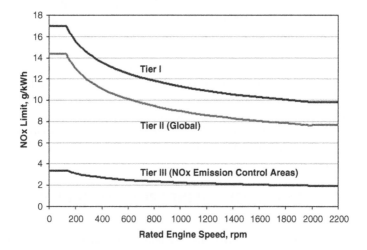

FIGURE 16.16 MARPOL Annex VI NO$_x$ emission limits.

At the time of writing (mid 2011) addition of not-to-exceed (NTE) testing requirements to the Tier III standards is being debated. NTE limits with a multiplier of 1.5 would be applicable to NO$_x$ emissions at any individual load point in the E2/E3 cycle.

One of the great weaknesses of the MARPOL rules is that large marine engines are tested using distillate diesel fuels, even though residual fuels, which are highly variable in content and virtually unregulated, are usually used in real-life operation.

In reading Table 16.8 it should be noted that:

- Engine torque is expressed as a percentage of the maximum available torque at a given engine speed
- Rated speed is the speed at which the OEM specifies the rated engine power
- Intermediate speed is the speed corresponding to the peak engine torque.

VEHICLE EVAPORATIVE EMISSIONS

Most vehicle emissions are produced by the combustion of hydrocarbon-based fuels within internal combustion engines; the exception that is important to the test engineer is the evaporative emissions of a complete vehicle.

TABLE 16.8 A List of ISO 8178 Test Cycles Often Referred To As the "Non-Road Steady Cycles" or NRSC

	Mode Number										
	1	2	3	4	5	6	7	8	9	10	11
Torque (%)	100	75	50	25	10	100	75	50	25	10	0
Speed	Rated speed					Intermediate speed					Low idle
Off-Road Vehicles											
Type C1	0.15	0.15	0.15	—	0.10	0.10	0.10	0.10	—	—	0.15
Type C2	—	—	—	0.06	—	0.02	0.05	0.32	0.30	0.10	0.15
Constant Speed											
Type D1	0.30	0.50	0.20	—	—	—	—	—	—	—	—
Type D2	0.05	0.25	0.30	0.30	0.10	—	—	—	—	—	—
Locomotives											
Type F	0.25	—	—	—	—	—	—	0.15	—	—	0.60
Utility, Lawn, and Garden											
Type G1	—	—	—	—	—	0.09	0.20	0.29	0.30	0.07	0.05
Type G2	0.09	0.20	0.29	0.30	0.07	—	—	—	—	—	0.05
Type G3	0.90	—	—	—	—	—	—	—	—	—	0.10

(Continued)

TABLE 16.8 A List of ISO 8178 Test Cycles Often Referred To As the "Non-Road Steady Cycles" or NRSC—cont'd

	Mode Number										
	1	2	3	4	5	6	7	8	9	10	11
Marine Application											
Type E1	0.08	0.11	—	—	—	—	0.19	0.32	—	—	0.30
Type E2	0.20	0.50	0.15	0.15	—	—	—	—	—	—	—
Marine Application Propeller Law											
Type E3, mode #	1		2		3		4				
Power (%)	100		75		50		25				
Speed (%)	100		91		80		63				
Weighting factor	0.2		0.5		0.15		0.15				
Type E4, mode #	1		2		3		4		5		
Power (%)	100		80		60		40		0		
Speed (%)	100		71.6		46.5		25.3		Idle		
Weighting factor	0.06		0.14		0.15		0.25		0.4		
Type E5, mode #	1		2		3		4		5		
Power (%)	100		75		50		25		0		
Speed (%)	100		91		80		63		Idle		
Weighting factor	0.08		0.13		0.17		0.32		0.3		

These emissions arise because the vehicle both at rest and in motion contains volatile compounds that evaporate differentially under changing solar and atmospheric conditions.

In the 1970s vehicles parked in the sun produced significant amounts of evaporative pollution. This pollution was quantified using a gas-tight chamber in which the vehicle was placed and subjected to typical diurnal temperature variations while exchanging the polluted air within the chamber with clean air from outside (volume compensation). The resulting test methodology and legislation requires such a chamber, commonly referred to as a SHED (sealed housing for evaporative determination). These are usually purchased as complete integrated systems that are designed to comply in every detail with the legislation.

Typically, the SHED consists of a "box" fitted with a vehicle access door capable of being closed with a gas-proof seal and lined with stainless steel, and constructed from materials that are chemically inert at the temperatures used in testing. This box or container, together with its external control cabinet, is invariably installed within a larger building and connected to the electrical and ventilation services of that building. A typical car SHED unit will have internal measurements of 6.5 m long by 3 m wide by 2.5 m high and will be equipped with systems to ensure the safe operation while handling potentially explosive gases. Legislation requires the test vehicle to be subjected to a diurnal temperature range of between 20 and 35 °C (98/69/EC) or 18.3 and 40.6 °C (CARB 2001), but some available units extend that range. Special, smaller than standard, SHED units for testing of motor cycles are available. There are also special units that allow refueling to be carried out within the sealed environment.

The withdrawal of polluted air for analysis and its replacement with treated air in the controlled environment of the SHED requires special control and some explosion-proof air handling equipment.

The vehicle subsystem directly involved with evaporative emissions includes the charcoal canister and its purge valve. The vast majority of in-service failures of the system (OBD-11 error code PO446) are caused by physical damage, saturation by overfilling, and leakage through breakage.

IMPLICATIONS OF CHOICE OF SITE ON LOW-LEVEL EMISSIONS TESTING

Test engineers not previously concerned with emissions testing should be aware of a number of special requirements that may not have the same importance elsewhere in other powertrain or vehicle testing facilities. The first of these is the subject of the test cell site.

The problem with measuring exhaust emission to the levels required to meet the standards set by SULEV legislation is that the level of pollution of the ambient air may be greater than that required to be measured at the vehicle exhaust.

Almost any trace of pollution of the air ingested by the engine or used to dilute the concentration of exhaust gas within the measuring process, from exhaust gas of other engines or fumes from industrial plant, e.g. paint or solvents, will compromise the results of emissions tests. Cross-contamination via ventilation ducts must be avoided in the design of the facility and it will be necessary to take into account the direction of prevailing winds in siting the cell. Other than siting the facility in an area having, so far as is possible, pristine air, there are three strategies available to help deal with SULEV level measurement:

- Reduction of the exhaust dilution ratio by having a heated CVS system and emissions system that prevents water condensation.
- Refinement or cleaning of the dilution air to reduce the background level of pollutants to <0.1 ppm. This supports a CVS system but is expensive and the plant tends to be large and complex.
- Use a partial flow system with "pure" (artificial) air for dilution. These bag mini-dilution systems do not comply with current CVS requirements and require incorporation of an exhaust flow-rate sensor.

Time will tell as to which of these methods become preferred and legislative practice, meanwhile investment in emission laboratories sited in areas with high levels of air pollution would not seem to be advisable.

In addition to the siting of emission laboratories, the following items and trends should be considered by engineers involved in the design and development of test facilities carrying out exhaust emission analysis:

- Since CO_2 reduction requires fuel consumption reduction, the measurement of fuel consumption will become more critical both in terms of mass fuel consumption accuracy and transient flow. Primary and secondary systems will have to be optimized to obtain the best control and measurement accuracy.
- Liquid fuel conditioning will become considerably more demanding. It will be necessary to take great care in the production, storage and handling of fuel, and in the avoidance of cross-contamination and deterioration. Control of fuel temperature to better than ± 1 °C at the entry to the engine fuel rail is already important and will continue to be so.
- It will become more important to condition combustion air, calling for the kind of systems described in Chapter 6. Temperature control of air will be required as a minimum, with humidity control becoming more common and requiring a high standard of system integration.
- Since much emissions testing will embrace cold starting and running at winter temperatures, there will be an increased need for cells capable of operating at temperatures of -10 °C or lower.
- The requirement to test engines fitted with the full vehicle exhaust system will increase the average cell floor area and call for exhaust dilution and extraction ducts.

- The running of increasingly complex emission test cycles calls, in many cases, for four-quadrant dynamometers and elaborate control systems.
- The requirement for testing and calibrating on-board diagnostic (OBD) vehicle systems will increase.

Finally, the delicate and complex nature of much emissions measurement instrumentation will call for standards of maintenance, cleanliness, and calibration at a level more often associated with medicine than with automotive engineering.

ABBREVIATIONS

AFR	Air/fuel ratio
CAFÉ	Clean Air for Europe (EEC program)
CI	Compression ignition (engine)
CNG	Compressed natural gas
CFO	Critical flow orifice
CTR	Continuously regenerating trap (diesel particle filter)
CVS	Constant-volume sampling
DI	Direct injection
DIR-x	Reference fuel with a total aromatic content of x wt%
DPF	Diesel particulate filter
DPM	Diesel particulate matter
EC	Elemental carbon
EPA	Environmental Protection Agency (US)
HD	Heavy duty
IDI	Indirect injection
LD	Light duty
MEXA™	Horiba analyzer range trade name
OC	Organic carbon (bound in hydrocarbon molecules)
PAC	Polyaromatic compounds
PAH	Polyaromatic hydrocarbons
PM 10	Particles of size below 10 μm
SOF	Soluble organic fraction
TEF	Toxic equivalence factor

REFERENCES

EEC L242. The Motor Vehicles (Type Approval) (Amendment) (No. 2) Regulations 1991 (http://www.legislation.gov.uk/si/si1991/Uksi_19912681_en_1.htm).

EPA CFR 40-86 (http://www.access.gpo.gov/nara/cfr/waisidx_04/40cfr86_04.html).

http://delphi.com/pdf/emissions/Delphi-Passenger-Car-Light-Duty-Truck-Emissions-Brochure-2010-2011.pdf

http://europa.eu/legislation_summaries/environment/air_pollution/l28186_en.htm

http://www.unece.org/trans/main/wp29/wp29regs1-20.html

FURTHER READING

BS 1747 Parts 1 to 13. Methods for Measurement of Air Pollution.

BS 4314 Part I. Infrared Gas Analysers for Industrial Use.

BS 5849. Method of Expression of Performance of Air Quality Infrared Analysers.

Horiba, Fundamentals of Exhaust Emissions Analysis and their Application, Horiba Instruments Ltd, Northampton, 1990.

USEFUL ADDRESSES

Environmental Protection Agency, 401 M Street SW, Washington, DC 20460, USA.

EU regs: http://www.dieselnet.com/standards/eu/ld.php

Key website for US regs: http://www.epa.gov/nvfel/testing/regulations.htm

National Environment Research Council, Polaris House, North Star Avenue, Swindon, Wilts, SN2 1EU.

Chapter 17

Chassis or Rolling-Road Dynamometers

Chapter Outline

Engine Testing. DOI: 10.1016/B978-0-08-096949-7.00017-0

INTRODUCTION

The continuing high volume of testing on chassis dynamometers is due not only to the need to certify vehicle exhaust emissions, but also to optimize the performance of the "sum of the parts" forming modern vehicles, by driving in controlled and repeatable conditions to test:

- Operation of on-board diagnostic (OBD) systems under loaded conditions and simulated faults
- Noise, vibration, and harshness (NVH)
- Drivability under transient load and braking conditions with varied driver profiles
- Performance under extreme climatic conditions, solar load, etc.
- Electromagnetic immunity and compatibility (EMC)
- Wind-tunnel performance (drag, side load stability, etc.)
- Power testing and vehicle cooling systems.

Compliance with ever more exacting statutory requirements governing both new vehicles, in order to gain type approval, and the after-market requires some form of testing on chassis dynamometers.

These requirements have called into existence a hierarchy of chassis dynamometers housed in test chambers of increasing complexity, including:

- Brake testers, ranging from simple to complex, installed in workshops
- In-service, small roller machines, for tuning and fault-finding installed in workshops
- End-of-line production rigs, housed in a booth installed with the factory building
- Emissions certification/testing installations installed in facilities
- Mileage accumulation facilities installed in a custom-built and open-sided facility
- Chassis dynamometers installed within anechoic and NVH and EMC test chambers
- Chassis dynamometers installed within climatic chambers of varied specification
- Independent or roller per wheel dynamometers ranging from workshop installations up to highly specialized R&D powertrain facilities
- Chassis dynamometers within wind tunnels (including "moving ground" or flat-track types) installed within very large custom-built facilities.

GENESIS OF THE ROLLING-ROAD DYNAMOMETER

The idea of running a complete vehicle under power while it was at rest was first conceived by railway locomotive engineers before being adopted by the road vehicle industry. As a matter of historical record the last steam locomotives built in the UK[1] were tested on multiple-axle units, with large eddy-current dynamometers connected to each driven axle fitted with rollers with rail-line profiles, the tractive force being measured by a mechanical linkage and spring balance. The locomotive on the test rig must have presented an awe-inspiring sight to an observer standing alongside when running at full power and a wheel speed of 90 mph.

Today the chassis dynamometer is used almost exclusively for road vehicles, although there are special machines designed for fork-lift and articulated off-road vehicles. The advantages to the designer and test engineer of having such facilities available are obvious; essentially they allow the static observation and measurement of the performance of the complete vehicle while it is operating within its full range of power and, in most respects, in motion.

Before about 1970 most machines were comparatively primitive "rolling roads", characterized by having rollers of rather small diameter, which inadequately simulated the tire contact conditions and rolling resistance experienced by the vehicle on the road. Such machines were fitted with various fairly crude arrangements for applying and measuring the torque resistance, while a single fixed flywheel was commonly coupled to the roller to give an approximate simulation of the tested vehicle's inertial mass.

The main impetus for development came with the rapid evolution of emissions testing in the 1970s. The diameter of the rollers was increased, to give more realistic traction conditions, while trunnion-mounted DC dynamometers with torque measurement by strain gauge load cells and more sophisticated control systems permitted more accurate simulation of road load. The machines were provided with a range of flywheels to give steps in the inertia; precise simulation of the vehicle mass was achieved by "trimming" the "iron inertia" by control of load contributed by the drive motor.

In recent years the development in digital control techniques and electrical drives has meant that incremental flywheels have been dispensed with in most emission testing designs and been replaced by single (2 × 2) or double (4 × 4) roll machines reliant on electrical simulation of all aspects of "road behavior".

THE ROAD LOAD EQUATION

The behavior of a vehicle under road conditions is described by the road load equation (RLE). It is a fundamental requirement of a true chassis dynamometer that it has to resist the torque being produced at the vehicle drive wheels in such a way as to simulate "real-life" resistance to vehicle motion.

The RLE is the formula that calculates the change in torque required with change of vehicle speed.

---

1. The National Railway Museum in York, England, holds both the locomotive "Evening Star" and the records of the test machine and displays the Heenan & Froude specification plate.

To simulate the real-life performance of a given vehicle, the RLE defines the traction or braking force that is called for under all straight-line driving conditions, which include:

- Steady travel at constant speed and coasting down on a level road
- Hill climbing and descent
- Acceleration, overrun, and braking
- Transitions between any of the above
- Effects of atmospheric resistance, load, towed load, tire pressure, etc.

The road load equation for a given vehicle defines the *tractive force* or *retarding force F* (newtons) that must be applied in order to achieve a specified response to these conditions. It is a function of the following parameters:

Vehicle specific:
Mass of vehicle	M (kg)
Components of rolling resistance	a_0 (N)
Speed-dependent resistance	$a_1 V$ (N)
Aerodynamic resistance	$a_2 V^2$ (N)

External:
Vehicle speed	V (m/s)
Road slope	θ (rad).

The usual form of the road load equation is:

$$F = a_0 + a_1 V + a_2 V^2 + M dV/dt + Mg \sin \theta \qquad (17.1)$$

where $M dV/dt$ = force to accelerate/brake vehicle and $Mg \sin \theta$ = hill climbing force.

More elaborate versions of the equation may take into account such factors as tire slip and cornering.

The practical importance of the road load equation lies in its application to the simulation of vehicle performance: it forms the link between performance on the road and performance in the test department.

To give a "feel" for the magnitudes involved, the following equation relates to a typical four-door saloon of moderate performance, laden weight 1600 kg, and C_d factor of 0.43:

$$F = 150 + 3V + 0.43V^2 + 1600 dV/dt + 1600 \times 9.81 \sin \theta$$

The various components that make up the vehicle drag are plotted in Figure 17.1: part (a) shows the level road performance and makes clear the preponderant influence of wind resistance.

The road load equation predicts a power demand at the road surface of 14.5 kW at 60 mph (96.6 kph), rising to 68.8 kW at 112 mph (180 kph) on a level road.

These demands are dwarfed by the demands made by hill climbing and acceleration. Thus, in Figure 17.1b, to climb a 15% slope at 60 mph calls for

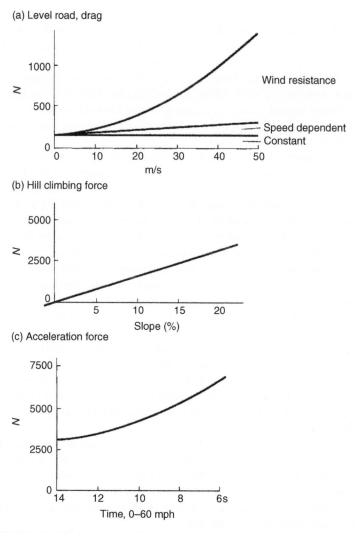

FIGURE 17.1 Vehicle drag, saloon of laden weight 1600 kg: (a) level road performance; (b) hill climbing; (c) acceleration force. Force is in newtons.

a total power input of $14.5 + 63.2 = 77.7$ kW, while in Figure 17.1c, to accelerate from 0 to 60 mph in 10 seconds at a constant acceleration calls for a maximum power input of $14.5 + 115 = 129.5$ kW.

CALIBRATION, COAST-DOWN, AND INERTIA SIMULATION

Like any other dynamometer, the torque measurement system of a chassis dynamometer, in this case also used to measure tractive force, must be regularly calibrated. Most machines use a strain gauge load cell that measures the torque reaction of the cradle-mounted drive motor, although the torque shaft or flange is also used. These measurement systems are calibrated in a similar way to that described for engine dynamometers in Chapter 10, with the possible detailed difference that, in the case of the chassis dynamometer, less midrange, if any, calibration weights are used during routine checks. Since the machinery is installed below ground level, in a pit, access to the load cell is often restricted; therefore, all good designs provide a calibration mechanism that allows the work to be done at cell floor level.

Speed measurement, on all but the after-market units, is done by an optical encoder. Before any test of a vehicle, or the dynamometer, is carried out, the machine must be run through a *warm-up routine* to thermally stabilize the whole system and reduce parasitic losses. Because of the time required to fully rig a vehicle undergoing emission development tests, most machines are provided with a mechanism that can lift the (fully rigged) vehicle up, so the wheels are clear of the rolls. In this way the warm-up routine can be run without disruption to the vehicle under test except for possible slackening of the restraint straps.

The control system of a chassis dynamometer may be checked by running a number of *coast-down* tests. Using the mass and road load equation for perhaps three vehicles representing the bottom, middle, and top of the range of vehicles to be run on the dynamometer, a plot of the theoretical coast-down graph of each is calculated between, typically, 130 kph (80 mph) and 16 kph (10 mph). The chassis dynamometer controller is then programmed with the details of the first test vehicle, run to a rotational speed equivalent to 135 kph, without any vehicle mounted, and allowed to coast down under control; the actual speed–time graph curve of the coast-down should exactly overlay that calculated.

To exactly calibrate a chassis dynamometer to a particular car, a similar technique can be adopted. This requires several actual coast-downs to be carried out by the test vehicle, at the correct loaded mass, on a straight level track in near still air. The averaged speed–time curve of these runs is then used to check and tune the control parameters of the dynamometer controller's RLE until the exact simulation is achieved.

Any differences between the mass inertia of the vehicle and that of the roller masses are compensated for through the motor control system, which must be sufficiently dynamic so that the machine's response to rapid vehicle accelerations are realistic.

The vehicle is essentially static in space when running on the rolls; therefore, it has no momentum. The "missing" momentum or mass inertia is generated in the rotating roller masses. As an example, the base inertia[2] of one

---

2. It is common practice to use the term "inertia" to mean the simulated or equivalent vehicle mass, rather than the moment of inertia of the rotating components. The inertia is thus quoted in kg rather than kgm^2.

particular single-axle, 125 kW, 48-inch roll dynamometer corresponds to a vehicle weight of 1369 kg (3000 lb) but, through control of its motor's torque input or absorption, it is capable of simulating the inertia of vehicles of weights between 454 kg (1000 lb) and 2722 kg (6000 lb).

It is a natural but incorrect perception, when driving a vehicle in a forward direction on a chassis dynamometer, to suppose that, if the vehicle restraint system suddenly failed, the vehicle would shoot forward. In fact, because the rolls have the momentum and they are rotating in the opposite direction to the wheels of the car, in those circumstances the car would shoot off the machine backwards (an even more worrying prospect?).

BRAKE SYSTEM TESTERS

Service garages and government test stations are equipped with quite simple but durable machines that are used for the statutory "in-service" testing of cars and commercial vehicles. They are installed in a shallow pit, and consist essentially of two pairs of rollers, typically of 170 mm diameter and having a grit-coated surface to give a high coefficient of adhesion with the vehicle tire. The rollers are driven by geared variable-speed motors and two types of test are performed. Either the rollers are driven at a low surface speed, typically 2–5 km/h, and the relation between brake pedal effort and braking force is measured, or the vehicle brakes are fully applied and torque from the rig motor increased and measured until the wheels slip (less recommended due to likely tire damage).

The testing of a modern ABS brake system requires a separately controlled roller for each wheel and a much more complicated interaction with the vehicle's braking control system and wheel speed transducers. Modern "end-of-line" brake test machines are used by vehicle manufacturers in a series of final vehicle check stations and also by major service depots; they therefore have features such as communication with the test vehicle control systems that are model specific.

TIRE-TESTING DYNAMOMETERS

Tires for every type of vehicle and airplane are tested by running them on a large-diameter steel drum; some are 5 meters in diameter and the vast majority have plain machined surfaces. The tire under test is fixed on its correct vehicle wheel, which may be fitted with its complete brake assembly. The test assembly is mounted on a rigid arm, fitted with strain gauges to measure imposed and resultant force vectors in all directions. The arm is hydraulically actuated and can press the wheel onto the roller with variable force to simulate realistic vehicle loading. The arm should also be able to turn the test wheel at an angle to the drum shaft axis to simulate "skewing" of the tire to the road surface.

The drum is accelerated to any required road speed by an electrical dynamometer whose control system is able to simulate the vehicle mass. Aircraft tire, wheel, and brake assemblies are tested on this type of test rig, which are capable of simulating the emergency landing and braking loads of a fully laden aircraft.

Such tire test rigs, even in the automotive sizes, can suffer from explosive tire failures and brake system fires so they require the highest possible standard of safety procedures and management relating to the exclusion of personnel from the rig area during test running.

ROLLING ROADS FOR IN-SERVICE TUNING AND ASSESSMENT

For a vehicle maintenance garage the installation of a rolling road represents a considerable investment, so to cater for this market a number of manufacturers produce complete packaged units, often based on nominally 8.5-inch (216 mm) diameter rollers, and requiring minimum subfloor excavation. This market has tended to be dominated by American suppliers, and grew rapidly when the US Environmental Protection Agency (EPA) called for annual emissions testing of vehicles, based on a chassis dynamometer cycle (IM 240).

Road load simulation capability at this level is usually limited to that possible with a choice of one or two flywheels and a comparatively simple control system. Most of the testing is concerned with fault diagnosis, possibly using OBD readers, brake balancing, and power checks. The tests have to be of short duration to avoid overheating of engine and tires.

In-service rolling roads for testing large trucks are confined almost exclusively to the USA. They are usually based on a single large roller capable of running both single- and double-axle tractor units. The power is usually absorbed by a portable water brake, such as that shown in Figure 4.2, which may also be dismounted from the roller and used for direct testing of truck engines, either between the chassis members or on an engine trolley, a useful feature in a large OEM agency, overhaul, and test facility.

ROLLING ROADS FOR END-OF-LINE (EOL) PRODUCTION TESTING

EOL chassis test rigs range from simple roller sets used for first start and basic function testing of two-wheel-drive cars, to multi-axle units with variable geometry of roll sets and ABS checking facility. EOL rigs require common special design features and any specification for such an installation should take them into account:

- The design and construction of the machine must minimize the possibility of damage from vehicle parts falling into the mechanism. It is quite common for small fixings, left in or on the vehicle during assembly, to shake down into the rolling road, where they can cause damage if there are narrow clearances or converging gaps between, for example, the rolls and lift-out beam, the wheelbase adjustment mechanism, or the sliding floor plate system.

- The vehicle must be able to enter and leave the rig quickly yet be safely restrained during the test. The usual configuration is for all driven vehicle wheels to run between two rollers. Between each pair of rolls there is a lift-out beam that allows the vehicle to enter and leave the rig. When the beam descends it lowers the wheels between the rolls and at the same time small "anti-climb-out" rollers swing up fore and aft of each wheel. At the end of the test, with the wheels at rest, the beam rises, the restraining rollers swing down, and the vehicle may be driven from the rig.

- To restrain the vehicle from slewing from side to side to a dangerous extent, specially shaped side rollers are positioned between the rolls at the extreme width of the machine, where they will make contact with the vehicle tire to prevent further movement. These rollers must be carefully designed and adjusted so that they don't cause tire damage; they are not recommended for use with vehicles fitted with low-profile tires due to possible "pinching" damage.

- Care should be taken with the operating procedures if "rumble strips" are to be used. These are raised strips or keys sometimes fitted to the rolls in line with the tire tracks, their purpose being to excite vibration to assist in the location of vehicle rattles. However, if the frequency of contact with the tire resonates with the natural frequency of the vehicle suspension, this can give rise to bouncing of the vehicle and even to its ejection from the rig. They can also create severe shock loads in the roll drive mechanism.

Pre-1990s rigs, still in use, that were designed to absorb power from more than one axle and be capable of adjustment of the center distance between axle, had shaft, chain, or belt connections between the front and rear roller sets; this complexity has been obviated by modern electronic speed synchronization of separate drive motors.

CHASSIS DYNAMOMETERS FOR EMISSIONS TESTING

The standard emissions tests developed in the USA in the 1970s were based on a rolling-road dynamometer developed by the Clayton Company and having twin rolls of 8.625 inch (220 mm) diameter. This machine became a de facto standard despite its limitations, the most serious of which was the small roll diameter, which resulted in tire contact conditions significantly different from those on the road.

Later models, which are still widely in use, have pairs of rollers of 500 mm diameter, connected by a toothed belt, and between which the vehicle tire sits. The roller sets are connected by a toothed belt to a set of declutchable flywheels that simulate steps of vehicle inertia. Any adjustment of simulated vehicle

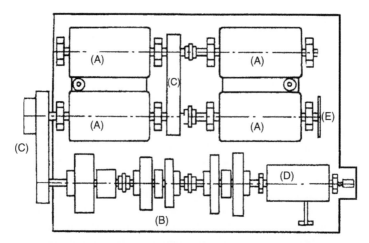

FIGURE 17.2   Schematic plan view of four-roller (2 × 2 = single driven axle) emission chassis dynamometer of late 1980s design. (A) Four 500-mm rollers. (B) Selectable flywheel set. (C) Belt drive. (D) DC motor fitted with shaft encoder. (E) Disk brake for vehicle loading. Between the rollers there is a vehicle loading and lift-out beam fitted with side-restraint rollers.

inertia between the flywheel mass increments is adjusted by the control system using a small (≤60 kW) four-quadrant DC dynamometer connected to the flywheel shaft, as shown diagrammatically in Figure 17.2.

The machine is designed for roller surface speeds of up to 160 km/h and the digital control system is capable of precise road load control, including gradient simulation. Such machines must be manufactured and installed to a very high standard to keep vibration, caused by variations of flywheel masses, to an acceptable level. It is possible to abuse them, notably by violent brake application that can induce drive vehicle wheel bounce and thus very high stresses in the roller and flywheel drive mechanism. Such a unit is shown in Figure 17.3.

When proven, high-power thyristor control systems became available, the US Environmental Protection Agency (EPA)-approved machines had a single pair of rollers of 48-inch diameter and 100% electronic inertia simulation. Most of the standard designs, now available from the major suppliers, are of this type and most are of a compact design having a double-ended AC drive motor with rollers mounted overhung on each stub shaft.

The performance figures quoted below are for a modern 48-inch roll chassis dynamometer designed for exhaust emission testing passenger cars to international industrial standards (in this case an AVL "ROADSIM" unit):

Nominal tractive force in motoring mode (per axle) 5870 N at $V \leq 92$ km/h
Nominal power in motoring mode (per axle) 150 kW at $V \geq 92$ km/h and < 189 km/h
Max. speed 200 km/h
Roller diameter 48 inches (1219.2 mm)

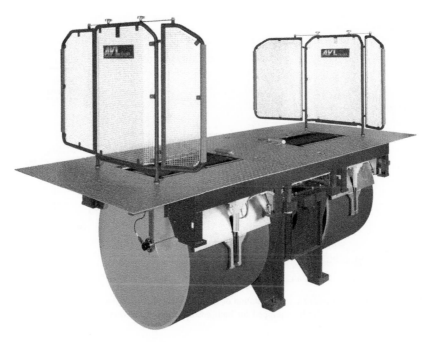

FIGURE 17.3 Single 48-inch roller, 150 kW AC chassis dynamometer package (AVL). *(Photo © AVL List GmbH.)*

Distance between outer roller edges 2300 mm
Inertia simulation range (2WD) 454–2500 kg
Inertia simulation range (4WD) 800–2500 kg
Max. axle load 2000 kg
Tolerance of tractive force measurement ±0.1% of full scale
Tolerance of speed measurement ≤0.02 km/h
Tolerance of tractive force control ≤0.2% of full scale
Tolerance of inertia and road load simulation ±1% of calculated value (but not better than force control).

This type of standard, pre-tested package enables chamber installation time to be minimized.

THE EMISSION DYNAMOMETER CELL AND ENVIRONS

As with engine test cells dealing with the testing of exhaust emissions, chassis dynamometers, particularly those associated with ultra-low emissions work (SULEV), should be built on sites that are out of the drift zone of automotive, agricultural, and industrial pollutants. All facilities have to be capable of running within the environmental conditions prescribed by EPA emission standards, namely −10 to +40 °C.

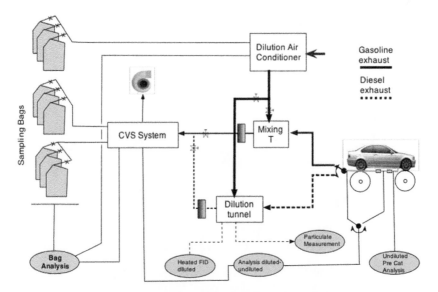

FIGURE 17.4 Diagrammatic representation of the major emission analysis equipment housed in a chassis dynamometer facilty equipped for both gasoline and diesel cars. Much of the valving detail is omitted for clarity.

Since it is more energy efficient to carry out several $-7\,°C$ cold-start tests while the chamber is cold, the cell designers have to decide either to have the cell built large enough to cold soak multiple cars or to have a smaller dynamometer cell and a separate refrigerated cold soak room for extra vehicles, perhaps using a refrigerated ISO container.

If both gasoline and diesel vehicles are to be tested in the same cell then the space required for the emissions equipment can be considerable and needs careful layout; Figure 17.4 shows diagrammatically the major plant to be housed. While the emission analyzers and even sample bag racks can be in a large control room, the AC or DC drive cabinets for the dynamometer must be in a separate, clean, well-ventilated room with a 15-meter or less, dedicated, subterranean cable route to the motors.

The blower needs to be mounted out of working sound range of the control room on the building roof while the housing of the sample and calibration gas bottles needs to be done according to the recommendations in Chapter 16.

MILEAGE ACCUMULATION FACILITIES

In the USA, prior to the adoption of Tier 1 emission standards, the EPA regulations required that the emissions of passenger cars and light vans had to be tested after 50,000 miles, accumulated either on a track or on a chassis dynamometer using a prescribed driving cycle known as the "AMA" (40 CFR 86, Appendix IV).

The cost and the physical strain of using human drivers on test tracks or public roads for driving vehicles the prescribed distances, in as little time as possible, are too high, so special chassis dynamometer systems were developed for running the specified sequences under automatic and robotic control, commonly over a period extending to 12 weeks.

Tier 1 emission standards, effective from the 1994 model year, included a doubling of the useful life period for light-duty vehicles from 50,000 to 100,000 miles. This raised concerns about the increased amount of time and associated costs needed to complete the required mileage and also concern that the AMA test cycle did not represent real-life usage. This resulted in the EPA allowing manufacturers to develop and submit their own strategies for proving the durability of their vehicles' emission control systems.

A number of manufacturers developed dynamometer driving cycles that were more severe than the AMA and that could reach 100,000 miles in a shorter amount of time. These are referred to as "whole-vehicle mileage accumulation" cycles.

Other manufacturers developed techniques for aging the catalytic converter and oxygen sensor on a test bench to the equivalent of the full useful life distance and then reinstalled them on the vehicle for emissions testing; this technique is called "bench aging".

Mileage accumulation dynamometers for passenger cars now tend to be of a very similar design as the emission test machines, being based on 48-inch or larger diameter, directly coupled to an AC motor in the range of 120–150 kW power rating.

In order to fully automate mileage accumulation tests, the vehicle has to be fitted with a robot driver (see "Robot Driver" section later in this chapter) and a system to allow automated fueling. Because the vehicle's own cooling system is stationary, a large motorized cooling fan, facing the front grille, is essential. This fan will be fitted with a duct to give a reasonable simulation of air flow, at the vehicle's front grille, under road conditions. The fan speed is usually controlled to match apparent vehicle speed up to about 130 km/h; above that speed the noise produced and power needed become too high to be practical on most sites. The fans and their discharge ducts have to be accurately positioned and very securely anchored to the ground.

Since mileage accumulation facilities are usually housed within an open-walled structure with just a roof and run 24 hours a day, they are "bad neighbors", and must be suitably shielded to prevent noise nuisance.

NVH AND EMC CHASSIS DYNAMOMETERS

The major features of anechoic cell design are discussed in Chapter 18. A critical requirement of an NVH facility is that the chassis dynamometer should itself create the minimum possible noise. The usual specification calls for the noise level to be measured by a microphone located 1 m above and 1 m from the centerline of the rolls. The specified sound level, when the rolls are rotating at a surface speed of 100 km/h, is usually ≤50 dBA.

To reduce the contribution from the dynamometer motor and its drive system, it is usually located outside the main chamber, in its own sound-proofed compartment, and connected to the rolls by way of a long shaft running through a transfer tube designed to minimize noise transfer. Note that the design of these shafts can present problems because of their unsupported length and the nature of the couplings at each end; lightweight tubular carbon-fiber shafts are sometimes used. The dynamometer motor will inevitably require forced ventilation and the ducting will require suitable location and treatment to avoid noise being transferred into the chamber.

Hydrodynamic bearings are commonly used in preference to rolling element, but they are not without their own problems as their theoretical advantages, in terms of reduced shaft noise, may not be realized in practice unless noise from the pressurized supply oil system is sufficiently attenuated.

In a well-designed chassis dynamometer the major source of noise will be the windage generated by the moving roll surface; this is not easily minimized at source or attenuated. Smooth surfaces and careful shrouding can reduce the noise generated by the roller end faces, but there is inevitably an inherently noisy jet of air generated where the roller surface emerges into the test chamber. If the roller has a roughened surface or is grit coated the problem is exacerbated. The noise spectrum generated by the emerging roller is influenced by the width of the gap and in some cases adjustment needs to be provided at this point.

The flooring over the dynamometer pit, usually of steel, sometimes aluminum plate, must be carefully designed and appropriately damped to avoid resonant vibrations.

A particular feature of NVH test cells is that the operators may require access to the underside of the vehicle for arranging sound-recording, photographic, and lighting equipment. This is usually accessed by way of a trench at least 1.8 m deep, lying between the vehicle wheels and covered by removable floor segments.

For running a vehicle under load within an EMC chamber (see Chapter 18), it often specified for the whole chassis dynamometer to be mounted on a turntable so that the running vehicle can be turned 360°, in relation to the electromagnetic beam of an emitter or antenna.

The chassis dynamometer of an EMC facility must not emit electromagnetic noise into the cell within the wavebands being used or investigated in the tests. Because of the wide spectrum of high-frequency emissions of a powerful IGBT-controlled AC dynamometer system, it is usual to specify DC motors for chassis dynamometers in EMC cells.

The subfloor layout of such complex and expensive cells has to be designed so that the thyristor-controlled DC drives and all other high-power electrical switching equipment is shielded by concrete and special attention is paid to the layout of cables according to the practices recommended in Chapter 5. To test the EMC of systems such as ABS braking and traction control and minimize electromagnetic noise from the test equipment, independent wheel dynamometers that are based on hydrostatic motor/pump circuits have been used.

SPECIAL FEATURES OF CHASSIS DYNAMOMETERS IN CLIMATIC CELLS

A typical modern 4 × 4 chassis dynamometer system built within a stainless-steel-lined climatic chamber is shown in Figure 17.5. Such machines intended to operate over the temperature range +40 to −10 °C are built from normal materials and apart from sensible precautions to deal with condensation they are standard machines. For operation at temperatures substantially below this range and particularly those working below −25 °C, certain special features and material use may be required.

Until the compact "motor between rollers" ("motor-in-middle" in the USA) design of dynamometer became common, the dynamometer motor in climatic cells was often isolated from the cold chamber and operated at normal temperatures. With the new designs, two quite different strategies can be adopted to prevent low temperatures causing temperature-related variability in the dynamometer system:

- The dynamometer pit can be kept at a constant temperature above or below that of the chamber by passing treated (factory) air through the pit by way of

FIGURE 17.5   Climatic cell equipped with 4 × 4 chassis dynamometer and full exhaust emissions analysis system. All cell surfaces are made from stainless steel. Control room window can just be seen in the left-hand wall beyond the mixing-tee module, which is mounted on castors. Tie-down pillars are in the foreground. Photo by author.

ducts cast in the floor. This strategy submits portions of the roll's surface to differential temperatures during the static cool-down phase and it is usual to place a cover over the exposed roll portion.

• The pit can be allowed to chill down to the cell temperature but the parts of the chassis dynamometer that are crucial to accurate and consistent performance, such as bearings and load cell, are trace heated.

As with tests at ambient temperature, the chassis dynamometer in a climatic cell is always run through a "warm-up" stabilization routine before testing takes place.

Whatever the layout, components exposed to temperatures below 25 °C, such as rolls and shaft, should be constructed of steel having adequate low-temperature strength, to avoid the risk of brittle fracture.

FLAT-TRACK CHASSIS DYNAMOMETERS

There are now a few different chassis dynamometer designs that are based on an endless steel belt tensioned between, and running around, two rollers horizontally disposed to each other. The more obvious industrial requirements for such dynamometer designs are threefold:

1. Testing of snowmobiles, which are themselves propelled through an endless belt in contract with the surface on which they are traveling.
2. Testing vehicles in wind tunnels (see Chapter 6), where it is required to accurately simulate the true relative motion between the underside of the vehicle and the road surface over which it is traveling, a situation not achieved with roller dynamometers, where the cell floor under the vehicle body, unlike the real-life situation, is stationary.
3. Tire testing where the flat track eliminates the problems related to differences in rolling resistance between running tires on rollers and on a flat surface (see "The Effect of Roll Diameter on Tire Contact Conditions" at the end of this chapter).

In the first two applications at least, the belt supporting the vehicle under test needs itself to be supported to prevent belt distortion and sag; the tension in the belt is insufficient to maintain a flat surface when loaded with the UUT. Some small units use a rack of small support rollers to provide this midspan support, but the larger and faster units use an inverted variation of the "air-float" system used to move heavy loads over smooth floor surfaces.

INDEPENDENT WHEEL DYNAMOMETERS

A limitation of the conventional rolling-road dynamometer, made up of one or more fixed cross-axles, is that it is unable to simulate cornering or differential wheel slip. With the universal adoption of electronically controlled traction and braking (ABS), there is a growing requirement for test beds that can simulate these conditions. There are two types of solution:

- *Four roll-set rolling roads.* These machines range from "end-of-line" rigs consisting of four sets of independent double-roller units, to complex development test rigs, some having steered articulation of each (wheel) roll unit. The production rigs permit the checking of on-board vehicle control systems, wheel-speed transducers, and system wiring by simulation of differential resistance and speeds of rotation. The development of hybrid vehicles has given impetus to the development of independent four-wheel chassis dynamometers because some designs are based on individual wheel drive motors. Electronically controlled transmission systems and limited slip differential (LSD) units on vehicles can also be tested on this type of dynamometer by the simulation of individual wheels losing traction.
- *Wheel substitution dynamometers.* A major problem faced by engineers carrying out NVH testing on a chassis dynamometer is that tire noise can dominate vehicle sound measurements. One answer is to absorb the power of each drive wheel with an individual dynamometer. The wheel hub is modified so that the vehicle can still sit on its tires, giving approximately the correct damping effect, while the tire contact and windage noise is eliminated. Alternatively, the wheels may be removed and the individual wheel dynamometers used to support the vehicle. The dynamometers should be four-quadrant machines, to simulate driving, coasting, and braking conditions. Hydrostatic, AC, and DC machines have been used for this type of application.

ARTICULATED CHASSIS DYNAMOMETERS

For QA and EOL testing of articulated, off-road vehicles, such as large front-end loaders, articulated chassis dynamometers have been built by using an inverted vehicle chassis frame, the axles and differentials of one of the tested vehicles. The rollers can be fitted in the place of the vehicle wheels and a dynamometer can be fitted to an extension of the drive shaft system. The floor plate system required to provide a safe surface in any axle position and to follow the movement of the swinging front rollers requires some ingenuity in the detailed design.

ROBOT DRIVERS OR SHIFTERS

For a human, driving a vehicle on a chassis dynamometer for any length of time can be a tedious task and their pattern of gear changing (shifting) will vary over that time, meaning that any one test will not be exactly repeatable. To drive through a fixed emission drive cycle requires the human driver to use a visual "driver's aid", which is very similar to an early version of a video game where the driver has to match the vehicle's speed and gear with a scrolling trace on a display screen. It is widely accepted that some drivers of emission cycles have "lower NO_x styles" than others, which is indicative of the variability of human driving characteristics even within a group of trained professionals.

A robotic driver provides a means of driving a vehicle installed on a chassis dynamometer, or testing a transmission system in a test cell, in a precisely repeatable sequence and manner. These two applications have produced three different types of robotic tools, which are:

- In-vehicle, floor mounted, with driver's seat removed. This type is normally used for mileage accumulation work where the long installation time required is negligible compared to the long test period and the need for absolute reliability.
- In-vehicle, seat mounted. This type is designed to have a short installation time and no modification of the car; they are normally used in emission testing for repetitive running of legislative drive cycles.
- Test cell, floor mounted, two- or three-axis, used in transmission testing.

Both in-cell gear shifters and in-vehicle drive robots are required to have the following common features:

- Appropriately quick set-up times and a built-in routine for "learning" the gear lever positions plus, in manual transmissions, learning and readjusting for change in the clutch "bite" position. Learning is normally carried out within a programmed teaching routine and involves manually shifting the gear lever with the robot attached and allowing the controller to map the x and y coordinates.
- Smooth and precise operation of the gear shift lever, of manual, semi-automatic, and fully automatic transmissions. Actuation in the x and y axes and free movement in the required arc of travel in the z axis are usual; powered operation in the z axis is much more complex and consequentially rarer. Most actuators are servo-motor-driven linear actuators. Pneumatic actuators give the best representation of the force applied by the human arm, since there is a smooth build-up of pressure over the synchronizer operation, but are reported to be more difficult to set up.
- Operation of the clutch and throttle pedals (in manual transmissions) precisely coordinated with operation of the gear change lever.
- Removal of all load and restraint on the gear shift lever when not shifting.
- Safety systems to prevent damage by overloading the lever in the case of non-engagement.
- High operational reliability, often over weeks of continuous operation.

The in-vehicle, seat-fixed designs have to overcome the problem of the compliance of their fixing and the constraints of the installation space. This means that their performance is usually less capable of extended operation;

they are also not capable of the most accurate force measurement or actuation in the "z" axis, none of which are practical constraints in running emission drive cycles. The problem of wear in the gearbox actuation and in the driveline clutch has to be overcome by regularly running "re-learn" routines during endurance work.

Because of their layout, most in-line, rear-wheel-drive, and 4×4 vehicles have gearboxes with shift mechanisms and actuators directly connected, but there is a significant increase in cable-operated gear change mechanisms that can create problems for robotic drivers. These cable designs are largely driven by NVH work that requires the decoupling of the transmission from the cabin and the use of softer engine mounts. The increased engine movement under load and cable stretch can cause the gear lever to change position during tests, which has created problems, such as the robot arm losing the gear knob position and the need for a "softer" actuation.

THE INSTALLATION OF CHASSIS DYNAMOMETERS

Whatever the type of chassis dynamometer the reader has to specify and install, there are some common planning and logistical problems that need to be taken into account. For single-roll machines, built within a custom-built chamber, the longitudinal space requirement may have to take account of front- and rear-wheel-drive vehicles plus room for tie-down mechanisms and the vehicle cooling fan. A typical layout is shown in Figure 17.6.

CHASSIS DYNAMOMETER CELLAR OR PIT DESIGN AND CONSTRUCTION DETAILS

Because of ease of vehicular access, many chassis dynamometers are built on the ground floor of a building and therefore need to be installed within a pit

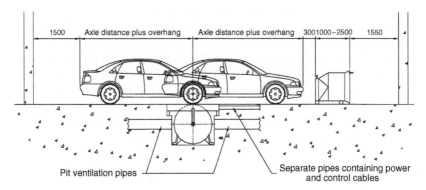

FIGURE 17.6 A single-roll-set (2×2) chassis dynamometer plus road speed following cooling fan, within a cell sized to test rear- and front-wheel-drive cars. *(AVL Zöllner.)*

made in that floor. However, some facilities have the chassis dynamometer's working floor above a large cellar with the machine sitting on a massive plinth within that space; this can also allow the working floor to be above the ground floor of the building. The chamber/cellar type of facility layout requires a large and expensive working floor (cellar ceiling) construction, but provides useful space to house dynamometer drives and other chamber services.

When a floor pit is required, its critical dimensions, including the exact specification of the pit and loading on the foundations into which a chassis dynamometer is to be installed, should be provided by the dynamometer's manufacturer.

The pit must be built to a standard of accuracy rather higher than is usual in some civil engineering practice. All pit-installed chassis dynamometers have a close dimensional relationship with the building in which they are installed and this requires that the pit construction has to meet three critical dimensional standards:

1. The effective depth of the pit in relation to the finished floor level of the test cell needs to be held to tight dimensional and level limits. Too deep is recoverable with steel packing shims, too shallow can be disastrous.
2. The lip of the pit has to be finished with edging steel; this has to interface accurately with the flooring plates that span the gap between the exposed rolls' surface and the building floor.
3. The centerline of the dynamometer has to be positioned and aligned both with the building datum and the vehicle hold-down structure or rails, which are cast into the cell floor. Only very limited movement and fine alignment can be achieved by the dynamometer installers because movement on the location pads will be very restricted.

Pit Flooding

Unless the chassis dynamometer is mounted above ground floor, its pit will form a sump into which liquids will drain. Spillage of vehicle liquids and cell washing will drain into the pit; therefore, a means of removing these liquids into a foul-liquid intercept drain is advisable. Flash flooding of the building due to an exceptional rainstorm or a broken water supply pipe have both caused expensive damage to chassis dynamometers. All dynamometer pits should be built with their floor, other than the feet support pads, sloping slightly towards a small sump fitted with a level alarm and within easy reach of the access ladder so that pumping out of any spilt liquids can be arranged.

Pit Depth

It is normal and sensible practice to make the pit floor lower than the datum dimension and then fix into the floor precisely leveled, steel "sole plates" at the required height minus 5–10 mm to allow some upward adjustment of the machine to the exact floor level. The final leveling can be done by some form of millwright's leveling pads, or more usually leveling screws and shim plates, as shown in Figure 17.7.

If the site has a water table that is at or above the full pit depth the whole excavation will have to be suitably "tanked" before final concrete casting to prevent groundwater seepage.

Subfloor cable and ventilation ducts, between the control room and the pit, and the drive cabinets and the pit, require pre-planning and close collaboration between the project engineer and builder; this task can complicate the task of creating a seal against groundwater leakage. Normally there will be the following subfloor duct systems, the majority of which are formed with single runs of smooth-bored plastic tubing:

- Signal cable ducts between control room and dynamometer pit
- Small power cable ducts between control room and dynamometer pit
- High-power drive cables between the drive cabinet room/space and the dynamometer pit
- Signal cable ducts between the drive cabinet room/space and the dynamometer pit
- Pit ventilation ducts that, at a minimum, will have to include a purge duct for removal of hydrocarbon vapors (see ATEX requirements in Chapter 4).

To allow later installation of cables, all tubes intended to contain wiring should be installed with strong rope run through and fixed at both ends.[3] After final commissioning the open tubes, which should always slope slightly downwards towards the cell, can be plugged with fire-stopping material.

On the wall of the test cell it is usual to fit a "breakout" box that allows transducers, microphones, or CAN-bus plugs to be used for special, vehicle-related, communications between control room and test unit.

Project managers of any major test facility building are advised to check the layout and take photographic records of all subfloor slab structures, service pipes, and steel work before concrete is poured.[4]

When, due to diminishing chamber access, the dynamometer has had to be located in its pit, and before the building has been completed and permanent electrical services are available, it may be vulnerable to pit flooding, condensation, and roll surface damage; action must be taken to guard against all eventualities.

---

3. When a "pull-through" rope has been forgotten, ferrets (*Mustela putorius furo*), normally used for rabbit hunting, have been used to pull a thin line through complex subfloor tubes, but pre-installation of rope-pulls is cheaper, if less exciting.

4. The author was once told by a steel fixer that his trade was comparable with that of a surgeon in that they both bury their mistakes, hence the advice to take detailed records of pre-pour structures.

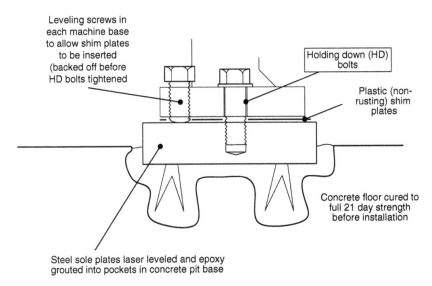

FIGURE 17.7   One of the leveled steel plates exactly leveled and cast in the pit floor to support the chassis dynamometer (one per dynamometer HD bolt location).

Dynamometer Flooring

In most large chassis dynamometer projects the metal cell floor, above the pit, and its support structure are provided by the dynamometer manufacturer. It is usual for the flooring to be centrally supported by the dynamometer structure and to contain access hatches for maintenance; these need to be interlocked with the control system in the same manner as is recommended for engine test cell doors. It will be clear to readers that, unless the pit shape is built to a quite precise shape and size, there will be considerable difficulty in cutting floor plates to suit the pit

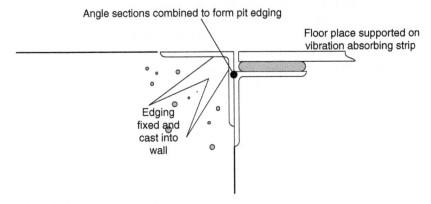

FIGURE 17.8   Section through one form of pit edging required to support pit flooring.

and dynamometer frame edges. Since errors made in floor plate and edging alignment are highly visible, it is strongly recommended that the civil contractor should supply the pit edging material and that it is cast into the pit walls under the supervision of the dynamometer supplier, before final edge grouting (see Figure 17.8). In some cases the pit edging assembly can be fixed in place using temporary jig beams to hold the rectangular shape true during grout pouring.

DESIGN AND INSTALLATION OF VARIABLE-GEOMETRY (4 × 4) DYNAMOMETERS

Many chassis dynamometers intended for testing four- or more wheel-drive vehicles are designed so that the inter-axle distance can be varied to accommodate a range of vehicles in a product family. The standard method adopted in 4 × 4 designs is for one, single-axle dynamometer module to be fixed within the building pit and for an identical second module to be mounted on rails so that it may be moved towards or away from the datum module. The inter-axle distance for light-duty vehicles (LDVs) will be in the range of 2–3.5 m, but for commercial vehicle designs the range of the wheelbases may be greater, requiring a commensurate increase in the roll-set movement. This range of movement is important in that it tends to determine the design needed to accommodate that section of cell flooring that is required to move with the traversing set of rolls. In the LDV designs where the total module movement is around 1.5 m, the moving plates can be based on a telescopic design with moving plates running over or under those fixed. Where the movement considerably exceeds 1.5 m, then the moving-floor designs may have to be based on a slatted sectional floor that runs down into the pit at either end, while being tensioned by counter weights, or as part of a tensioned cable loop. All these designs require good housekeeping and maintenance standards to avoid problems of jamming in operation.

The traversing roll-set is normally moved by two or more, electrically powered, lead screw mechanisms, coupled to a linear position transducer so that any predetermined position can be selected from the control room. For operational and maintenance reasons it is recommended that some positional indication, vehicle specific or actual, inter-axle, centerline distance is marked on the operating floor.

LDV units are usually traversed on flat machined surface rails having lubricated foot pads fitted with sweeper strips to prevent dirt entrainment; this is very similar to long-established machine tool practice and has to be installed with a similar degree of accuracy.

Large commercial vehicle rigs often use crane traversing technology, with crane rails installed on cast ledges within the pit wall design and the axle modules running within "crane beams" running on wheels, one of which on each side are connected by a common, electrically powered, shaft (to prevent "crabbing").

DRIVE CABINET HOUSING

The same precautions relating to the integration of high-powered AC or DC drives covered in Chapters 5 and 10 relate to those fitted to chassis dynamometers. The drive and control cabinets for high-power chassis dynamometers may be over 6 m long, 2.2 m high, and although only 600–800 mm deep will require space for front and sometimes rear access.

OFFLOADING AND POSITIONING CHASSIS DYNAMOMETER UNITS

Even small chassis dynamometer machines, that are delivered in one unit, will require lifting equipment to offload, transport, and position it in the pit. In all cases it is usually necessary to provide temporary floor strengthening to take the loads of the transport unit plus the machine being moved.

Forty-eight-inch roll machines or larger units may have to be positioned in already built or partly completed chambers. The installation will often require specialist equipment and contractors to lift and maneuver the machine into a building area, of limited headroom and limited floor access, as in the case of lined climatic chambers. It is a project phase that is used as an example in the project management section of Chapter 1, which is recommended reading to those readers in charge of the task for the first time.

TIRE BURST DETECTORS

On all mileage accumulation rigs and others undertaking prolonged automatic (robot-driven) test sequences, there should be some form of detector that can safely shut down the whole system in the event of a tire deflating. The most common form is a limit switch mounted on a floor stand with a long probe running under the car at its longitudinal midpoint, adjusted so that any drop of the vehicle's normal running body height will be detected; there needs to be one such device fitted on each side of the vehicle.

LOADING AND EMERGENCY BRAKES

It must be possible to lock the rolls to permit loading and unloading of the vehicle. These consist of either disk brakes fitted to the roll shafts or brake pads applied to the inside surface of one or both rollers. In normal operation these brakes have to be of sufficient power to resist the torques associated with driving the vehicle on and off the rig. Pneumatic or hydraulically powered, they are controlled both by the operational controls (vehicle loading) or by the safety instrumentation (EM stop), and are designed to be normally "on" and to require active switching to be off (machine operational).

It is not considered good practice to rely on these brakes to bring the rig to rest in the case of an emergency: they would normally only be used in this case when there had been a major electrical power failure.

VEHICLE RESTRAINTS

The test vehicle must be adequately restrained against fore and aft motion under the tractive forces generated and against slewing or sideways movements caused by the wheels being set at an angle to the rig's longitudinal axis. Any restraint used must not impose loads on the vehicle that will cause body distortion or uncharacteristic tire distortion.

Vehicle restraint is less of a problem on the twin-roller units as used for short-duration test work, such as the low-power end-of-line test rigs or emission rigs shown in Figure 17.2. These rigs are fitted with a combined, pneumatically powered, mechanism consisting of a "lift-out" beam that is positioned between the pair of rollers, and lateral-restraint roller mechanisms. When the vehicle drives onto the rig the lift-out beam is up and the vehicle is stopped with the drive wheels centered above them; as the beam descends, the vehicle wheels are dropped between the dynamometer's rolls and a frame comes up either side of the vehicle wheel carrying a pair of small free-running rollers (about 50 mm); these prevent the wheel from climbing out of the rolls during braking or acceleration. At the end of the test the beam comes up, lifting the vehicle, and the anti-climb-out rollers drop down, out of the vehicle path. No other restraint is required providing the driver doesn't accelerate and brake violently.

Where full restraint systems are used, the cell floor must be provided with strong anchorage points. There are many different designs of vehicle restraint equipment; the chain or loading strap fixed to vehicle tow points are the most common but may impose unrealistic forces to the vehicle structure if over-tensioned.

Three different types of strap or chain restraint system may be distinguished:

- A vehicle with rear-wheel drive is the easiest type to restrain. The front wheels, sitting on the solid cell floor, can be prevented from moving by fore and aft chocks linked across the wheel by a tie bar. The rear end may be prevented from slewing by two high-strength straps with integral tensioning devices; these should be arranged in a crossover configuration with the floor fixing points outboard and to the rear of the vehicle.
- Front-wheel-drive vehicles need careful restraint since, with the rolls running at speed, any movement of the steering mechanism can lead to violent yawing of the vehicle. Restraints may be similar to those described for rear-wheel drives, but with the straps at the front and chocks at the rear. Human drivers require practice with the handling characteristics of the vehicle on the rolls, while for robotic operation the steering wheel should be locked; otherwise disturbances such as a burst tire could have serious

consequences. Tire pressures are usually set above the normal road pressure level to minimize heating; both pressures and wear levels should be carefully equalized to prevent differential wheel diameters giving differential wheel speeds, thus affecting transmission noise and power absorption.

- Four-wheel-drive vehicles usual rely on crossover strapping at both ends, but details depend on fixing points built into the vehicle.

When using straps it is good practice to tie the rear end first and then to drive the vehicle slowly ahead with the steering wheel loosely held. In this way the vehicle should find its natural position on the rolls and can be restrained in this position, giving the minimum of tire scrubbing and heating.

The other type of vehicle restraint used is sometimes called a "sled" restraint and attaches to a vehicle's tow bar or a specially fitted structure. These restraints may take the form of a rigid, floor-fixed, steel pillar fitted with a restraint arm that is allowed to slide up and down as the car moves, but prevents lateral and fore or aft movement.

GUARDING AND SAFETY

Primary operational safety of chassis dynamometer operators is achieved by restricting human access to machinery while it is rotating. Access doors or hatches to the pit, or critical areas of the cellar, should be interlocked with the EM stop system, and latched EM stop buttons positioned in plant areas remote from the control room, such as the drive room and cell extremities.

The small exposed segment of dynamometer rolls, on which the tires rotate, is the most obvious hazard to operators and drivers when the rig is in motion and personnel guards must always be fitted. Such guards are a standard part of the dynamometer system for good safety reasons, but perhaps also because it is the only operational piece of the manufacturer's equipment that is visible to the visitor after installation and therefore will usually be fitted with the company logo. Roll cover plates, for use when no vehicle is installed, are also advisable to prevent surface damage during maintenance periods.

ROLL SURFACE TREATMENT

In the past, most twin-roll and many single-roll machines had a normal finish-machined steel surface and no special treatment was applied. The roll surface of modern single-roll machines may now be sprayed with a fine-grained tungsten carbide coating, which gives better grip and creates a more natural the tire-roller noise but causes greater tire attrition, unless wheel alignment is of a high standard.

Brake testers and some production rigs always have a high-friction surface, which again can give rise to severe tire damage if skidding under high load occurs.

ROAD SHELLS AND BUMP STRIPS

For development work and particularly for NVH development, it is sometimes desirable to have a simulated road surface attached to the rolls. Usually the required surface is a simulation of quite coarse-stoned asphalt and is achieved by use of a "road shell" enclosing the machine's steel roll. Road shell design and manufacture is difficult; they should ideally be made up in four or more segments of differing lengths and with junction gaps that are helically cut to prevent the gaps creating tire noise with a regular "beat".

The most usual techniques for producing road surfaces for attachment to chassis dynamometer rolls are:

- Detachable cast aluminum alloy road shells, made in segments that may be bolted to the outside diameter of the rolls. The surface usually consists of parallel-sided pits that give an approximation to a road surface.
- Detachable fiberglass road shells having an accurate molding of a true road surface. These shells are usually thicker than the aluminum type and in both cases the cell's floor plates must be adjustable in order to accommodate the increased roll diameter.
- A permanently fixed road surface made up of actual stones bonded into a rubber belt, which is, in turn, bonded to the roll surface.

Most road shells are not capable of running at anywhere near maximum rig speed because they are difficult to fix and to balance; neither will they be capable of transmitting full acceleration torque. It is therefore necessary to provide safety interlocks so that the computerized speed and torque limits are set to appropriate lower levels when shells are in use. Since road shells also change the effective rolling radius and the base inertia of the dynamometer, this requires appropriate changes in the control and data processing software parameters.

Potholes in damaged road surfaces and increasing use of "traffic calming" strips at the end of high-speed road sections are sometimes simulated by fitting bump strips onto the roll surface at diametrically opposite positions; again, safety interlocks are required.

DRIVER'S AIDS

The control room needs to be in two-way audio contact with the human driver in a test vehicle on a chassis dynamometer. The driver will normally wear a headset with microphone of the type used by rally drivers with which there is a permanently open, two-way voice channel with the control desk. The driver will also be observed by the control room operator, through CCTV, and may be sent operational instructions visually, through the external display screen of the driver's aid (DA).

A major function of a DA screen is to give the driver instructions relating to standardized test sequences, such as production test programs or emissions test drive cycles. This kind of VDU display is often graphical in form, showing the speed demanded and the actual speed achieved. In the case of emissions tests the test profile must be followed within defined limits, so it is usual to include an error-checking routine in the software to avoid wasted test time.

FIRE SUPPRESSION

For a more general treatment of the subject of fire suppression, see Chapter 4. The following relates specifically to the avoidance and treatment of fires in chassis dynamometer cells.

The first sensible precaution, in order to reduce the fire load in the cell, is to have the minimum required fuel in the test vehicles for the tests being run. Even with the cell's large frontal fan, the cooling air flow around a vehicle, mounted on a chassis dynamometer, will be different from that experienced on the open road. When running on a chassis dynamometer the underfloor exhaust systems can become very hot and the fire risk from this source increases with increasing power absorption. All vehicle test facilities should be equipped with substantial hand-held or hand-operated fire extinguishers, and all staff should be trained and practiced in their use.

A fixed fire suppression system is more difficult to design than in an engine test cell, both because of the larger volume of vehicle test cells and because access to the seat of the fire is difficult since it may well be within the vehicle's engine bay.

Conventional factory water sprinkler systems are not recommended because of the high level of consequential damage if dynamometer pit flooding occurred.

Water fog suppression systems, which can be arranged to include discharge nozzles mounted beneath the vehicle and thus near the potential seat of fires, are to be recommended. Automatic gas-based systems of the type used in some engine test cells are not recommended in vehicle cells in view of cell volume and, in most cases, the need to ensure that the driver has been evacuated before they are used.

Another fire protection method, for vehicles rigged for long-duration tests, is to fit them with a system of the type designed for rally cars; this enables the driver or control room to flood the engine compartment with foam extinguishant.

There must be a clear and unimpeded escape route for any test driver and the impairment in vision, caused by steam or smoke, must be taken into account by the risk assessment, particularly in the case of anechoic cells, where the escape door positions must not be camouflaged within the coned surface.

THE EFFECT OF ROLL DIAMETER ON TIRE CONTACT CONDITIONS

The bulk of the rolling resistance is the consequence of hysteresis losses in the material of the tire and this gives rise directly to heating of the tire. There is an obvious difference in running a tire on a flat road and running it on rollers of different diameters.

A widely accepted formulation describing the effect of the relative radii of tire and roll is:

$$F_{xr} = F_x\left(1 + \frac{r}{R}\right)^{1/2} \tag{17.2}$$

where F_{xr} = rolling resistance against drum, F_x = rolling resistance on flat road, r = radius of tire, and R = radius of drum.

Figure 17.9a shows the situation diagrammatically and Figure 17.9b the corresponding relation between the rolling resistances: this shows that rolling resistance (and hence hysteresis loss) increases linearly with the ratio r/R. For tire and roller of equal diameter, the rolling resistance is $1.414 \times$ resistance on a flat road, while for a tire three times the roller diameter, easily possible on a brake tester, the resistance is doubled. A typical value for the "coefficient of friction" (rolling resistance/load) would be 1% for a flat road.

To indicate the magnitudes involved, a tire bearing a load of 300 kg running at 17 m/s (40 mph) could be expected to experience a heating load of about 500 W on a flat road, increased to 1 kW when running on a roller of one-third its diameter. Clearly the heating effects associated with small-diameter rolls are not negligible.

Tires used even for a short time on rolling roads may be damaged by heating and distortion effects. Some dynamometer systems are fitted with tire-cooling

(a)

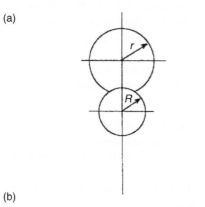

FIGURE 17.9 (a) Relationship between tire and chassis dynamometer roller diameters. (b) Linear relationship between rolling resistance and the tire/roll ratio.

(b)

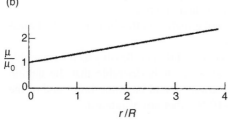

systems to reduce tire damage; however, all tires used for any but short-duration tests should be specially marked and changed before the vehicle is allowed on public roads.

LIMITATIONS OF THE CHASSIS DYNAMOMETER

Like all devices intended to simulate a phenomenon, the chassis dynamometer has its limitations, which are easily overlooked. It should be obvious that a vehicle, restrained by elastic ties and delivering or receiving power through contact with a rotating drum, is not dynamically identical with the same vehicle in its normal state as a free body traversing a fixed surface. During motion at constant speed the differences are minimized, and largely arise from the absence of air flow and the limitation of simulated motion to a straight line. Once acceleration and braking are involved, however, the vehicle motions in the two states are quite different. To give a simple example, a vehicle on the road is subjected to braking forces on all wheels, whether driven or not, and these give rise to a couple about the center of gravity, which causes a transfer of load from the rear wheels to the front, with consequent pitching of the body. On a chassis dynamometer, however, the braking force is applied only to the driven wheels, while the forward deceleration force acting through the center of gravity is absent, and replaced by tension forces in the front-end vehicle restraints (remember: braking on a rolling road tends to throw the vehicle backwards).

It will be clear that the pattern of forces acting on the vehicle, and its consequent motions, are quite different in the two cases. These differences make it difficult to investigate some aspects of vehicle ride and NVH testing on the chassis dynamometer. A particular area in which the simulation differs most fundamentally from the "real-life" situation concerns all aspects of driveline oscillation, with its associated judder or "shuffle".

The author and his original co-author had occasion to study this problem in some detail and, while the analysis is too extensive to be repeated here, one or two of their conclusions may be of interest:

- The commonly adopted arrangement, whereby the roll inertia is made equal to that of the vehicle, gives a response to such disturbances as are induced by driveline oscillations, i.e. judder, that are significantly removed from on-road behavior.
- For reasonably accurate simulation of phenomena such as judder, roll inertia should be at least five times vehicle inertia.
- Electronically simulated inertia is not effective in this instance; actual mass is necessary.
- The test vehicle should be anchored as lightly and flexibly as possible; this is not an insignificant requirement, since it is desirable that the natural frequency of the vehicle on the restraint should be at the lower end of the range of frequencies, typically 5–10 Hz, that are of interest.

It is strongly recommended that investigations of vehicle behavior under conditions of driveline oscillation should proceed with caution if it is intended to involve running on a chassis dynamometer.

SOLAR HEAT LOAD TESTING IN CHASSIS DYNAMOMETER ENVIRONMENTAL CELLS

Subjecting whole vehicles and vehicle modules to simulated solar light energy and heat load, using banks of lamps that emit light within the solar wavelength spectrum, has been practiced for at least the last 25 years in order to test emissions, the efficiency of car cooling systems and material durability, color fastness, etc. Legislative emphasis on vehicle testing designed to reduce fuel consumption has meant that the power required to run vehicle air-conditioning systems has come under close scrutiny. The EPA SC03 standard defines a supplementary test procedure for passenger cars that has to be carried out in a test facility capable of providing the environment given in Table 17.1.

All legislation covering solar simulation will define standards for the spectral content and spatial uniformity; the first criterion will be the specification of the lamp units used, which should meet the EPA spectral energy distribution given in Table 17.2.

For the EPA SC03 test the radiant energy must be uniform, within ±15% over a 0.5 meter grid at the centerline of the vehicle at the base of the window

TABLE 17.1

Facility Parameter	Parameter Specification
Air temperature	35 °C (95 °C)
Relative humidity	40% (100 grains or 0.0648 grams of water per pound of dry air)
Solar heat load	850 W/m²
Vehicle cooling	Air flow from external vehicle cooling fan proportional to vehicle speed on chassis dynamometer (max speed 55 mph, average speed 21.55 mph)

TABLE 17.2

	Wavelength Band (nm)			
	<320	320–400	400–780	>780
% of total spectrum	0	0–7	45–55	35–53

screen. Such spatial uniformity has to be achieved by the design of the lighting array and the frame in which the array is supported. The solar array frame fitted in a multipurpose chassis dynamometer cell is normally a planar design with a target area of perhaps 6 m long × 2.5m wide; it will have to be supported from the cell roof by a mechanism that allows it to be lowered to operating height corresponding to the vehicle and similarly raised, when not in use, out of the way of normal cell operation. The lamps of commercially available solar arrays are capable of intensity variation typically of between 600 and 1100 W/m^2. The electrical power requirements for full vehicle arrays can be considerable, typically around 80 kW.

Anechoic Test Cells: NVH and EMC Testing

INTRODUCTION

Anechoic means without an echo. To investigate the quantitative and qualitative characteristics of noise, or any other wave energy, from a source, it is important not to have an echo of the source emission corrupting the signal measurement. Such reflected energy waves can have additive or subtractive (canceling) effects on the original transmission.

Engine Testing. DOI: 10.1016/B978-0-08-096949-7.00018-2

The everyday concept of an "echo" is the bounce-back of a sound wave, but the same idea holds good for electromagnetic wave energy with frequencies above the audible wavebands. In the automotive engine, powertrain, and component test industries, anechoic cells are designed specifically for either electromagnetic compatibly testing (EMC[1]) working with frequencies in the megahertz and gigahertz ranges, or noise, vibration, and harshness (NVH) testing, working in the range of human hearing, around 16 Hz to 20,000 Hz. Both technologies are able to carry out some forms of testing on a flat road in the wide-open air, which is inherently a free field over a reflecting plane. Much of the legislative "drive-by noise test" (also called "pass-by") work is carried out on a specially surfaced and instrumented, open-air track section (certified to ISO 362). However, the ideal situation of a still day in the flat desert of Arizona is not available to most test engineers so, for EMC and NVH work on a static unit under test they require test cells that simulate those anechoic conditions.

Both EMC and NVH test cells exist in two different forms:

1. *Semi-anechoic* (in the USA, commonly and correctly "hemi-anechoic"), in which walls and ceiling are lined with sound-absorbent materials while the floor is reflective. This space simulates the situation where the source, often an engine or vehicle, is located in a wide open space but resting on a reflective horizontal plane representing a road.
2. *Full or fully anechoic*, in which all surfaces including the floor are nonreflective (energy absorbent); such cells tend to be used for component testing rather than road vehicle or engine work.

However, there are some important differences between EMC and NVH anechoic cells, both in the operational and construction features; they are not dual-purpose facilities.

PART 1. NVH TESTING AND ACOUSTIC TEST FACILITIES

Both "noise at work" and vehicle, type approval, legislation that limits the noise produced by power units and new motor vehicles have been progressively tightened over the last 30 years. However, with the increase in traffic flows over the whole 24-hour period, traffic noise in particular is perceived to have increased despite the noise emissions in tests of individual powertrains and vehicles being significantly reduced[2] over the same period.

The development and wider use of electrically powered hybrid vehicles has brought about large reductions in engine noise at slow speeds, even to the

---

1. EMC is commonly used as a generic term when used to describe test cells, and in this book covers both electromagnetic interference (EMI) and electromagnetic susceptibility (EMS) unless specifically stated.

2. EU 2010 figures indicated that noise from individual cars has, since 1970, been reduced by 85% and the noise from trucks by 90%.

point that it has caused serious safety concerns for organizations representing visually impaired people. The testing of noise sources other than the engine, in the powertrain, and from tires running on road surfaces is becoming even more important due, in part, to the widespread noise pollution from fast auto-routes (freeways).

In automotive engineering, NVH testing is divided between meeting type approval requirements and the vehicle refinement process; in non-automotive engine development it is more concerned with health and safety (H&S) of operators and machinery health monitoring through vibration analysis.

First, some basic definitions of terms as used in powertrain and automotive engineering:

- Noise is usually considered as unwanted sound. This is distinct from required sound (driver feedback) and desired sound (from vehicle sound system, exhaust note, etc.).
- Vibration is the oscillation of the UUT that is typically, but not exclusively, felt by the operator and passengers rather than being heard.
- Harshness is used to describe the severity and discomfort associated with unwanted sound or vibration, particularly from short-duration events (e.g. ABS operation).

The Anechoic Cell Structure

BS EN ISO 3740:2001, entitled "Acoustics. Determination of Sound Power Levels of Noise Sources. Guidelines for the Use of Basic Standards", is the set of standards concerned with methods of determining sound power levels of noise sources, and they mention certain aspects of anechoic cell design, including:

- Desirable shape and volume of the cell
- Desirable absorption coefficient of surfaces
- Specification of absorptive treatment
- Guidance regarding avoidance of unwanted sound reflections.

A key part of the technical specification of an anechoic cell is known as the *"cut-off" frequency*; this is the frequency below which the rate of sound level decay in the chamber no longer replicates the outdoors or, more precisely, a "free-field" environment. The lower the cut-off frequency, the better its simulation of a true free field. The maximum cut-off frequency normally required for automotive anechoic cells (both engine and vehicle) is around 120 Hz, but many modern facilities are built to have cut-offs as low as 60 Hz.

The acceptance test of the anechoic characteristic of a cell is carried out using a broadband generator placed at the geometric center of the cell. The level of sound decay is measured by a microphone physically moved along chords tensioned between the noise source and the high-level corners of the cell. The decay measured is compared with the 6 dB distance-doubling characteristic of a free field.

The task of certification of an anechoic cell and the equipment required is a specialist field requiring the services of qualified contractors.

There are certain other points that should be remembered by the nonspecialist if he or she is required to take responsibility for setting up a test facility of this kind, the first of which is the geographical location of the facility.

The structure of an anechoic cell should be isolated from any environment in which high levels of airborne noise or ground vibration are generated, such as production plant, nearby public roads, or railways. Many acoustic anechoic cells built in less than ideal locations take the form of a "building within a building" in which the inner structure is isolated by dampers from external vibration.

The walls and ceiling of the chamber have to absorb sound emitted by the UUT and this is usually achieved by completely lining them with pyramidal shapes, often made of a dense foam, that attenuates sound by both scattering and absorption. The total length (height) from the surface mounting base of the cones may be some 700 mm and behind the mounting frame may be an air gap of some 300 mm to the cell wall. Therefore, the effective volume of such spaces is considerably reduced from the full enclosed space; the smaller the cell volume, the more marked the reduction.

Access door frames and furniture can affect the acoustic performance adversely, by creating hard reflective surfaces, so must be carefully incorporated into the overall lining. Windows should not be required for the same reasons and are invariably made unnecessary by the use of color CCTV cameras.

Safety note concerning doors: Along with fitting the legally required, illuminated, emergency exit notices above cell exits, the acoustic lining covering exit door latches or operating buttons should be color marked to make their location and the opening direction obvious.

Treatment and Location of Services for Anechoic Cells

Since so much care has to be taken in the creation of a quiet cell environment, it is obvious that any noise created by the fluid and ventilation support systems has to be minimized and attenuated, so far as is possible.

Ventilation

The ventilation of anechoic cells designed for the testing of running engines presents particular problems, because of "wind noise". However since full-power running in such cells is normally of a short duration and the precise control of inlet air temperature is not so critical to the sound recording, the full thermodynamic rating of the cell (as described in Chapters 3 and 6) is not usually required.

The technique adopted for ventilation systems is to have an air flow through the cell of adequate volume, entering and leaving at as low a velocity as possible to reduce nozzle and wind noise. This minimization of air velocity into and out of the cell is achieved by making the inlet and discharge ducting as large as possible.

A common design is to make the last 500 mm above floor level of the whole end-wall act as the ventilation air inlet and for the air to exit through ductwork behind the cones of the upper corner sections of the opposite end of the cell.

Shared services of any sort between anechoic cells, including ventilation systems, should be avoided because they will act as inter-space transmission paths for noise.

In anechoic engine test cells the dynamometer must be physically separate and isolated acoustically from the cell proper. Since the engine under test should be in the horizontal center of the anechoic cell space, such an arrangement requires an overlong coupling shaft system, the design of which will present problems discussed in Chapter 11.

In many designs, such as that shown in Figure 18.1, the shaft system is split into two sections, with an intermediate bearing mounted on a support, or pedestal block, just inside the anechoic chamber. The shaft section running through the wall, between the dynamometer and the pedestal bearing, will run within a sound-absorbing "silencer tube", designed to minimize transmission of noise from the dynamometer room. The pedestal, which needs to be a massive (nonreverberatory) construction, will often be used to house, or support, transducer and service connection boxes.

Fluid Services

The cooling water piping and control valves can create variable noise patterns, at unpredictable times; therefore, as much of the system as possible is kept outside the cell and the pipework run under floor level as much as possible. The distance between the temperature control devices and the engine usually means that the thermal inertia of the system is too high to have good transient response but, as with the ventilation air flow, this is rarely critical to the type of recordings being made during NVH testing. However, pumped circulation of the coolant fluid is often needed to reduce the temperature control time-lag. Sound transmission through steel pipes needs to be minimized by decoupling with nonmetallic sections.

Fuel systems have to be fitted with all the isolation valves and interlocks described for other cells in Chapter 8, but in the case of acoustic anechoic cells the valves are often fitted below the bottom row of the lining cones.

Engine Exhaust Piping

Most types of acoustic cell linings can absorb liquids and may be inflammable, so running a hot engine exhaust tube through such linings increases the level of fire hazard. Two solutions are commonly used to run exhausts out of the cell:

1. The engine exhaust is taken out through the floor into a cellar space, which may house the seismic block. The floor-fixed transit tubes need

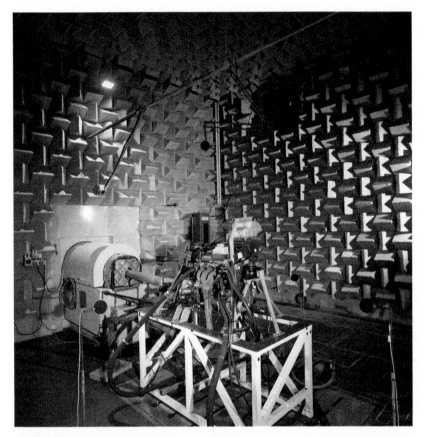

FIGURE 18.1 A semi-anechoic NVH test cell for engine refinement work. Note the two-piece shaft system is connected to a dynamometer installed in the adjoining services room system that includes a pedestal bearing mounted on a plinth, around which are mounted various engine system support and transducer connection boxes. The engine is mounted on a hollow frame and the bedplate mounted on a substantial seismic block. *(Photo supplied courtesy of MIRA Ltd.)*

to have a nonrigid decoupling section plus top and bottom caps for installation when not in use, as shown diagrammatically in part section in Figure 18.2.

2. The exhaust is taken out in a double-skinned tube that is jacketed with circulating cooling water. These systems must be fitted with a water flow switch in the EM stop chain in order to prevent both overheating and the possibility of the outlet being shut off or blocked, which would lead to a steam explosion.[3]

---

3. This is not a theoretical hazard and the author has experienced the result of such an event.

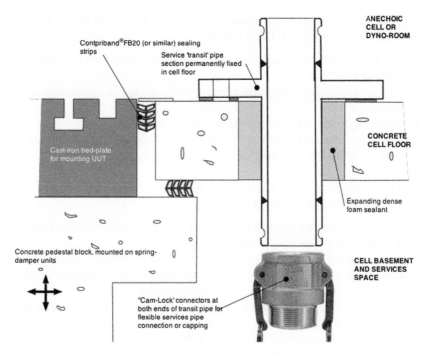

FIGURE 18.2 A part section through the floor of an anechoic cell or its dynamometer room and a cellar space below, which houses the support plinth and services. The floor is fitted with transition tubes that allow the temporary connection of services or plugging when not in use, between the cell and service space. Sketch by the author from an actual site design.

Engine Mounting

The engine in an anechoic cell must be raised more than is usual above floor level, typically to give a gap of 1.0–1.2 meters from floor to shaft centerline, to allow microphones to be located at a distance below it. This calls for tall, nonstandard, engine-mounting arrangements that are frequently pre-rigged on a skeletal pallet frame (without any reverberatory drip tray).

Chassis Dynamometers

These dynamometers, for both NVH and EMC cells, are covered in Chapter 17.

Instrumentation

Acoustic research makes use of highly sensitive instruments and correspondingly sensitive measurement channels. The design of the signal cabling system must be carefully considered at the design stage to avoid interference and a multitude of trailing leads in the cell.

The modules of instrumentation commonly used in NVH test work and mounted on or near the UUT in a cell are:

- Single-axis and triaxial accelerometers, filtered and unfiltered, having millivolt outputs
- Charge amplifier accelerometers similar to those used in combustion analysis work
- Externally polarized precision condenser microphones
- Signal pre-amplifiers that are mounted close to the transducers and boost their signal to the recording and sound analysis equipment housed in the control room
- Multimicrophone array stands or individual microphone stands or hanging cables (some visible in Figure 18.1).

Accelerometers may be attached to the UUT either by an adhesive or by a fixing stud, drilled and tapped into the engine or gearbox body.

In the control room, the engine and cell control system is usually less dominant in position and function than in performance cells, being replaced by multichannel amplifiers and sound analysis equipment.

NVH test work on full vehicles may use a "binaural head" to record the sound within the vehicle as it is perceived by a human driver.

Modern acoustic analysis techniques such as transfer path analysis (TPA), which helps engineers trace and differentiate the structural and airborne paths taken from source to the user's ear, can require substantial computing power. Such work is normally carried out away from the cell control room as a post-processing exercise. The advance of digital recording and processing has tended to mean that test cells have low levels of shaft turning and spend most of their time having the UUT prepared for short multistage test sequences, the recorded sounds from which can be analyzed and digitally altered without further running. In spite of all the advanced modeling and analysis tools and software, there is still a place in the vehicle refinement process for the subjective judgment of human users, and all major test facilities will have access to a studio where a "jury", made up of a selection of the target user group, is asked to judge the level of acceptability of recorded and post-processed sounds.

Vehicle Noise: Measurement Practice

A number of regulations, both national and international, lay down permitted vehicle noise levels and the methods by which they are to be measured. In Europe it is specified that A-weighted sound pressure level (SPL) should be measured during vehicle passage at a distance of 7.5 m from the centerline of the vehicle path. The EEC rules (Directive 70/157/EEC) on the permissible sound level and the exhaust system of motor vehicles covers all motor vehicles with a maximum speed of more than 25 km/h. The limits range from 74 dBA for motor cars to 80 dBA for high-powered goods vehicles.

Many motor sport venues worldwide have introduced their own noise limits for vehicles using their tracks, often in the range 95–98 dBA, measured at a fixed percentage of rated, maximum, engine speed and at a set distance from the exhaust, relying on the use of hand-held meters operated by scrutineers. Unless care is taken to avoid sound reflection from surroundings, the results of such tests can be highly misleading, nor do they presently measure noise levels coming from engine induction systems. Some racetracks are now fitted with permanent, complex "drive-by" monitoring systems, which allow the authorities to measure peak and average noise levels to keep within the local limits of noise pollution.

Fire Detection and Suppression in Anechoic Cells

The detection and quenching systems described in Part 2 of Chapter 4 are relevant to all acoustic anechoic chambers.

In acoustic cells in particular, automatic smoke and flame detectors are important because the operator has no direct vision of the cell and may be attending to a very localized area of the cell and UUT through a CCTV image.

Since ventilation air flow, and therefore air-cooling flows, will be lower in anechoic cells than in other normal powertrain cells, there is a real risk of local overheating, and serious fires in anechoic engine cells have occurred. In EMC cells the danger of localized electrical fires may be higher than normal. In all cases the rule is to reduce the potential and discretionary fire load by restricting the amount of fuel in the cell.

In choosing quenching systems, special consideration must be given to vehicle cells where the volume of the cell is very large compared with the vehicle, which is the only common source of a fire and which would require very large volumes of any gaseous suppressant. The engine bay cannot be rigged with a rally car extinguishing system because of the likely effect on engine noise, so as advised in Chapter 17 the fitting of a water mist system, with nozzles below and at the sides of the vehicle position, is recommended.

PART 2. ELECTROMAGNETIC COMPATIBILITY (EMC) TEST FACILITIES

The Task

The physical and functional separation of a vehicle's power and control subsystems, together with the selection and routing of the interconnecting looms combined into one vehicle system, is referred to as the vehicle's E/E (electrical/electronic) architecture.

Each module and connecting loom of this architectural assemblage has to be tested, as individual modules and as an assembled system, in order to:

- measure, in order to reduce, the emissions produced by a module or component (PCB) during its full range of operation (EMC);

- measure, in order to increase, its level of immunity to externally generated electromagnetic radiation (EMI);
- measure a module's or component's immunity to damage by electrostatic discharge in order to reduce its vulnerability;
- measure the EMC of the complete vehicle system, in its electromagnetic environment, under running conditions.

The electromagnetic environment that has to be considered when designing these tests is not only that generated by the vehicle system itself, but also the surroundings in which it may operate. Examples are when the vehicle is in proximity to radar beams at airports or transformer installations at power stations.

Along with the EMC of the vehicle's fixed modules, we also have to consider the function of devices used in or near vehicles that are not part of the vehicle, such as removable GPS systems, mobile phones, CB radios, and biomedical devices such as heart pacemakers, all of which need to be used, without causing or suffering from significant degradation in performance, at random locations within or close to the vehicles.

The new generation of electric and hybrid vehicles is producing a whole new source of radio frequency (RF) noise at higher energy levels than their predecessors. Safety-critical modules such as airbags are multiplying, as are wireless devices, such as forward ranging, collision avoidance radar, and tire pressure transducers.

Thus, the design and implementation of EMC test regimes is becoming ever more complex and is employing ever more specialist facilities and staff.

Coupling Mechanisms Between EMI Source and "Victim"

Terms such as "the electromagnetic environment" and "RF noise" lead one to believe that almost all electromagnetic interference is due to radiation and that most EMC testing is done in anechoic chambers in which vehicles are "zapped" with powerful beams of energy at radio frequencies.

It is true that in EMC immunity tests vehicles and components are exposed to beams, in an anechoic cell, having frequencies typically between 10 kHz and 18 GHz, at field strengths of 100–200 V/m, and their emissions may be checked over a frequency range of perhaps 20 Hz up to 170 GHz at set distances, typically 1, 3, and 10 m. However, much of the work of EMC test engineers is involved at the level of PCB and wiring loom design, where the coupling between noise source and its victim is frequently by one of the other coupling mechanisms shown in Figure 18.3. The common results of the various coupling mechanisms give rise, in the vehicle, to phenomena such as voltage drop-outs, fluctuations or continuous disturbance, all of which have to be simulated and any problems thus caused solved, by redesign or shielding.

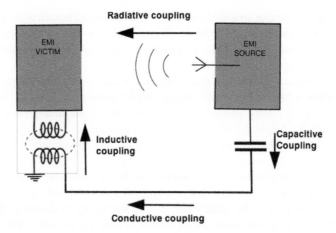

FIGURE 18.3   Diagram showing the paths of the four "noise" coupling mechanisms between the EM noise source and the victim.

The *permitted performance levels* will depend on the criticality of the subsystem. All vehicle manufacturers grade systems by "class", which range from those that are a convenience up to those that are safety critical. Then the systems are graded for their *function performance status* when subjected to EMI within test limits. Typically, these will be variations on the following grading:

1. The system will function as designed during and after exposure to the disturbance (fully immune).
2. Temporary functional deviation imperceptible to the operator. The system will not suffer permanent damage or loss in I/O parametric values or permanent reduction in functionality.
3. The system may deviate from its design performance within set limits, or revert to a fail-safe mode during an exposure to the disturbance providing it does not affect the safe operation of the vehicle. It should return to normal operation automatically and safely when the disturbance is removed.
4. As in 3 above, but operator action, such as cycling through ignition key functions or replacing a fuse, may be required to return the function to normal after the disturbance is removed.

None of the systems that are permitted reduction of functionality may have a "knock-on" effect on systems that can affect vehicle safety.

Pulse Interference or Electrostatic Discharge (ESD)

Before major modules or vehicles are subjected to full EMC testing, it is usual practice to test their components and subassemblies for their immunity from damage, or reduction in functionality, caused by electrostatic discharge.

Materials that make up a modern vehicle and its clothed human users vary in their ability to generate and hold an electrical charge through triboelectric[4] charge generation. The antistatic property of a material defines its ability to resist being charged triboelectrically.

During normal vehicle use there can be operations, such as slipping the clutch, when a high triboelectric charge can be generated. Actions, such as refueling from a plastic gasoline container that has been sliding on a trunk carpet for hours, might create safety-critical situations.

To deal with such scenarios the test engineer has to anticipate, test, and mitigate.

Electronic components and modules can be at risk from ESD during installation and maintenance, which is why work areas and tools are chosen to have high "resistivity", the quality that defines the ability to dissipate charge. The human body appears near the top of the + tribological series; therefore, workers involved in the construction of PCBs and their testing have to wear earthing wrist straps to prevent a charge building up.

There is an international ESD association that coordinates test methods and all matters relating to the protection of equipment from the effects of ESD. For example, ESD Association Standard S5.1 is entitled "Human Body Model (HBM) Electrostatic Discharge Sensitivity Testing" and it covers a procedure for performing ESD susceptibility tests.

The most common ESD test is specified in IEC 61000-4-2, which includes the specification of a standardized pulse generator. The component under test is placed on a wooden table topped with a horizontal coupling plane that is connected to the ground reference plane through a cable that includes bleed resistors of 470k at each end. The component is insulated from the coupling plane by a thin plastic sheet and the pulses are applied at points judged to be appropriate by an experienced tester.

The test voltages used in most harmonized standards are 4 kV for direct probe contact and 8 kV for air discharge (spark), but many original equipment manufacturer (OEM) and military tests use voltages of 15 or 25 kV for modules that are directly accessible from outside the vehicle, such as keyless entry devices or diagnostic system plug sockets.

Legislation

The world of EMC legislation is no less complex than the DoE and testing problems needing to be solved. It is very easy for the nonspecialist to become confused in an area where both legislation and technology are developing so quickly. The wide field of EMC legislation is where legislative bodies controlling radio communications, such as the Federal Communications Commission (FCC) in the USA, overlap with the work carried out by

---

4. "Tribo" comes from the Greek for "rubbing".

committees in groups dealing with vehicle, machinery, and marine safety, and even military defense.

International standards similarly fall under the control of two standards organizations:

- International Organization for Standardization (ISO) Working Group TC 22/SC 3/WG 3 covering electrical interference
- Special International Committee for Radio Interference (CISPR, which is part of the IEC), Subcommittee D covering automotive products.

In the USA the American National Standards Institute (ANSI) has delegated the EMC standards writing task to the Society of American Engineers (SAE). The SAE in turn has spread the task between two committees:

- The *SAE EMI* standards committee covers *immunity* from electromagnetic interference of automotive systems
- The *SAE EMR* standards committee covers *emissions* radiated from vehicle components that may cause interference to other devices.

While there is general agreement between the standards listed in Table 18.1, there is no comprehensive agreement between the SAE and ISO on test methods.

To add complexity to an already complicated situation, the major US OEMs—Ford, GM, and Daimler-Chrysler—have produced their own global and comprehensive EMC standards and test methods that, while they are

TABLE 18.1

International Standard	SAE Standard	Standard Header Details
CISPR 12:2009	SAE J551-2	Vehicles, boats, and internal combustion engines Radio disturbance characteristics Limits and methods of measurement for the protection of *off-board* receivers
CISPR 25:2008	SAE J551-4 and J1113-41	Vehicles, boats, and internal combustion engines Radio disturbance characteristics Limits and methods of measurement for the protection of *on-board* receivers
ISO 7637-2:2011	SAE J1113-42	Road vehicles Electrical disturbances from conduction and coupling, methods and procedures

generally harmonized with, and referenced to, the SAE test methods, all contain detailed variations. The Ford document (EMC-CS-2009.1), which is average to suppliers is 121 pages long and covers every aspect of EMC test methodology required by Ford and their Tier 1 suppliers.

As explained at the beginning of Chapter 1, any vehicle or automotive electrical module manufacturer has to obtain type approval from a test facility accredited to ISO 17025 who will be able to both design and run the required EMC test regimes to obtain the type approval.

Type Approval "E" and "e" Marking

The e-mark is an EU mark and is part of the vehicle homologation requirements for approved vehicles and vehicle components sold into the EU (see Figure 18.4a). It is the type approval mark required by devices related to all safety-relevant functionality of the vehicle. Such devices will have to be e-mark certified by a certifying authority; manufacturers cannot self-declare and affix an e-mark, unlike most CE marking, where manufacturers' self-declaration is the norm.

Since late 2002, all electronic devices intended for use in vehicles, including after-market electronics products, are required to obtain formal type approval for products before placing them on the market.

The E-mark is a United Nations mark for approved vehicles and vehicle components sold into the EU and some other countries under UN-ECE Regulation 10.02 (see Figure 18.4b).

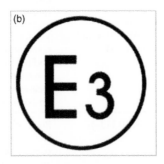

FIGURE 18.4 (a) The "e" mark, issued by a certifying authority followed by the EU country code number for the UK. (b) The "E" mark followed by the country code, in this case for Italy.

Both marks are affixed to approved products with other numbers relating to the legislation under which the approval has been gained.

EMC Test Facilities

Open-air test sites, or OATS, are the reference sites in most standards. They are especially useful for emissions testing of large equipment systems such as aircraft.

Anechoic RF Cells

As discussed earlier in this chapter, acoustic anechoic cells have to be isolated from external noise and vibration; for the same reasons of test integrity, anechoic cells working in the radio frequency (RF) waveband have to be insulated and isolated from electromagnetic noise coming from outside the test field. The test cell structure is built to form a Faraday cage, the external skin being formed from perforated sheet metal in the case of large cells, or solid metal sheet in the case of small-component test units.

All support structures and fittings used in EMC cells should be made of nonconductive wood or plastic.

Large automotive EMC cells may contain a turntable that allows the test vehicle to be turned in a directional energy beam or present a different aspect to a receiver. In the most complex facility a chassis dynamometer is mounted within a turntable; this allows a vehicle, running under load, to be presented at 360° to the energy beam or receiver. In such installations DC drives are used rather than AC, which can produce high levels of RF noise; the DC drive cabinets are shielded below ground by concrete. Such large chambers typically have dimensions, internal to the lining, of 20 m × 13 m × 8 m high.

For radiated emission measurement, antennas are used as transducers. The standard that defines the requirements for antennas to be used in EMC measurement is CISPR publication 16-1:1993, "Specification for Radio Disturbance and Immunity Measuring Apparatus and Methods, Part 1."

CISPR 16 lays down the requirements that have to be met by fully and semi-anechoic chambers designed for EMC testing and often have a subdesignation relating to the distance of separation between the UUT, more commonly called equipment under test (EUT) in EMC work, and the antenna. Thus, chambers may be designated CISPR 16 10m, or CISPR 16 3m. Since all the performance of all types of anechoic lining declines as the angle of incidence of the waves deviates from normal (90°), "the bigger the better" for chambers of a particular CISPR separation designation.

EMC Anechoic Cell Linings

The lining materials of EMC anechoic chambers are designed to absorb radio wave and microwave energies, and are referred to as RAM (radar absorbing material). RAM is made from two distinctly different types of material, although both types are commonly met in different areas of the same chamber:

1. Closed-cell polyurethane or polystyrene foam molded into a steep pyramidal shape and "dosed" (impregnated) with carbon, then painted with a fire-resistant paint. The principle of this type of absorber is to mimic free space by "progressive impedance"; that is, absorbing the radio or microwave by internal reflection through a resistive structure that has a progressively larger shape, during which the wave energy is being converted into heat. The absorbing performance is directly related to the wavelength of the energy and the figure quoted (in dB) relates to waves with a normal angle of incidence.

2. Ferrite tiles, which are usually made as 100 mm × 100 mm × 6 mm thick pieces with a central fixing hole. They are made from sintered iron/nickel material, which is ground to a precise shape so that large wall sections can be precisely assembled. Since they are of about the same density as steel, the chamber support frame and roof structure needs to be suitably substantial and they have to be attached to a backing layer. In some forms the tiles are supplied in panels ready assembled with a tuned dielectric backing layer. Ferrite linings work by "magnetic permeability" and operate well at lower frequencies, where the wavelength is too long for pyramidal absorbers to work effectively.

3. Hybrid linings and mixed-cell linings. The attenuating performance of ferrite tiles is modified by placing wedge- or pyramid-shaped absorbers in front of them, and hybrid linings are often used where the chamber needs to operate at frequencies above 1000 MHz. Mixed-cell linings of both ferrite tiles and pyramids are used in many large automotive cells with the patches of pyramid or hybrid lining in large patches at a normal incidence to the vehicle under test and the ferrite tiles over the remainder of the surface. See Figure 18.5.

Automotive chambers run by accredited organizations have chambers that typically have specifications around the following:

Immunity testing:	10 kHz to 18 GHz with a field strength of 200 V/m
Emissions:	10 kHz to 40 GHz at 1 m or 3 m ranges.

Reverberation Chamber for EMC Testing

A reverberation chamber can be used for a more limited range of automotive EMC testing than a full anechoic chamber, but they are often capable of working at much higher RF field energy levels; their use is covered by IEC 61000-4-21.

They normally take the form of a rectangular enclosure, with perfectly conducting walls, inside which the electromagnetic field distribution is characterized by a standing-wave pattern created by the reflection from the walls. However, it is possible to change the standing-wave pattern within the cavity

FIGURE 18.5 The mixed, pyramidal, and ferrite tile absorption lining on the end-wall of a large automotive EMC semi-anechoic test cell. Chamber dimensions: 20 m × 11 m × 7 m high. *(Photo courtesy of MIRA Ltd.)*

with "stirrers", which take the form of zigzagged reflectors mounted on a rotating shaft; these effectively change the boundary of the chamber inside a reverberation chamber. There are cost advantages of using a reverberation chamber:

- The high conductivity allows for high field power levels to be generated from moderate levels of input power.
- The chamber construction costs are lower than an anechoic chamber because no expensive and heavy absorbing lining is needed.
- The reverberation chamber provides a realistic simulation for electronics functioning within a cavity like an electrical cabinet.

The largest and highest powered chambers are used for military work, while automotive versions tend to be used only in subassembly testing and have internal volumes of around 60 m^3 with a door that is a size capable of allowing a 2.2-m-tall electrical cabinet to be installed.

Health and Safety in RF Cells

There are health and safety (H&S) risks to all closed anechoic test chambers, the interiors of which can be disorienting and have exit routes that may not be obvious to the inexperienced user. Managers and operators should always be aware that first-time visitors may suffer from mild or even severe claustrophobia and be aware of the need to get the sufferer into the open air quickly.

The health effects of exposure to high levels of RF energy have been the subject of much discussion for years; health scares led to rules covering the leakage from microwave ovens and still lead to concerns about mobile telephone mast locations today. One of the major sets of guidelines is that issued by the International Council on Non-Ionizing Radiation Protection (ICNIRP).

It is not disputed that there are two health effects of exposure to high-energy RF fields, which are heating of the human body and RF shocks and burns (electrostimulation) resulting from direct contact with an RF radiator such as an antenna.

Although some scientists believe that exposure to low-level RF fields for long periods of time can result in other harmful effects, none of these nonthermal effects are incorporated in major standards or guidelines.

The key H&S requirement of EMC cells is the control of access and ensuring that unsupervised, solo work is not carried out in any chamber, be it in an open or closed state.

Each test facility has to develop and strictly enforce their own rules, based upon published guidelines and according to the type of work being carried out, which might range between checking car radio reception during engine cranking, to testing military equipment for immunity against the energy pulse of a nuclear weapon.

FURTHER READING

M.J. Crocker, Handbook of Noise and Vibration Control, John Wiley, 200 ISBN-13: 978-0471395997.

IEC 61000-4-21 ed. 2.0. Electromagnetic Compatibility (EMC), Part 4-21: Testing and Measurement Techniques—Reverberation Chamber Test Methods.

T. Williams, EMC for Product Designers, fourth ed., Newnes, 2006. ISBN-13: 978-0750681704.

Photographs of EMC vehicle cell and NVH engine cell supplied by MIRA, UK; http://www.mira.co.uk/Facilities/

The Pursuit and Definition of Accuracy

INTRODUCTION AND TERMS

This chapter covers what is perhaps the most important subject in the book and while it is primarily aimed at university students, it may be useful to the general readership. Some of the original text carried over from the first edition in 1992 may appear to some readers as being "old fashioned". Some readers will not be acquainted with the Bourdon pressure gauges mentioned in the text, but these are devices in which the effect of the internal forces being measured can be seen to produce physical movement in the instrument linkage. It is the disconnection from such close acquaintance with the forces and temperatures being measured, induced by remote, "black box" instrumentation, that is an important justification for this chapter.

Modern instrumentation, data manipulation, simulation, and "modeling" have tended to obscure questions of accuracy and to give an illusion of

Engine Testing. DOI: 10.1016/B978-0-08-096949-7.00019-4

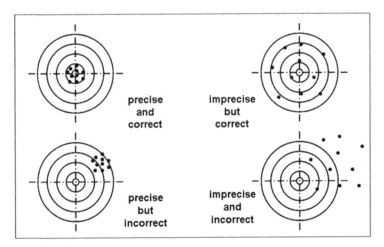

FIGURE 19.1   A "target" diagram that seeks to disambiguate the meaning of the terms precise, imprecise, correct, and incorrect. *(From the Guide to the Expression of Uncertainty in Measurement (GUM).)*

precision to experimental results that is often totally unjustified. It is a repeated statement of the obvious that the purpose of engine testing is to produce data, and that the value of that data, particularly when transformed into "information", depends critically on its accuracy.

Measurement accuracy of individual instruments is only part of the total combined measurement uncertainties, a subject that is addressed in the Guide to the Expression of Uncertainty in Measurement (GUM) and ISO 3534-1:2006 (see Figure 19.1).

Along with accuracy and precision, instruments will have an inherent measurement resolution, which is the smallest change in the physical quantity that produces a change in the measurement. Appropriate accuracy in the test process presents the biggest challenge facing the test engineer, who, along with a complete understanding of the unit under test (UUT), must master the following skills:

- Experience in the correct application of instruments
- Knowledge of methods of calibration and an awareness of the different kinds of error to which instruments are subject, both individually and collectively (as part of a complex system)
- A critical understanding of the relative merits and limitations of different methods of measurement and their applicability to different experimental situations
- An understanding of the differences between true and observed values of experimental quantities
- An appreciation that repeatability and accuracy are separate and distinctly different attributes of data

- An appreciation of the effects on raw data of the signal processing chain
- An appreciation of the changes caused by variations in climatic conditions and in the support services of the test laboratory
- An appreciation of data storage and retrieval, and relational database systems.

This is a very wide range of skills, only to be acquired by experience.

The first essential is to acquire, as a habit of mind, a sceptical attitude to all experimental observations: all instruments tend to be liars.

ANALOG VERSUS DIGITAL

There was a time in the working life of the author, and perhaps some readers, when virtually all instrument displays in the engine test cell were made up of individual analog dials, but as technology and the need for more accuracy in display developed, the individual digital panel meter became common, then followed by the tabulated numbers displayed on visual display units (VDUs). The inherent problems of analog gauges (limited resolution, parallax, moving parts, etc.) were eliminated by a digital display.

However, the digital display of a value gives an illusion of accuracy that may be totally unjustified. The temptation to believe that a reading of, say, 97.12 °C is to be relied upon to the second decimal place is very strong, but in the absence of convincing proof it must be resisted. An analog indicator connected to a thermocouple cannot, in most cases, be read to closer than 1 °C and the act of reading such an indicator is likely to bring to mind the many sources of inaccuracy in the whole temperature measuring system. Replace the analog indicator by a digital instrument reading to 0.1 °C and it is easy to forget that all the sources of error are still present: the fact that the readout can now discriminate to 0.05 °C does not mean that the overall accuracy of the measuring system is producing a reading within comparable limits.

Analog display of data has a different message to give to the observer to that of a digital display, because it shows how far the measured point is from other values above and below it, along with the rate and direction of change. This difference between the ways that the human brain processes digital and analog displays of numerical information is far from subtle[1] and should be considered carefully by control equipment designers; when rate of change or proximity to

---

1. It is interesting to note that engineers old enough to remember carrying out calculations using a slide rule can well appreciate the key "advantage" of being presented with data in an analog form—in one glance one is able to see not only the calculated answer, but also how near or far that answer is from other values. A simple but revealing example is the calculation of a pipe size for a given fluid flow; the digital calculator might show the diameter needs to be 55.38 mm (2.18″) while the slide rule gives the same answer and also "shows" the nearest standard pipe size, which is 50 mm (2″), but the next size up, and a better choice, would be 65 mm (2.5″).

a critical value needs to be observed, the analog display is easier for most staff to use intuitively.

EXAMPLE: MEASUREMENT OF EXHAUST TEMPERATURE OF A DIESEL ENGINE

Temperature measurements are some of the most difficult to make accurately, and the following classic example illustrates many of the pitfalls encountered in the pursuit of accuracy. Figure 19.2 shows a typical situation: the determination of the exhaust temperature of a diesel engine. The instrument chosen is a vapor pressure thermometer, such as is suitable for temperatures of up to 600 °C and commonly installed in the individual exhaust ports of medium-speed diesel engines. It comprises a steel bulb, immersed in the gas of which the temperature is to be measured, and connected by a long tube to a Bourdon gauge, which senses the vapor pressure but is calibrated in temperature.

Let us consider the various errors to which this system is subject.

The observer may induce a parallax error by viewing the needle at an angle that is not straight on. Sensing errors are associated with the interface between the system on which the measurements are to be made and the instruments responsible for those measurements. In the present case there are a number of sources of sensing error. First, the bulb of the temperature indicator can "see" the walls of the exhaust pipe, and these are inevitably at a lower temperature than that of the gas flowing in the pipe. It follows that the temperature of the bulb must be less than the temperature of the gas. This error can be reduced but not eliminated by shielding the bulb or by employing a "suction pyrometer". A further source of error arises from heat conduction from the bulb to the support, as a result of which there is a continuous flow of heat from the

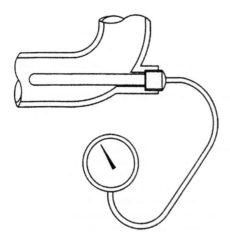

FIGURE 19.2 Measurement of exhaust gas temperature by vapor pressure thermometer.

exhaust gas to the bulb and no equality of temperature between them is possible.

A more intractable sensing error arises from the circumstances that the flow of gas in the exhaust pipe is constant neither in pressure, velocity, nor temperature. Pulses of gas, originating at the opening of the exhaust valves in individual cylinders, alternate with periods of slower flow, while the exhaust will also be to some extent diluted by scavenge air carried over from the inlet. The thermometer bulb is thus required to average the temperature of a flow that is highly variable both in velocity and in temperature, and it is unlikely in the extreme that the actual reading will represent a true average.

A more subtle error arises from the nature of exhaust gas. Combustion will have taken place, resulting in the creation of the exhaust gas from a mixture of air and fuel, perhaps only a few hundredths of a second before the attempt is made to measure its temperature. This combustion may be still incomplete, the effects of dissociation arising during the combustion process may not have worked themselves out, and it is even possible that the distribution of energy between the different modes of vibration of the molecules of exhaust gas will not have reached its equilibrium value; as a consequence, it may not be possible even in principle to define the exhaust temperature exactly.

We can deal with some of these sensing errors by replacing the steel bulb of Figure 19.2 by a slender thermocouple surrounded by several concentric screens (Figure 19.3) and further improve accuracy with a suction pyrometer (Figure 19.4) in which a sample of the exhaust gas is drawn past the sensor at uniform velocity by external suction. This deals with the "averaging" problem but still leaves the question regarding the definition of the exhaust temperature open.

The example includes a Bourdon-type pressure gauge and these may be particularly prone to a variety of classic instrument errors that can occur in any system. Two of the commonest are zero error and calibration error. The zero

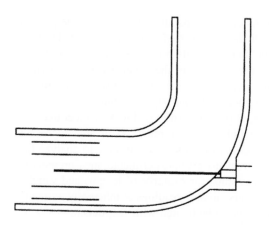

FIGURE 19.3 Measurement of exhaust gas temperature by shielded thermocouple.

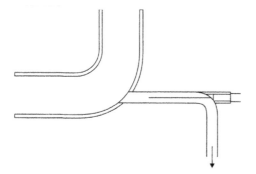

FIGURE 19.4 Measurement of exhaust gas temperature by suction pyrometer.

error is present if the pointer does not return precisely to the zero graduation when the gauge is subjected to zero or atmospheric pressure.

Calibration errors are of two forms: a regular disproportion between the instrument indication and the true value of the measured quantity, and errors that vary in a nonlinear manner with the measured quantity. This kind of fault may be eliminated or allowed for by calibrating the instrument, in the case of a pressure gauge by means of a dead-weight tester. These are examples of systematic errors.

In addition, the pressure gauge may suffer from random errors arising from friction and backlash in the mechanism. These errors affect the repeatability of the readings.

The sensitivity of an instrument may be defined as the smallest change in applied signal that may be detected; in the case of a pressure gauge it is affected particularly by friction and backlash in the mechanism.

The precision of an instrument is defined in terms of the smallest difference in reading that may be observed. Typically, it is possible to estimate readings to within one-tenth of the space between graduations, provided the reading is steady, but if it is necessary to average a fluctuating reading the precision may be much reduced.

Finally, one must consider the effect of installation errors. In the present case these may arise if the bulb is not inserted with the correct depth of immersion in the exhaust gas or, as is quite often the case, if it is installed in a pocket and is not subjected to the full flow of the exhaust gas. If a thermocouple probe that is physically smaller that the bulb is used, some of the vagaries of gas flow temperature measurement may be greater.

A consideration of this catalog of possible errors will make it clear that it is unlikely that the reading of the indicator will reflect with any degree of exactness the temperature of the gas in the pipe. It is possible to analyze the various sources of error likely to affect any given experimental measurement in this way, and while some measurements, for example of lengths and weight, require a less complex analysis, others, notably readings of inherently unsteady properties such as flow velocity, need to be treated with skepticism.

A hallmark of the experienced experimenter is that, as a matter of habit and training, he questions the accuracy and credibility of every experimental observation.

A similar critical analysis should be made of all instrumentation. This example has been dealt with in some detail to illustrate the large number of factors that must be taken into account whatever instrumentation, digital or analog, is being used.

A detailed description of the methods used in the calculation of uncertainty is given in Ref. [1] at the end of this chapter.

SOME GENERAL PRINCIPLES

1. While cumulative measurements are generally more accurate than rate measurements, when taken over a period of fluctuating engine powers, they don't indicate where high and low values are recorded. Examples are the measurement of speed by counting revolutions over a period of time and the measurement of fuel consumption by recording the time taken to consume a given volume or mass.

2. Be wary of instantaneous readings. Few processes are perfectly steady. This is particularly true of flow phenomena, as will be clear if an attempt is made to observe pressure or velocity head in a nominally steady gas flow. A water manometer fluctuates continually and at a range of frequencies depending on the scale of the various irregularities in the flow. No two successive power cycles in a spark-ignition engine are identical. If it is necessary to take a reading instantaneously it is better to take a number in quick succession and statistically analyze them.

3. Be wary of signal damping. Often it is entirely sensible, but sometimes leads to serious misrepresentation of the true nature of a fluctuating value.

4. A related problem: "time slope effects". If one is making a number of different observations over a period of nominally steady-state operation, make sure that there is no drift in performance over this period.

5. Individual measurement events must be recorded in the correct order for the correct causal deductions to be made; the more complex the data streams and number of sources, the more difficult this becomes.

6. The closer one can come to an "absolute" method of measurement, e.g. pressure by water or mercury manometer or dead-weight tester, force and mass by dead weights (plus knowledge of the local value of "g"; see Chapter 10), the less the likelihood of error.

7. In instrumentation, as in engineering generally, simplicity is a virtue. Each elaboration is a potential source of error.

DEFINITION OF TERMS RELATING TO ACCURACY

This is a particularly "gray" area [2–4]. The statement "this instrument is accurate to $\pm 1\%$" is in fact entirely meaningless in the absence of further

TABLE 19.1 Definitions of Various Terms that are Used, Often Incorrectly, in any Discussion of Accuracy

Error	The difference between the value of a measurement as indicated by an instrument and the absolute or true value.
Sensing error	An error arising because of the failure of an instrument to sense the true value of the quantity being measured.
Systematic error	An error to which all the readings made by a given instrument are subject; examples of systematic errors are zero errors, calibration errors, and nonlinearity.
Random error	Errors of an unpredictable kind. Random errors are due to such causes as electrical supply "spikes" or friction and backlash in mechanisms.
Observer error	Errors due to the failure of the observer to read the instrument correctly, or to record what he has observed correctly.
Repeatability	A measure of the scatter of successive readings of the same quantity.
Sensitivity	The smallest change in the quantity being measured that can be detected by an instrument.
Precision	The smallest difference in instrument reading that it is possible to observe.
Average error	Take a large number of readings of a particular quantity, average these readings to give a mean value, and calculate the difference between each reading and the mean. The average of these differences is the average error; roughly half the readings will differ from the mean by more than the average error and roughly half by less than the average error.
Correlation error (as defined in this book and by common practice in the test industry)	The difference between the value of a measurement made of the same parameter (under the same claimed test conditions) in two different cells or at two different sites.

definition. Table 19.1 lists the terms commonly used in any discussion of accuracy.

STATEMENTS REGARDING ACCURACY: A CRITICAL EXAMINATION

We must distinguish between the accuracy of an observation and the claimed accuracy of the instrument that is used to make the observation.

Our starting point must be the true value of the quantity to be measured (Figure 19.5a).

The following are a few examples of such quantities associated with engine testing:

- Fuel flow rate
- Air flow rate
- Output torque
- Output speed
- Pressures in the engine cylinder
- Pressures in the inlet and exhaust tracts
- Gas temperatures
- Analysis of exhaust composition.

By their nature, none of these quantities is perfectly steady, and decisions as to the sampling period are critical.

Sensing Errors

These are particularly difficult to evaluate and can really only be assessed by a systematic comparison of the results of different methods of measuring the same quantity.

Sensing errors can be very substantial. In the case discussed above regarding measurement of exhaust temperature, the author has observed errors of over 70 °C while balancing fuel pumps on large diesel engines (Figure 19.5b).

Typical sources of sensing error include:

1. Mismatch between temperature of a gas or liquid stream and the temperature of the sensor
2. Mismatch between true variation of pressure and that sensed by a transducer at the end of a connecting passage
3. Inappropriate damping or filtration applied within the instrument's measuring chain
4. Failure of a transducer to give a true average value of a fluctuating quantity
5. Air and vapor present in fuel systems
6. Time-lag of sensors under transient conditions.

Note that the accuracy claimed by its manufacturer does not usually include an allowance for sensing errors. These are the responsibility of the user.

Systematic Instrument Errors

Typical systematic errors (Figure 19.5c) include:

1. Zero errors—the instrument does not read zero when the value of the quantity observed is zero

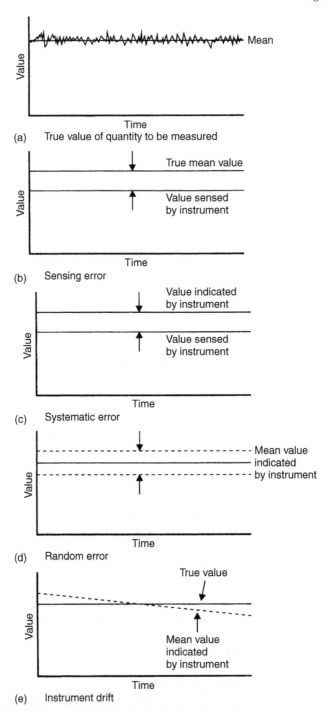

FIGURE 19.5 Various types of instrumentation error. (a) True value of quantity to be measured plotted against time—on the same time scale. (b) Sensing error. (c) Systematic error. (d) Random error. (e) Instrument drift.

2. Scaling errors—the instrument reads systematically high or low
3. Nonlinearity—the relation between the true value of the quantity and the indicated value is not exactly in proportion; if the proportion of error is plotted against each measurement over full scale, the graph is nonlinear
4. Dimensional errors—for example, the length of a dynamometer torque arm may not be precisely correct.

Random Instrument Errors

Random errors (Figure 19.5d) include:

1. Effects of stiction in mechanical linkages
2. Effects of friction, for example in dynamometer trunnion bearings
3. Effects of vibration or its absence
4. Effects of electromagnetic interference (see Chapter 5).

Instrument Drift

Instrument drift (Figure 19.5e) is a slow change in the calibration of the instrument as the result of:

1. Changes in instrument temperature or a difference in temperature across it
2. Effects of vibration and fatigue
3. Fouling of the sensor, blocking of passages, etc.
4. Inherent long-term lack of stability.

It will be clear that it is not easy to give a realistic figure for the accuracy of an observation.

INSTRUMENTAL ACCURACY: MANUFACTURERS' CLAIMS

One of the significant variables in all powertrain testing is ambient (atmospheric) conditions, as sensed in the test cell; therefore, any instrument specification should qualify the conditions under which the measurements were made. Manufacturers often state instruments are corrected for "standard conditions" without specifying them, leading to confusion and errors. International Standard Metric Conditions for natural gas and similar fluids are 288.15 K and 101.325 kPa.

With these considerations in mind, it is possible to look more critically at the statement "the instrument is accurate to within $\pm 1\%$ of full-scale reading". At least two different interpretations are possible:

1. No reading will differ from the true value by more than 1% of full scale. This implies that the sum of the systematic errors of the instrument, plus the largest random error to be expected, will not exceed this limit.
2. The average error is not more than 1% of full scale. This implies that about half of the readings of the instrument will differ from the average by less

than 1% full scale and half by more than this amount. However, the definition implies that systematic errors are negligible: we are only looking at random errors.

Neither of these definitions is satisfactory. In fact, the question can only be dealt with satisfactorily on the basis of the mathematical theory of errors.

UNCERTAINTY

While it is seldom mentioned in statements of the accuracy of a particular instrument or measurement, the concept of uncertainty is central to any meaningful discussion of accuracy. Uncertainty is a property of a measurement, not of an instrument. The uncertainty of a measurement is defined as the range within which the true value is likely to lie, at a stated level of probability.

It is noteworthy that in most scientific disciplines data, when presented graphically, has each data point represented by a cruciform indicating the degree of uncertainty or error band; this practice is rarely seen in engine test data and helps to confirm the unjustified illusion of exactness.

The level of probability, also known as the confidence level, most often used in industry is 95%. If the confidence level is 95% there is a 19:1 chance that a single measurement differs from the true value by less than the uncertainty and one chance in 20 that it lies outside these limits.

If we make a very large number of measurements of the same quantity and plot the number of measurements lying within successive intervals, we will probably obtain a distribution of the form sketched in Figure 19.6. The corresponding theoretical curve is known as a normal or Gaussian distribution (Figure 19.7). This curve is derived from first principles on the assumption that the value of any "event" or measurement is the result of a large number of independent causes (random sources of error).

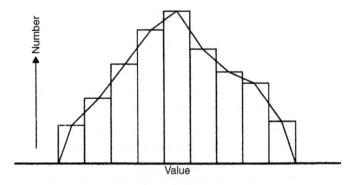

FIGURE 19.6   A frequency distribution.

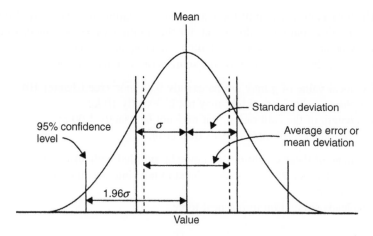

FIGURE 19.7 Normal or Gaussian distribution.

The normal distribution has a number of properties shown in Figure 19.7:

- The *mean value* is simply the average value of all the measurements.
- The *deviation* of any given measurement is the difference between that measurement and the mean value.
- The σ^2 (sigma^2) is equal to the sum of the squares of all the individual deviations divided by the number of observations.
- The *standard deviation* σ is the square root of the variance.

The standard deviation characterizes the degree of "scatter" in the measurements and has a number of important properties. In particular, the 95% confidence level corresponds to a value sigma = 1.96. Ninety-five percent of the measurements will lie within these limits and the remaining 5% in the "tails" at each end of the distribution.

In many cases the "accuracy" of an instrument as quoted merely describes the average value of the deviation, i.e. if a large number of measurements are made, about half will differ from the true or mean value by more than this amount and about half by less. Mean deviation = 0.8σ approximately.

However, this treatment only deals with the random errors; the systematic errors still remain. To give a simple example, consider the usual procedure for checking the calibration of a dynamometer torque transducer. A calibration arm, length 1.00 m, carries a knife-edge assembly to which a dead weight of 10.00 kg is applied. The load is applied and removed 20 times and the amplifier output recorded. This is found to range from 4.935 to 4.982 V, with a mean value 4.9602 V.

At first sight we could feel a considerable degree of confidence in this mean value and derive a calibration constant:

$$k = \frac{4.9602}{10 \times 9.81 \times 1.0} = 0.050563 \text{ V/Nm}$$

The 95% confidence limit for a single torque reading may be derived from the 20 amplifier output readings and, for the limiting values assumed, would probably be about ± 0.024 V, or $\pm 0.48\%$, an acceptable value.

There are, however, four possible sources of systematic error:

- The local value of g may not be exactly 9.81 m/s^2 (see Chapter 10)
- The mass of the dead weight may not be exactly 10 kg
- The length of the calibration arm may not be exactly 1.00 m
- The voltmeter used may have its own error.

In fact, none of these conditions can ever be fulfilled with absolute exactness. We must widen our 95% confidence band to take into account these probably unknown errors.

This leads straight on to our next topic.

TRACEABILITY

In a properly organized test department, careful thought must be given to this matter. In many small departments the attention given to periodic recalibration is minimal. In many more, while some effort may be made to check calibrations from time to time, the fact that the internal standards used for these checks may have their own errors is overlooked. To be confident in the validity of one's own calibrations, these standards must themselves be checked from time to time.

Traceability refers to this process. The "traceability ladder" for the dead weight referred to in the above example might look like this:

- The international kilogram (Paris)
- The British copy (NPL)
- National secondary standards (NPL)
- Local standard weight (portable)
- Our own standard weight.

COMBINATION OF ERRORS

Most derived quantities of interest to the experimenter are the product of several measurements, each of them subject to error. Consider, for example, the measurement of specific fuel consumption of an engine. The various factors involved and typical 95% confidence limits are as follows:

- Torque $\pm 0.5\%$
- Number of revolutions to consume measured mass of fuel $\pm 0.25\%$
- Actual mass of fuel $\pm 0.3\%$.

The theory of errors indicates that in such a case, in which each factor is involved to the power 1, the confidence limit of the result is equal to the square root of the sum of the squares of the various factors, i.e. in the present case the

95% confidence limit for the calculated specific fuel consumption is $\sqrt{0.5^2 + 0.25^2 + 0.3^2} = \pm 0.58\%$.

THE NUMBER OF SIGNIFICANT FIGURES TO BE QUOTED

One of the commonest errors in reporting the results of experimental work is to quote a number of significant figures, some of which are totally meaningless: the illusion of accuracy once more, often engendered by the misuse of a common spreadsheet format. This temptation has been vastly increased by the now universal employment of digital readouts.

Consider a measurement of specific fuel consumption, readings as follows:

Engine torque	110.3 Nm
Number of revolutions	6753
Mass of fuel	0.25 kg.

Specific consumption equals:

$$\frac{3.6 \times 10^6 \times 0.25}{2 \times \pi \times 6753 \times 110.3} = 0.1923 \text{ kg/kWh}$$

Let us assume that each of the three observations has the 95% confidence limit specified in the previous paragraph. Then the specific fuel consumption has a confidence limit of 0.58%, i.e. between 0.1912 and 0.1934 kg/kWh. Clearly it is meaningless to quote the specific fuel consumption to closer than 0.192 kg/kWh.

Particularly in development work, where the aim is often to detect small improvements, this question must be kept in mind.

ABSOLUTE AND RELATIVE ACCURACY

A topic linked with the last one. In a great deal of engine test work—perhaps the majority—we are interested in measuring relative changes and the effect of modifications. The absolute values of a parameter before and after the modification may be of less importance; it may not be necessary to concern ourselves with the absolute accuracy and traceability of the instrumentation. Sensitivity and precision may be of much greater importance. Some test engineers have unhealthy respect for the "repeatability" of results in a given test cell, while not considering that this can be almost completely disconnected from accuracy of results.

THE COST OF ACCURACY: A FINAL CONSIDERATION

An engineer with responsibility for the choice of instrumentation should be aware of the danger of mismatch between what he thinks he requires and what is actually necessary for an adequate job to be done. The cost of an instrument to perform a particular task can vary by an order of magnitude; in most cases the main variable is the level of accuracy offered and this should be compared with the accuracy that is really necessary for the job in hand.

REFERENCES

[1] Evaluation of measurement data, An introduction to the Guide to the Expression of Uncertainty in Measurement and related documents, GUM aka JCGM 104 (2009). Available at: http://www.bipm.org/utils/common/documents/jcgm/JCGM_104_2009_E.pdf.

[2] A.T.J. Hayward, Repeatability and Accuracy, Mechanical Engineering Publications, London, 1977. ISBN-13: 978-0852983515.

[3] C.F. Dietrich, Uncertainty, Calibration and Probability, second ed., Taylor & Francis, 1973. ISBN-13: 978-0750300605.

[4] P.J. Campion et al., A Code of Practice for the Detailed Statement of Accuracy, HMSO, 1973. ISBN-13: 978-0950449661.

FURTHER READING

BS 4889. Method for Specifying the Performance of Electrical and Electronic Measuring Equipment.

BS 5497 Part 1. Guide for the Determination of Repeatability and Reproducibility for a Standard Test Method by Inter-Laboratory Tests.

BS 7118 Parts 1 and 2. Linear and Non-Linear Calibration Relationships.

Tribology, Fuel, and Lubrication Testing

INTRODUCTION

Because of the increasingly important role played by lubricants in exhaust emissions, the ever-increasing expectations of drivers for long inter-service periods and the reliability of their vehicles, major powertrain manufacturers of the world now regard engine and transmission lubricants as important powertrain components, rather than simply a liquid lubricant. In this respect the general public, and the publicity to which they are subjected, probably lag behind this perception.

The complex web of international standards covering lubricating oils is supported by specialist laboratories, analyzing the chemical and physical characteristics of oil products before and after strictly prescribed usage, which

Engine Testing. DOI: 10.1016/B978-0-08-096949-7.00020-0

is carried out both on component test rigs and in engine test cells. This book will mention only a few of the very numerous standard tests that are in constant use for assessment and routine confirmation of fuel and lubricant properties. The development, updating, and monitoring of these standard tests, not to mention the cost in fuel, components, and work hours, represent a significant cost to engine and lubricant manufacturers.

While "fuels and lubes" (F&L) testing continues to be a specialized branch of the engine test industry carried out in certified laboratories, most major original equipment manufacturer (OEMs) are increasingly producing their own lubrication test specifications, based on their own engine and transmission tests, while still following the strict methodologies of the international standards organizations.

New engine technologies, such as direct injection and exhaust gas after-treatment, have created new problems, or increased old problems, for the lubrication oil industry, examples of which are increased fuel dilution of lubricating oil and poisoning of catalytic exhaust gas devices. There is also a continuous development of lubricants required to provide longer periods between vehicle oil changes and of fuels producing cleaner engine components and less pollutants in the exhaust stream.

CALORIFIC VALUE OF FUELS

The calorific value of a fuel is defined in terms of the amount of heat liberated when a fuel is burned completely in a calorimeter. Detailed methods and definitions are given in Ref. [1], but for the present purpose the following is sufficient.

Since all hydrocarbon fuels produce water as a product of combustion, part of these products (the exhaust gas in the case of an internal combustion engine) consists of steam. If, as is the case in a calorimeter, the products of combustion are cooled to ambient temperature, this steam condenses, and in doing so gives up its latent heat. The corresponding measure of heat liberated is known as the higher or gross calorific value (also known as gross specific energy). If no account is taken of this latent heat we have the lower or net calorific value (also known as net specific energy). Since there is no possibility of an internal combustion engine making use of the latent heat, it is normal practice to define performance in terms of the lower calorific value C_1. Table 20.1 shows values of the lower calorific value and density for some typical fuels.

GASEOUS FUELS

These fuels, which have favorable emissions characteristics, are used by vehicles in areas where the infrastructure exists and they are maintaining their importance as a fuel in power generation engines. Natural gas (NG) is also

TABLE 20.1 Properties of Five Liquid Fuels

Fuel	Lower Calorific Value (MJ/kg)	Stoichiometric Air/Fuel Ratio	Density (kg/l)
Gasoline	43.8	14.6	0.74
Gas oil	42.5	14.8	0.84
Methanol	19.9	6.46	0.729
Ethanol	27.2	8.94	0.79
Heavy fuel oils	39.7	—	0.88–0.98

sometimes described as compressed natural gas (CNG) and, when transported in bulk at very low temperature, as liquefied natural gas (LNG).

Natural gas consists mainly of methane but, having evolved from organic deposits, invariably contains some higher hydrocarbons and traces of N_2 and CO_2. Composition varies considerably from field to field and the reader involved in work on natural gas should establish the particulars of the gas with which he is concerned. North Sea Gas has become a standard gaseous fuel in the UK and its approximate properties are shown in Table 20.2.

Liquefied Petroleum Gas (LPG or LP Gas)

LPG is a product of the distillation process of crude oil or a condensate from wet natural gas. It consists largely of propane and, unlike natural gas, can be stored in liquid form at moderate pressures. In view of its good environmental

TABLE 20.2 Approximate Properties of Typical Gaseous Fuels

Natural (North Sea) Gas		Liquified Petroleum Gas (LPG)	
Methane, CH_4	93.3%	Propane, C_3H_8	90%
Higher hydrocarbons	4.6%	Butane, C_4H_{10}	5%
N_2 plus CO_2	2.1%	Unsaturates	5%
Lower calorific value	48.0 MJ/kg		46.3 MJ/kg
Stoichiometric air/fuel ratio	14.5:1		15.7:1
Approximate density at 0 °C	0.79 kg/m^3		2.0 kg/m^3

It should be noted that some gas suppliers quote the higher rather than the lower calorific value, which in the case of methane, with its high H/C ratio, can make a difference of almost 10%.

properties (low unburned hydrocarbon emissions, low CO, virtually no particulate emissions, and no sulfur) it is a favorable vehicle engine fuel, attracting lower tax rates in many countries. NO_x emissions tend to be higher than for gasoline, owing to its high combustion temperature. It has a high octane number, RON 110, which permits higher compression ratios. Approximate properties are given in Table 20.2.

LUBRICANT CLASSIFICATION AND CERTIFICATION

Up to the early 1980s, the US military played a major role in the development of fuel and lubricant specifications, but since that time and with the proliferation worldwide of interested parties, engine manufacturers, oil companies and the chemical industry, a degree of harmonization of test methods has been achieved. Any student of the subject will have to be aware of the abbreviations, acronyms, and names of the important governing or standard-setting organizations listed in alphabetical order below:

ACEA	The European Automobile Manufacturers Association
ACT	Additives Technical Committee (Technical Committee of Petroleum Additive Manufacturers in Europe)
ASTM	The American Society for Testing of Materials
ATIEL	Association Technique de l'Industrie Européenne des Lubrifiants
BSI	British Standards Institution
CEC	Coordinating European Council—for the Development of Performance Tests for Transportation Fuels, Lubricants, and Other Fluids
COFRAC	The French Committee for Accreditation
CONCAWE	The oil companies' European association for Environment and H&S in refining and distribution
DIN	Deutsches Institut für Normung; the German Institute for Standardization, that country's ISO member body
EELQMS	European Engine Lubricants Quality Management System
ERC	(ATC's) European Registration Center
IP	Institute of Petroleum
ISO	International Organization for Standardization, the international standard-setting body composed of representatives from national standards bodies
TDG	Test Development Group (coordinated under CEC)

The CEC has support groups working under their Secretariat:

- Statistical Development Group, charged with assuring the quality of test results
- Rating Group, charged with ensuring consistent fuel and lubricant ratings across the industry
- Reference Fuels; the CEC maintains a suite of reference fuels supplied for use by the TDG test groups to ensure consistency
- Reference Lubricants Reference Oils; as with reference fuels, reference oils are supplied to TDG test groups to aid test result correlation ("round robin" testing) and the development of new tests.

In Europe the ACEA details lubrication oil specifications backed by the EELQMS standardized tests and performance criteria; these have to be met by any product sold to and by their members. Along with the ACEA tests, individual member companies have their own test sequences and processes. The nomenclature of the ACEA process is rather complex; in descending hierarchy, oils are named by sequence, class, then category, plus a year of implementation designation.

The current ACEA sequences and classes are: ACEA A/B (combined) for gasoline and light-duty diesel engines, and ACEA C for gasoline and diesel engines having advanced after-treatment (catalyst compatible).

The category indicates the type of application the lubricant is intended for within its class; there are currently four in the general sequence, designated A1/B1, A3/B3, A3/B4 and A5/B5, and three in the catalyst-compatible oils, C1 (low SAPS (sulfated ash, phosphorus, and sulfur (SAPS) oils), C2 (for vehicles with TWC and DPF), and C3 (high-performance gasoline or diesels).

There are four classes in the oils designed for heavy-duty diesels, designated E2 (general purpose, normal drain intervals), E4 (highly rated engines meeting Euro 3 and 4), E6 (highly rated engines running on low SAPS fuel), and E7 (highly rated engines with EGR and SCR).

The year of implementation number is placed after the class designation thus: $C1_{-04}$. Using these classes to set the limits of laboratory and engine-based tests, the ACEA produces tables indicating the minimum allowable quality of oils of the various classes under a system of self-assessment to EELQMS.

TRIBOLOGY

The science of tribology, launched as a separate discipline with a new name by the Jost Report of 1966, is concerned with lubrication and wear; both are central to IC engine operation. It was pointed out some years ago that the difference between a brand new automotive powertrain and one that was totally worn out was the loss of perhaps 100 grams of material at critical points in the mechanisms. So far as friction is concerned, the great majority of the power developed by an automobile engine is ultimately dissipated as friction (not all wastefully!). Even a large slow-speed diesel engine is unlikely to achieve a mechanical efficiency exceeding 85%, most of the losses being due to friction.

As the evolution of materials and powertrain technology continues, different aspects of lubrication and wear call for concentrated attention. Some problems, such as excessive bore wear and bore polishing in diesel engines, have been largely overcome and sunk into the background. Others emerge as engine technologies (such as common-rail diesel) develop, and as operating conditions become more demanding or emission legislation sets new parameters. The following are the current (2011) development problems of tribology-related concern:

- The effect on exhaust emissions of lubrication oil constituents
- Interactions between fuel and lubricant
- Poisoning of exhaust treatment systems by lubrication oil constituents
- Valve train and injector wear using RNT methods (see below)
- Development of synthetic and "stay-in-grade" or "fill-for-life" lubricants.

The pursuit of higher efficiency and reduced friction losses is a continuing activity in all engine development departments.

Bench Tests

A wide range of standard (non-engine) tests are written into standards such as those of the ASTM, DIN, CEC, and IP.

These tests for oils can range from chemical analysis such as ASTM D1319-10 (aromatics and olefins content), through physical laboratory tests such as ASTM D2386 (freezing point), to special rig tests such as DIN 51 819-3, which uses a roller bearing test rig. Similarly, there are series of tests specifically for greases and fuel analysis.

The principle of operation of many of the lubrication oil test machines has not changed for many years and even though the test conditions they impose differ from any to be found in a real engine, their value is their ability to reliably reproduce the classic mechanical wear and/or failure conditions.

The number of analytical and diagnostic tools available to test laboratories has increased significantly in the last two decades, and ranges from laser measuring devices able to differentiate into the realms of nanotechnology, through use of radio-tracing techniques of wear particles, to electronic scanning microscopes.

Laboratories supporting tribology research carry out a wide variety of work, including:

- Visual and microscopic examination
- Wear debris quantification and analysis
- Surface roughness measurement
- Metallographical sections and etching
- Microhardness measurement
- Scanning electron and scanning tunneling electron microscopy
- Infrared and Raman spectroscopy.

After each engine test, whether it is an oil- or fuel-related test sequence, the whole engine or that part of it of specific interest will be stripped under laboratory conditions and examined. In addition to the surface examinations listed above that look at discoloration, corrosion, and wear, fuel tests will examine such effects as the reduction in flow through injector nozzles and ring sticking.

OIL CHARACTERISTICS

Viscosity is a measure of the resistance of a fluid to deform under shear stress. It is commonly perceived as resistance to pouring. It is measured by a viscometer, of which there are various designs commonly based on the time taken to flow through an orifice or a capillary tube when at a standard temperature. The SAE "multigrade" rating known to most motorists has a nomenclature based on flow through an orifice at 0 and 210 °F, with "W" (standing for winter) after the number for low-temperature rating and a plain number for the high-temperature rating, as in SAE10W40.

Dynamic viscosity: The SI physical unit of dynamic viscosity (μ) is the Pascal-second (Pa s), which is identical to $1 \text{ kg m}^{-1} \text{ s}^{-1}$. The physical unit for dynamic viscosity in the centimeter gram second system of units (cgs) is the *poise* (P), named after Jean Poiseuille. It is more commonly expressed, particularly in ASTM standards, as centipoise (cP). The centipoise of water is almost unity (1.0020 cP) at 20 °C:

1 poise $= 100$ centipoise $= 1 \text{ g cm}^{-1}\text{s}^{-1} = 0.1$ Pa s
1 centipoise $= 1$ mPa s.

Kinematic viscosity is the ratio of the viscous force to the inertial force or fluid density ρ:

$$v = \frac{\mu}{\rho}$$

Kinematic viscosity has SI units of m^2s^{-1}. The physical unit for kinematic viscosity is the *stokes* (St), named after George Stokes. It is sometimes expressed in terms of *centistokes* (cS or cSt); 1 stokes $= 100$ centistokes $= 1 \text{ cm}^2\text{s}^{-1} = 0.0001 \text{ m}^2\text{s}^{-1}$.

Viscosity index (or VI) is a lubricating oil quality indicator, an arbitrary measure for the change of kinematic viscosity with temperature.

Total base number (TBN) of oil is the measure of the alkaline reserve, or the ability of the oil to neutralize acids from combustion tested to ASTM D2896 by a laboratory method known as potentiometric perchloric acid titration. Depletion of the TBN in service results in acid corrosion and fouling within the engine.

FUEL CONTAMINATION OF LUBRICATION OIL

One of the serious side-effects of the pre- and post-injection strategies in direct injection gasoline and diesel engines has been seen as an increase in fuel dilution of lubricating oil, which can lead to complete engine failures. This is a problem not only for the lubrication specialist who has to measure the effect, but for test engineers carrying out any long-duration testing of modern, downsized DI engines. The latter group need to carry out offline checks of the engine's oil using commercially available oil condition test kits. Modern real-time methods for monitoring fuel contamination use minute amounts of a tracer fluid that can be detected by laser excitation [2].

REFERENCE FUELS AND LUBRICANTS

Any laboratory carrying out fuels and lube testing will have to store and handle correctly reference fuels and oils. Many component inspection results after a running engine test will be based on comparative judgments based on the use of these reference liquids, which are expensive and which may have limited storage life unless kept in optimum conditions. Test limits for such criteria as piston cleanliness or sludge accumulation in some engine-based lubrication tests will be based on reference lubricant results that are commonly designated with a CEC reference such as in the case of CEC RL191 (15W-40).

DESIGNATED ENGINES AND TEST REGIMES IN FUEL AND LUBE TESTING

While the science of tribology studies the phenomena associated with wear and lubrication, the classification and specification of all fuels and crankcase lubricants worldwide are based upon standardized engine tests.

There is a complete section of the engine test industry attached to the oil industry using internationally specified engines in exactly determined configurations to run strict test protocols to certify engine fuels and lubricants. The use of a prescribed engine builds redundancy into the test system and the need for regular updating of the certification process because the availability of any engine is limited, as is its suitability to represent the current technology that the fuels and lube industry has to support. Examples of just three of the engines commonly designated in tests are:

- Mercedes M-111: used in the fuel economy test CEC-L-54-T-96 (C1$_{-04}$, C2$_{-04}$, C3$_{-04}$).
- Mack T-8: uses Mack E7-350 six-cylinder commercial diesel engine (11.93 L). This engine test measures an oil's ability to resist viscosity increases due to excessive soot formation. ASTM 5967 (E4$_{-99}$, E6$_{-04}$, E7$_{-04}$).
- VW 1.6 l TC: the procedure listed as CEC L-46-T-93 uses a TC intercooled Volkswagen diesel engine (1.6 L) to measure an engine oil's ability to resist ring sticking and to maintain piston cleanliness under high-temperature conditions.

Typically a fuel and lube test stand will be configured specifically for a designated engine that will remain in place for prolonged periods, only being removed, in whole or part, for component inspection after testing. The engine will be mounted on a base plate, connected to an eddy-current dynamometer, and rigged with fuel and/or oil control systems, instrumented exactly as required by regulation. The running of the test sequence, the manipulation of the test results, and the post-test examination of engine parts, such as valves and injectors, are similarly tightly prescribed.

The major disadvantages of such tests are the cost and time involved and the difficulty, where measurements of such factors as surface wear, deposits, oil consumption, and friction are concerned, of obtaining consistent results. The measurements can be much affected by variations in engine build, and components, also of fuel in the case of lubrication tests or of lubrication oil in the case of fuel tests; therefore, most fired tests are run with strictly prescribed reference fluids. There are a few internationally recognized specialist test laboratories that provide the service of running such tests and each of the major engine manufacturers have their own special test cells running both public-domain tests sequences and their own tests on new engine designs.

Motored transmission, engine or "part engine" tests are used for assessing specific components such as valve trains. These tests do not so much represent real-life conditions but rather provide a repeatable test sequence that gives a good insight into the problems of wear and lubrication within a particular powertrain module. By the successive removal of components from a motored engine, a good estimate can be made of the contribution made to frictional losses by the different components (see Figure 20.1).

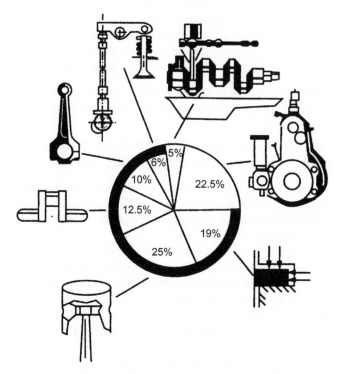

FIGURE 20.1 Typical percentile distribution of friction losses in an internal combustion engine.

RNT (Radionuclide Technique) Wear Tests

Engine component wear measurement within tribology research is increasingly based on the radionuclide technique (RNT), sometimes referred to as "Radionuclide Technique in Mechanical Engineering" (RTM). It is a way of determining the tribological performance of a system continuously, at high resolution and in real time. Test-piece preparation uses a method called thin layer activation (TLA), which requires that the iron or steel alloy component,[1] whose wear rate is to be investigated, is radioactively "labeled" within a cyclotron device. During the process a homogeneous, radioactive, surface layer of an isotope of cobalt (^{56}Co or ^{57}Co) is deposited, with a thickness that is adjusted, according to the expected wear measurement depth, between 20 μm and 0.2 mm. While the mechanical and chemical properties of the components are not affected by TLA, the minute particles resulting from wear during running of an engine's or transmission's functional test may be detected by one of two possible methods:

- Concentration measurement method (CMM). The radioactive wear particles, suspended in the lubricating oil, are pumped through the flow chamber with the detector; the count rate of the detector is proportional to the total wear.
- Thin layer difference (TLD) method. There is a linear relationship between the measured, characteristic, gamma radiation intensity from the labeled component and the depth of the irradiated zone. The radiation is measured by a detector that is mounted externally to the engine, making the method suitable for use in some in-service situations such as large marine engines.

RNT techniques have moved from pure research and are now used in a number of OEM "in-house" wear tests such as the VW 650-hour RNT test.

Example of (Non-RNT) Engine Tests Involving Tribology: Measurement of Compression Ring Oil Film Thickness

A feature of many tribological situations is the extreme thinness of the oil films present, usually to be measured in microns, and various techniques have been developed for measuring these very small clearances, the latest of which uses ultrasonic reflection [3].

Figure 20.2 shows measured oil film thicknesses between top ring and cylinder liner near top dead center in a large turbocharged diesel engine. There is a wide variation in film thickness for the different strokes, and an effective breakdown during the firing stroke. It is notable that the thinner oil, SAE 10W/30, gives better protection than a thicker one. This was thought to be the result of higher oil flow to this critical area with the thinner oil and is confirmed by the

---

1. TLA cannot be applied to synthetic (nonmetallic) materials.

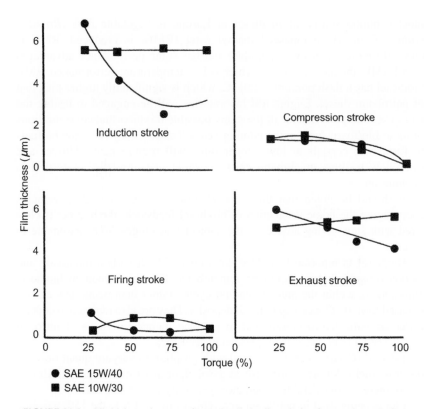

FIGURE 20.2 Oil thickness between ring and cylinder liner during four-stroke operation.

generally lower rates of wear of aluminum from the piston and chromium from the ring surfaces.

BIOFUELS (see chapter 7 concerning storage of biofuels)

To reduce dependency on fossil fuel oil stocks, there are a number of global initiatives intended to increase the use of oil derived from renewable sources. The European Commission introduced directive 2003/30/CE that proposed increasing to 5.75% the proportion of energy delivered by biofuels by 2010. However, justifiable political and media attention will continue, during the life of this book, to be focused on the use of biofuels and the issues related to the effect of their use on food security, food prices, effects on biodiversity, and perceived CO_2 emission advantages over fossil fuels; these are criteria for use well outside the professional role of the test engineer.

Biodiesel in Europe has to conform to the EN14214 specification and can be derived from a range of animal and vegetable oils; the most common forms are fatty acid methyl ester (FAME) and fatty acid ethyl ester (FAEE). The most common source of biodiesel in Europe is vegetable rape (*Brassica napus*), from which rapeseed methyl ester (RME) is produced. RME is currently more expensive to produce but has some performance advantages over FAME, the most critical of which is low-temperature performance. RME biodiesel has a flash point of >150 °C, which is significantly higher than that of petroleum diesel. Engine test laboratories will be engaged in testing the emissions and performance of the many possible mixtures and concentrations of these bio/fossil fuels in the coming years. The tendency for some biofuels to form wax crystals at low temperatures will require new additives and filter flow testing[2] in climatic cells; a whole new direction of testing is opening up.

It should be noted that there are worldwide environmental movements encouraging the home production of biodiesel feedstock; the internet is cluttered with sites giving advice on the subject that ranges from reasonable to lunatic.

Biodiesel is reported to reduce emissions of both carbon monoxide and carbon dioxide, but it should be remembered that the carbon in biodiesel emissions is within the animal "carbon cycle" rather than being new carbon released from that locked up in fossil deposits. They are also reported to contain fewer aromatic hydrocarbons, but to produce higher emissions of nitrogen oxides (NO_x) than diesel from petroleum feedstock.

Ethanol (C_2H_6O) can be blended with gasoline in varying quantities; the resulting fuel is known in the USA as gasohol, or as gasoline type C in Brazil, where there is considerable experience, generally use of E20 and E25.

E85 is being used in public service vehicles in the USA, the 15% gasoline being required to overcome cold starting problems experienced with higher ethanol mixtures.

Methanol (CH_3OH) can also be used as a gasoline additive but is highly toxic and less easy to produce from biological sources than ethanol.

The subject of biofuels will continue to be a contentious issue as land is used for its production rather than for food crops and tropical jungle is cleared for the same purpose.

REFERENCES

[1] BS 7420. Guide for Determination of Calorific Values of Solid, Liquid and Gaseous Fuels (1991).
[2] Rapid In Situ Measurement of Fuel Dilution of Oil in a Diesel Engine using Laser-Induced Fluorescence Spectroscopy. SAE Paper Number 2007-01-4108.
[3] R.S. Dwyer-Joyce, D. Green, P. Harper, R. Lewis, S. Balakrishnan, P.D. King, H. Rahjenat, The Measurement of Oil Film Thickness in Piston Ring and Skirt Contacts by an Ultrasonic Means, SAE, 2005.

---

2. Cold filter plugged point (CFPP) is a standard test requirement for winter grade fuels.

FURTHER READING

A.J. Caines, R.F. Haycock, Automotive Lubricants Reference Book, second ed., Wiley-Blackwell, 2004. ISBN-13: 978-0768012514.

T. Mang, W. Dresel (Eds), Lubricants and Lubrication, Wiley-VCH, 2005. ISBN-13: 978-3527314973.

G.E. Totten, The Handbook of Lubrication and Tribology: Theory and Practice (Handbook of Lubrication, Theory & Practice of Tribology), second ed., CRC Press, 2006. ISBN-13: 978-0849320958.

The authority responsible for fuel and lube tests in Europe is:

The Coordinating European Council for the Development of Performance Tests for Lubricants and Engine Fuels (CEC).

A useful website concerning CEC tests and standards is: http://www.cectests.org/

Chapter 21

Thermal Efficiency, Measurement of Heat, and Mechanical Losses

INTRODUCTION

A measure of the performance of an internal combustion engine is the proportion of the heat of combustion of the fuel that is turned into useful work at the engine flywheel. The thermal efficiency at full load of internal combustion engines ranges from about 20% for small gasoline engines up to more than 50% for large slow-running diesel engines, the latter being the most efficient means currently

Engine Testing. DOI: 10.1016/B978-0-08-096949-7.00021-2

available of turning the heat of combustion of fuel into mechanical power. Thermodynamics allows us, within certain limitations, to predict the theoretical maximum thermal efficiency of an internal combustion engine. The proportion of the heat of combustion that is not converted into useful work appears elsewhere: in the exhaust gases, in the cooling medium, and as convection and radiation from the hot surfaces of the engine. There may in addition be appreciable losses in the form of unburned or late-burning fuel. It is important to be able to evaluate these various losses. Of particular interest are losses from the hot gas in the cylinder to the containing surfaces, since these directly affect the indicated power of the engine. The so-called "adiabatic engine" seeks to minimize these particular losses. Some of the heat carried away in the exhaust gas may be converted into useful work in a turbine or, in large diesel engines, used for such purposes as steam generation or the production of hot water.

IDEAL STANDARD CYCLES: EFFECT OF COMPRESSION RATIO

Many theoretical cycles for the internal combustion engine have been proposed, some of them taking into account such factors as the exact course of the combustion process, the variation of the specific heat of air with temperature, and the effects of dissociation of the products of combustion at high temperature. However, all these cycles merely modify the predictions of the cycle we shall be considering, generally in the direction of reduced attainable efficiency, without much changing the general picture. For a detailed discussion, see Heywood [1].

The air standard cycle, also known as the Otto cycle, is shown in Figure 21.1. It consists of four processes, forming a complete cycle, and is based on the following assumptions:

1. The working fluid throughout the cycle is air, and this is treated as a perfect gas.
2. The compression process 1–2 and the expansion process 3–4 are both treated as frictionless and adiabatic (without heat loss).

FIGURE 21.1 Air standard cycle (see text).

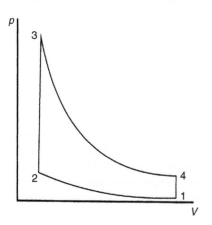

3. In place of heat addition by internal combustion, phase 2–3 of the process is represented by the addition of heat from an external source, the volume remaining constant.
4. The exhaust process is replaced by cooling at constant volume to the initial temperature.

It will be evident that these conditions differ considerably from those encountered in an engine; nevertheless, the thermodynamic analysis, which is quite simple, gives useful indications regarding the performance to be expected from an internal combustion engine, in particular with regard to the influence of compression ratio. The air standard cycle efficiency is given by:

$$\eta_{as} = 1 - \frac{1}{R^{\gamma-1}} \tag{21.1a}$$

The course of events in an engine cylinder departs from this theoretical pattern in the following main respects:

1. Heat is lost to the cylinder walls, reducing the work necessary to compress the air, the rise in temperature and pressure during combustion, and the work performed during expansion.
2. Combustion, particularly in the diesel engine, does not take place at constant volume, resulting in a rounding of the top of the diagram (point 3) and a reduction in power. A better standard of reference for the diesel engine is the limited pressure cycle (Figure 21.2), for which the efficiency is given by the expression:

$$\eta = 1 - \frac{1}{R^{\gamma-1}} \left[\frac{\alpha\beta^{\gamma} - 1}{\alpha\gamma(\beta - 1) + \alpha - 1} \right] \tag{21.1b}$$

where

$$\alpha = \frac{p3}{p2a} \qquad \beta = \frac{V3b}{V3a}$$

FIGURE 21.2 Limited pressure cycle.

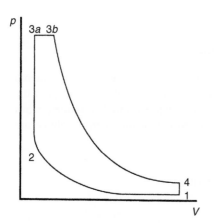

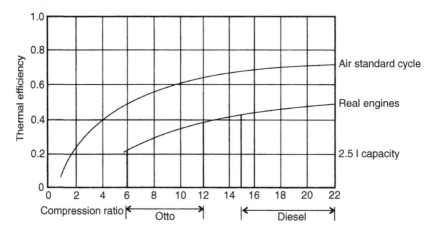

FIGURE 21.3 Variation of air standard cycle efficiency with compression ratio.

This reduces to equation (21.1a) when $\beta = 1$.

3. The properties of air, and of the products of combustion, do not correspond to those of an ideal gas, resulting in a smaller power output than predicted.

4. The gas exchange process is ignored in the standard cycle.

Figure 21.3 shows the variation of air standard cycle efficiency with compression ratio, and shows the range of this ratio for spark-ignition and diesel engines. It is clearly desirable to use as high a compression ratio as possible. Also shown is the approximate indicated thermal efficiency to be expected from gasoline and diesel engines of 2.5 liter swept volume. Larger engines, and in particular large slow-speed diesel engines, can achieve significantly higher efficiencies, mainly because heat losses from the cylinder contents become less in proportion as the size of the individual cylinders increases.

THE ENERGY BALANCE OF AN INTERNAL COMBUSTION ENGINE

The distribution of energy in an internal combustion engine is best considered in terms of the steady flow energy equation, combined with the concept of the control volume. In Figure 21.4 an engine is shown, surrounded by the control surface. The various flows of energy into and out of the control volume are shown.

In:

- Fuel, with its associated heat of combustion
- Air, consumed by the engine

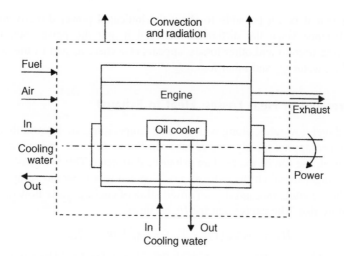

FIGURE 21.4 Control volume of an IC engine showing energy flows.

Out:

- Power developed by the engine
- Exhaust gas
- Heat to cooling water or air
- Convection and radiation to the surroundings.

The steady flow energy equation gives the relationship between these quantities, and is usually expressed in kilowatts:

$$H_1 = P_s + (H_2 - H_3) + Q_1 + Q_2 \tag{21.2}$$

in which the various terms have the following meanings:

H_1 = combustion energy of fuel = $\dot{m}_f C_L \times 10^3$
P_s = power output of engine
H_2 = enthalpy of exhaust gas[1] = $(\dot{m}_f + \dot{m}_a)C_p T_e$
H_3 = enthalpy of inlet air = $\dot{m}_a C_p T_a$
Q_1 = heat to cooling water[2] = $\dot{m}_w C_w (T_{2w} - T_{1w})$
Q_2 = convection and radiation.

This assumes that the specific heat of the exhaust gas, the mass of which is the sum of the masses of air and fuel supplied to the engine, is equal to that of air. This is not strictly true, but permits an approximate calculation to be made if the temperature of the exhaust gas is measured (exact measurement of exhaust temperature is no simple matter, see Chapter 20).

---

1. For a detailed description of enthalpy see, for example, Ref. [2].

2. This may also include heat transferred to the lubricating oil and subsequently to the cooling water via an oil cooler.

Note that it is not possible to show the indicated power directly in this energy balance since the difference between it and the power output P_s, representing friction and other losses, appears elsewhere as part of the heat to the cooling water Q_1 and other losses Q_2.

Measurement of Heat Losses: Heat to Exhaust

If air and fuel flow rates, along with exhaust temperature, are known, this may be calculated approximately (see H_2 above). For an accurate measurement of exhaust heat, use can be made of an exhaust calorimeter. This is a gas-to-water heat exchanger in which the exhaust gas is cooled to a moderate temperature and the heat content measured from observation of cooling water flow rate and temperature rise. The expression for H_2 becomes:

$$H_2 = \dot{m}_c C_w (T_{2c} - T_{1c}) + (\dot{m}_f + \dot{m}_a) C_p T_{co} \qquad (21.3)$$

The rate of flow of cooling water through the calorimeter should be regulated so that the temperature of the gas leaving the calorimeter, T_{co}, does not fall below about 60 °C (333 K). This is approximately the dew point temperature for exhaust gas: at lower temperatures the steam in the exhaust will start to condense, giving up its latent heat (see section on "Calorific Value of Fuels" in Chapter 20).

Sample Calculation: Analysis of an Engine Test

Table 21.1 is an analysis, based on equation (21.2), of one test point in a sequence of tests on a vehicle engine. Then, we note that:

Specific heat of air at constant pressure $C_p = 100$ KJ/kg·K
Specific heat of water $C_w = 4.18$ kJ/kg·K
$H_1 = 0.00287 \times 41.87 \times 10^3 = 120.2$ kW
$P_s = 36.8$ kW
$H_2 = (0.00287 + 0.04176) \times 1.00 \times 1066 = 47.6$ kW
$H_3 = 0.04176 \times 1.00 \times 292 = 12.2$ kW
$H_2 - H_3 = 35.4$ kW
$Q_1 = 0.123 \times 4.18(72.8 - 9.2) = 32.7$ kW
Q_2 (by difference) $= 15.3$ kW.

We may now draw up an energy balance (quantities in kilowatts):

Heat of combustion H_1	120.2	Power output P_s	36.8 (30.6%)
		Exhaust ($H_2 - H_3$)	35.4 (29.5%)
		Other losses Q	15.3 (12.7%)
	120.2		120.2 (100%)

The thermal efficiency of the engine is:

$$\eta_{th} = \frac{P_s}{H_1} = 0.306$$

TABLE 21.1 Energy Balance of a Gasoline Engine at Full Throttle (Four-Cylinder, Four-Stroke Engine, Swept Volume 1.7 Liter)

Engine speed	3125 rev/min
Power output	$P_s = 36.8$ kW
Fuel consumption rate	$\dot{m}_f = 0.00287$ kg/s
Air consumption rate	$\dot{m}_a = 0.04176$ kg/s
Lower calorific value of fuel	$C_L = 41.87 \times 10^6$ J/kg
Exhaust temperature	$T_c = 1066$ K (793 °C)
Cooling water flow	$\dot{m}_w = 0.123$ kg/s
Cooling water inlet temperature*	$T_{1w} = 9.2$ °C
Cooling water outlet temperature*	$T_{2w} = 72.8$ °C
Inlet air temperature	$T_a = 292$ K (19 °C)

* The engine was fitted with a heat exchanger. These are the temperatures of the primary cooling water flow to the exchanger.

The compression ratio of the engine $R = 8.5$, giving:

$$\eta_{as} = 1 - \frac{1}{8.5^{1.4-1}} = 0.575$$

The mechanical efficiency of this engine at full throttle was approximately 0.80, giving an indicated thermal efficiency of:

$$\frac{0.306}{0.8} = 0.3825$$

This is approximately two-thirds of the air standard efficiency.

Sample Calculation: Exhaust Calorimeter

In the test analyzed in Table 21.1 the heat content of the exhaust was also measured by an exhaust calorimeter with the following result:

Cooling water flow, $\dot{m}_c = 0.139$ kg/s
Cooling water inlet temperature, $T_{1c} = 9.2$ °C
Cooling water outlet temperature, $T_{2c} = 63.4$ °C
Exhaust temperature leaving calorimeter, $T_{co} = 355$ K (82 °C).

Then from equation (21.3):

$$H_2 = 0.139 \times 4.18(63.4 - 9.2) + (0.00287 + 0.04176) \times 1.00 \times 355$$
$$= 47.3 \text{ kW}$$

$H_3 = 12.2$ kW, as before, giving heat to exhaust $H_2 - H_3 = 35.1$ kW. This shows a satisfactory agreement with the approximate value derived from the exhaust temperature and air and fuel flow rates.

Energy Balances: Typical Values

Figure 21.5 shows full power energy balances for several typical engines. Results for the gasoline engine analyzed above are shown in (a). Table 21.2 shows energy balances in terms of heat losses per unit power output for various engine types. This will be found useful when designing such test cell services as cooling water and ventilation. They are expressed in terms of kW/kW power output.

These proportions depend on the thermal efficiency of the engine, and are only an approximate guide.

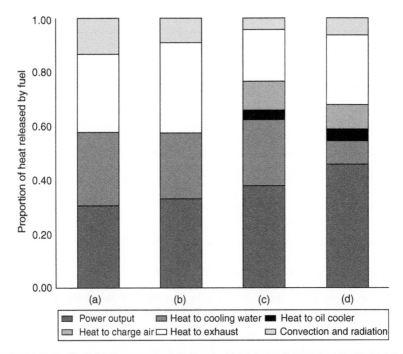

FIGURE 21.5   Typical full-power energy balances: (a) 1.7 L gasoline engine (1998); (b) 2.5 L naturally aspirated diesel engine; (c) 200 kW medium-speed turbocharged marine diesel engine; (d) 7.6 MW combined heat and power unit.

TABLE 21.2 An Approximate Energy Balance (kW per kW Power Output) of Three Engine Types

	Automotive Gasoline	Automotive Diesel	Medium Speed
Power output	1.0	1.0	1.0
Heat to cooling water	0.9	0.7	0.4
Heat to oil cooler			0.05
Heat to exhaust	0.9	0.7	0.65
Convection and radiation	0.2	0.2	0.15
Total	3.0	2.6	2.2

The Role of Indicated Power in the Energy Balance

It should be observed that the indicated power output of the engine does not appear in any of our formulations of the energy balance. There is a good reason for this: the difference between indicated and brake power represents the friction losses in the engine and the power required to drive the auxiliaries, and it is impossible to allocate these between the various heat losses in the balance. Most of the friction losses between piston and cylinder will appear in the cooling water; bearing losses and the power required to drive the oil pump will appear mostly in the oil cooler while water pump losses will appear directly in the cooling water. An exact analysis is problematic and beyond the scope of this book.

Chapter 3 deals with the concept of the engine test cell as a thermodynamic system and the recommendation is made that at an early stage in designing a new test cell an estimate should be made of the various flows: fuel, air, water, electricity, heat, and energy into and out of the cell. In such cases it is usually necessary to make some assumptions as to the full power performance of the largest engine to be tested. This is possible on the basis of information given earlier and in this chapter and an example follows.

Prediction of Energy Balance

Taking the example of a 250 kW turbocharged diesel engine at full power. Assume specific fuel consumption = 0.21 kg/kWh (LCV 40.6 MJ/kg; thermal efficiency 0.42). Then, following the general recommendations of Table 21.2, we may make an estimate, summarized in Table 21.3.

TABLE 21.3 Energy Balance, 250 kW Turbocharged Diesel Engine

In		Out	
Fuel	592 kW	Power	250 kW (42.2%)
		Heat to water	110 kW (18.6%)
		Heat to oil cooler	15 kW (2.5%)
		Heat to exhaust	117 kW (29.9%)
		Convection and radiation	40 kW (6.8%)
Total	592 kW		592 kW

The corresponding thermodynamic system is shown in Figure 21.6. This also shows rates of flow of fuel, air and cooling water, based on the following estimates:

Fuel flow
Assume fuel density 0.9 kg/liter

$$\text{Fuel flow} = \frac{250 \times 0.21}{0.9} = 5811/\text{hour} \ (52.5 \ \text{kg/hour})$$

Induction air flow
Assume full load air/fuel ratio = 25:1
Air flow = $250 \times 0.21 \times 25 = 1312.5$ kg/h
Taking air density as 1.2 kg/m^3
Air flow = 1094 m^3/h (0.30 m^3/s, 10.7 ft^3/s)

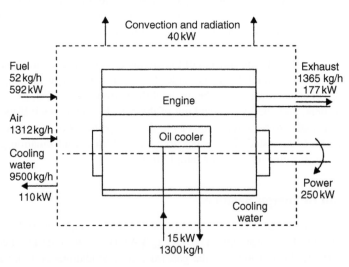

FIGURE 21.6 Control volume, 250 kW diesel engine showing energy and fluid flows.

Cooling water flow

Assume a temperature rise of 10 °C through the jacket and oil cooler. Then, since the specific heat of water = 4.18 kJ/kg °C and 1 kWh = 3600 kJ:

$$\text{Flow to jacket} + \text{oil cooler} = \frac{125 \times 60}{4.18 \times 10} = 180 \text{ kg/min (180 l/min)}$$

Exhaust flow

Sum of fuel flow + induction air flow = 1312.5 + 52.5 = 1365 kg/h.

Energy Balance for Turbocharger

In the case of turbocharged engines it is useful to separate the energy flows to and from the turbocharger [3] and associated air cooler from those associated with the complete engine. In this way an energy balance may be drawn up, covering the following:

- Exhaust gas entering the turbine from engine cylinders
- Induction air entering the compressor
- Cooling water entering the air cooler
- Exhaust leaving the turbine
- Induction air leaving the cooler
- Cooling water leaving the cooler.

A separate control volume contained within the control volume for the complete engine may be defined and an energy balance drawn up.

Summary of Stages in Calculation of Energy Balance From an Engine Test

1. Obtain information on the lower calorific value of the fuel.
2. From a knowledge of the compression ratio of the engine, calculate air standard cycle efficiency as a yardstick of performance.
3. For one or a number of test points measure:
   a. fuel consumption rate
   b. air flow rate to engine
   c. exhaust temperature
   d. power output
   e. cooling water inlet and outlet temperatures and flow rate.
4. Calculate the various terms of the energy balance and the thermal efficiency from this data.

Prediction of Energy Balance for a Given Engine

1. Record type of engine and rated power output.
2. Calculate fuel flow rate from known or assumed specific fuel consumption.
3. Draw up energy balance using the guidelines given in Table 21.2.

MEASUREMENT OF MECHANICAL LOSSES IN ENGINES

It is a curious fact that, in the long run, all the power developed by all the road vehicle engines in the world is dissipated as friction: either mechanical friction in the engine and transmission, rolling resistance between vehicle and road, or wind resistance. Mechanical efficiency, a measure of friction losses in the engine, is thus an important topic in engine development and therefore engine testing. It may exceed 80% at high power outputs but is generally lower and is zero when the engine is idling. Under mixed driving conditions for a passenger vehicle, between one-third and one-half of the power developed in the cylinders is dissipated either as mechanical friction in the engine, in driving the auxiliaries such as alternator and power-steering pump, or as pumping losses in the induction and exhaust tracts. Since the improvement of mechanical efficiency is such an important goal to engine and lubricant manufacturers, an accurate measure of mechanical losses is of prime importance. In fact, the precise measurement of these losses is a particularly difficult problem, to which no completely satisfactory solution exists.

The starting point in any investigation of mechanical losses should ideally be a precise knowledge of the power developed in the engine cylinder. This is represented by the indicator diagram (Figure 21.7), which shows the relation between the pressure of the gas in the cylinder and the piston stroke or swept volume (also see Chapter 15).

For a four-stroke engine account must be taken of both the positive area A_1, representing the work done on the piston during the compression and expansion strokes, and the negative area A_2, representing the "pumping losses", the work performed by the piston in expelling the exhaust gases and drawing the fresh charge into the cylinder. In the case of an engine fitted with a mechanical

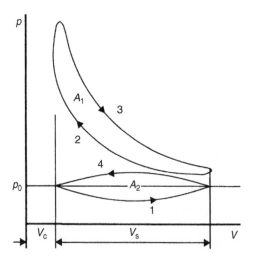

FIGURE 21.7 Indicator diagram, four-stroke engine. (1) Induction. (2) Compression. (3) Expansion. (4) Exhaust.

supercharger or an exhaust gas turbocharger, there are additional exchanges of energy between the exhaust gas and the fresh charge, but this does not invalidate the indicator diagram as a measure of power developed in the cylinder. The universally accepted measure of indicated power is the indicated mean effective pressure (i.m.e.p.). This represents the mean positive pressure exerted on the piston during the working strokes after allowing for the negative pressure represented by the pumping losses. The relation between i.m.e.p. and indicated power of the engine is given by the expression:

$$P_i = \frac{\bar{p}_i V_s n}{60K} \times 10^{-1} \text{ kW} \tag{21.4}$$

Here $n/60K$ represents the number of power strokes per second in each cylinder. The useful power output of the engine may be represented by the brake mean effective pressure:

$$P = \frac{\bar{p}_b V_s n}{60K} \times 10^{-1} \text{ kW} \tag{21.5}$$

The mechanical losses in the engine plus power to drive auxiliaries are represented by $(P_i - P)$ and may be represented by the friction mean effective pressure:

$$\bar{p}_f = \bar{p}_i - \bar{p}_b$$

It may not be realistic, for several experimental reasons, to place much faith in measurements of indicated power based on the indicator diagram. Several other techniques, each having their limitations, will be discussed and the critical determination of top dead center (TDC) is covered in detail in Chapter 15.

Motoring Tests

One method of estimating mechanical losses involves running the engine under stable temperature conditions and connected to a four-quadrant dynamometer. Ignition or fuel injection are then cut and the quickest possible measurement made of the power necessary to motor the engine at the same speed.

Sources of error include:

- Under nonfiring conditions, the cylinder pressure is greatly reduced, with a consequent reduction in friction losses between piston rings, cylinder skirt and cylinder liner, and in the running gear.
- The cylinder wall temperature falls very rapidly as soon as combustion ceases, with a consequent increase in viscous drag that may to some extent compensate for the above effect.
- Pumping losses are generally much changed in the absence of combustion.

Many detailed studies of engine friction under motored conditions have been reported in the literature (see Refs [4,5]). These usually involved the progressive removal of various components: camshaft and valve train, oil and fuel pumps, water pump, generator, seals, etc., in order to determine the contribution made by each element.

The Morse Test

In this test, the engine is run under steady conditions and ignition or injection is cut off in each cylinder in turn: it is of course only applicable to multicylinder engines.

On cutting out a cylinder, the dynamometer is rapidly adjusted to restore the engine speed and the reduction in power measured. This is assumed to be equal to the indicated power contributed by the nonfiring cylinder. The process is repeated for all cylinders and the sum of the reductions in power is taken to be a measure of the indicated power of the engine.

A modification of the Morse test [5] makes use of electronically controlled unit injectors, allowing the cylinders to be disabled in different ways and at different frequencies, thus keeping temperatures and operating conditions as near normal as possible.

The Morse test is subject, though to a lesser extent, to the sources of error described for the motoring test.

The Willan Line Method

This is applicable only to unthrottled compression ignition engines. It is a matter of observation that a curve of fuel consumption rate against torque or b.m.e.p. at constant speed plots quite accurately as a straight line up to about 75% of full power (Figure 21.8). This suggests that for the straight-line part of the characteristic, equal increments of fuel produce equal increments of power; combustion efficiency is constant.

At zero power output from the engine, all the fuel burned is expended in overcoming the mechanical losses in the engine, and it is a reasonable inference that an extrapolation of the Willan line to zero fuel consumption gives a measure of the friction losses in the engine.

Strictly speaking, the method only allows an estimate to be made of mechanical losses under no-load conditions. When developing power the losses in the engine will undoubtedly be greater.

SUMMARY

The four standard methods of estimating mechanical losses in an engine and its auxiliaries have been briefly described. No great accuracy can be claimed for any of these methods and it is instructive to apply as many of them as possible and compare the results. Measurement of mechanical losses in an engine is still something of an "art". While no method can be claimed to give a precise

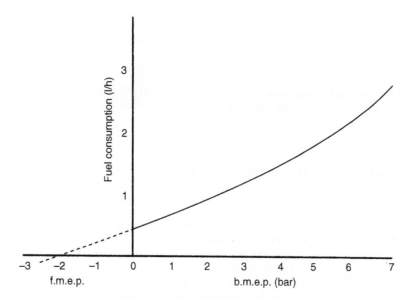

FIGURE 21.8 Willan's line for a diesel engine.

absolute value for mechanical losses they are, of course, quite effective in monitoring the influence of specific changes made to a particular engine.

NOTATION

Indicated mean effected pressure (i.m.e.p.)	$\bar{p}_i$ (bar)
Brake mean effected pressure (b.m.e.p.)	$\bar{p}_b$ (bar)
Friction mean effected pressure (f.m.e.p.)	$\bar{p}_f$ (bar)
Swept volume of engine	V_s (liter)
Engine speed	n (rev/min)
Constant: 1 for two-stroke, 2 for four-stroke engines	K
Indicated power	P_i (kW)
Engine power output	P (kW)
Thermal efficiency	
Lower calorific value	C_L (MJ/kg)
Compression ratio	R
Ratio of specific heats of air	$\gamma\ (= 1.4)$
Air standard efficiency	η_{as}
Mass flow rate of fuel	$\dot{m}_f$ (kg/s)
Mass flow rate of inlet air	$\dot{m}_a$ (kg/s)
Power output of engine	P_s (kW)
Specific heat of air at constant pressure	C_p (kJ/kg·K)
Specific heat of water	C_w (kJ/kg·K)
Ambient temperature	T_a (K)
Exhaust temperature	T_e (K)
Cooling water inlet temperature	T_{1w} (K)

Cooling water outlet temperature	T_{2w} (K)
Mass flow rate of cooling water	$\dot{m}_w$ (kg/s)
For exhaust calorimeter	
Mass flow rate of cooling water	$\dot{m}_c$ (kg/s)
Cooling water inlet temperature	T_{1c} (K)
Cooling water outlet temperature	T_{2c} (K)
Temperature of exhaust leaving calorimeter	T_{co} (K)

REFERENCES AND FURTHER READING

[1] J. Heywood, Internal Combustion Engine Fundamentals, McGraw-Hill, Maidenhead, 1988. ISBN-13: 978-0070286375.

[2] T.D. Eastop, A. McConkey, Applied Thermodynamics for Engineering Technologists, Longman, London, 1993. ISBN-13: 978-0582091931.

[3] N. Watson, M.S. Janota, Turbocharging the Internal Combustion Engine, Wiley-Interscience, New York, 1982. ISBN-13: 978-0333242902.

[4] F.A. Martin, Friction in internal combustion engine bearings, I. Mech. E. (1985), paper C67/85.

[5] S.N.M. Haines, S.A. Shields, The determination of diesel engine friction characteristics by electronic cylinder disablement, Proc. I. Mech. E. Part A 203 (A2) (1989) 129–138.

Martyr's Laws of Engineering Project Management

This list of empirically developed "laws", relevant to the process of building a test facility or any multidisciplinary laboratory, started life as something of a joke. For some years I had, when occasions demanded it, admonished my colleagues to "remember Martyr's second law" (see below). When, later in my working life, I was employed to carry out formal teaching I was asked to also quote the first and any subsequent laws. Thus it was that, in order to create a lighthearted lecture to fill any gap in schedules, I created the following list that, rather than being used as an occasional schedule filler, became a foundation lecture and was finally turned into a printed tee-shirt, the ultimate accolade!

First Law:

No complex project should be allowed to proceed without a clear specification that can be understood by all participants.

The vast majority of project failures, particularly those initiated by government committees, are caused by breaches of this law.

Second Law:

Paper is cheaper to change than concrete.

Meaning that spending time and money getting a project right at the start is cheaper than putting it right at the end.

Third Law:

Sometimes the best solution to a problem is not to have it.

Always consider avoiding a difficult problem (technical, logistical, or managerial) by using an alternative strategy. This has quite deep meaning and some research in the concept of Ockham's Razor[1] is recommended.

---

1. "Plurality should not be posited beyond necessity"—even the English translation from the original Latin is beautiful.

Fourth Law:

Not even the most talented customer (or student) can break the Laws of Physics—but it may not stop them trying.

It is not uncommon for customers of research instruments and facilities to include in their specification functions that are physically impossible (or financially not viable). This gives a problem to the company quoting the work, which is always at risk of losing the job to the competitor who agrees the (impossible) specification, and later fails to perform.

Fifth Law:

CHANGE is an inherent part of any project; if you don't manage it, then you will be the victim of it.

Well-run projects always have a "change request" system built into their procedures that allows changes to the specification or program to be formally requested and considered, with clear vision of the cost and time implications (which can be either positive or negative).

Sixth Law:

Nine women can't produce a baby in one month.

An old cliché but one designed to remind us that there are some tasks that take as long as they take and that throwing more labor at the job, or shouting loudly, does not change the time scale. Concrete curing time to full strength is a good example.

Seventh Law:

The triangle of "cost–quality–time" is constant for each and all projects, although the priority of each will be unique. In real life, if two are fixed the third is variable.

This has been disputed but then proved too many times in my life to discuss further!

Eighth Law:

With absolute compliance to procedures and having blind faith in the process it is possible to go to hell under perfect control.

All the major engineering projects in my life had suffered at some time from some sort of crisis, be they caused by weather events, a dock strike, a truck being driven into a canal, or the financial collapse of a subcontractor, none of which were resolved by strict adherence to rules, procedures or processes, rather the opposite.

This law caused much debate in groups of students who had been taught to believe that following the rules will naturally produce the intended outcome. However, there are many well-documented cases where projects and companies have come to a disastrous end because the management of the process was considered more important than the achievement of the specified outcome.

Ninth Law:

Late delivery of test instrumentation to a finished site is better, for everybody, than an "on time" delivery into an unfinished site.

This was written specifically for powertrain test facility projects and is based on repeated experiences of expensive equipment being put into buildings (that were running behind project schedules) and that were still occupied by inappropriate trades, such as tiling or plastering. Precision instruments do not survive well in such conditions, so the result is that installation might be on program but premature failures and the consequential "blame game" inevitably follows.

Tenth Law:

The project process should not be seen as linear but rather as a circle, with end results fed back to the beginning.

Well-organized and open post-project reviews, that include all the major parties, can be valuable learning experiences both individually and corporately.

Finally:

Don't forget Murphy's Law: "If it is possible for a fault to occur, it will occur—usually at the worst possible time."

Vigilant, technically competent, commercially empowered, site and project management is the best insurance for any project in minimizing the chances of "Murphy occurrences".

Index

Printed and bound by CPI Group (UK) Ltd, Croydon, CR0 4YY

03/10/2024

01040414-0011